CAMBRIDGE MONOGRAPHS ON PARTICLE PHYSICS, NUCLEAR PHYSICS AND COSMOLOGY
16

General Editors: T. Ericson, P. V. Landshoff

ELECTRON SCATTERING FOR NUCLEAR AND NUCLEON STRUCTURE

JOHN DIRK WALECKA

College of William and Mary

CAMBRIDGE
UNIVERSITY PRESS

CAMBRIDGE
UNIVERSITY PRESS

Shaftesbury Road, Cambridge CB2 8EA, United Kingdom

One Liberty Plaza, 20th Floor, New York, NY 10006, USA

477 Williamstown Road, Port Melbourne, VIC 3207, Australia

314–321, 3rd Floor, Plot 3, Splendor Forum, Jasola District Centre, New Delhi – 110025, India

103 Penang Road, #05–06/07, Visioncrest Commercial, Singapore 238467

Cambridge University Press is part of Cambridge University Press & Assessment,
a department of the University of Cambridge.

We share the University's mission to contribute to society through the pursuit of
education, learning and research at the highest international levels of excellence.

www.cambridge.org
Information on this title: www.cambridge.org/9781009290579

DOI: 10.1017/9781009290616

First published 2002
Reissued as OA 2022

A catalogue record for this publication is available from the British Library.

ISBN 978-1-009-29057-9 Hardback
ISBN 978-1-009-29059-3 Paperback

Contents

Preface

In the summer of 1986 I left Stanford University, after 26 years on the faculty, to assume the job of Scientific Director at the Continuous Electron Beam Accelerator Facility (CEBAF) now known as the Thomas Jefferson National Accelerator Facility (TJNAF). This facility, funded by the Department of Energy and located in Newport News, Virginia provides a high-energy, high-intensity, high-duty-factor electron accelerator for studying the internal structure of nuclei and nucleons. It has long been a top priority for the field of nuclear physics in the United States. Each year I gave a physics lecture series at the site. The initial series on electron scattering was based on a set of lectures I had given at Argonne National Laboratory in the winter of 1982–1983. As Scientific Director, I was continually called upon to make presentations on this topic. This book is based both on the lecture series on electron scattering, and on the many presentations I have given on this subject over the years.

The scattering of high-energy electrons from nuclear and nucleon targets essentially provides a microscope for examining the structure of these tiny objects. The best evidence we have on what nuclei and nucleons actually look like comes from electron scattering. An intense continuous electron beam with well-defined energy provides a powerful tool for structure investigations. Inclusive experiments, where only the final electron is detected, examine static and transition charge and current densities in the target. Coincidence experiments, where other particles are detected together with the scattered electron, provide valuable additional information.

In electron scattering experiments where the momentum of the initial and final electron are well-defined, a virtual quantum of electromagnetic radiation is produced which interacts with the target. The energy of this quantum is determined by the energy transfer from the electron, and the momentum of the quantum from the momentum transfer. The electromagnetic interaction is well-understood; the interaction is with the local, static and dynamic charge and current densities. The scattering cross section is determined by the four-dimensional Fourier transform of these quantities. For a given energy transfer to the target, one can vary the three-momentum transfer by varying the momentum vector of the final electron. One then maps out the Fourier transform of the spatial densities, and by inversion of the Fourier transform, one determines the spatial distribution of the densities themselves. The wavelength with which the target is examined is inversely proportional to the three-momentum transfer. In electromagnetic studies in nuclear physics one focuses on how matter is put together from its constituents and on distance scales $\sim 10\,\mathrm{fm}$ to $\sim 0.1\,\mathrm{fm}$ where $1\mathrm{fm} = 10^{-13}\,\mathrm{cm}$. Particle physics concentrates on finer and finer details of the substructure of matter with experiments at high energy which in turn explore much shorter distances. To carry out such studies, one needs electron accelerators of hundreds of MeV to many GeV.

A theoretical description of the nuclear and nucleon targets is required to interpret the experiments. The appropriate description employed depends on the distance scale at which one examines the target. Imagine that one looks at the earth from space. The appropriate quantities used to describe these observations, the appropriate *degrees of freedom*, are macroscopic ones, the location and shape of continents, oceans, clouds, etc. When one gets closer, finer details emerge, trees, houses, cars, people, and these must be included in the description. At the microscopic level of observations, it is the atomic and subatomic description which is relevant. It is thus self-evident that

> *The appropriate set of degrees of freedom depends on the distance scale at which we probe the system.*

At the macroscopic level, one describes nuclei in terms of properties such as size, shape, charge, and binding energy. Further refinement describes, for example, the spatial distribution of the charge. A finer and more detailed description is obtained using nucleons, protons and neutrons, as the degrees of freedom. The *traditional* approach to nuclear physics starts from structureless nucleons interacting through static two-body potentials fitted to two-body scattering and bound-state data. These two-body potentials are then inserted in the non-relativistic many-body Schrödinger equation and that equation is solved in some approximation — it can be solved exactly for few-body systems using modern computing techniques. Electromagnetic and weak currents are then constructed from the properties of free nucleons and used to probe the structure of the nuclear system.

Although this traditional approach to nuclear physics has had a great many successes, it is clearly inadequate for an understanding of the nuclear system on a more microscopic level. A more appropriate set of degrees of freedom then consists of the *hadrons*, the strongly-interacting mesons and baryons, where baryon number, a strictly conserved quantity, counts the number of nucleons that now exhibit internal structure and dynamics. There are many arguments that one can give in support of this picture. For example, the long-range part of all modern two-nucleon potentials consists of the exchange of mesons including π with $(J^\pi, T) = (0^-, 1), \sigma(0^+, 0), \omega(1^-, 0)$, and $\rho(1^-, 1)$. We know that at long range the force between two nucleons comes from meson exchange. Moreover, the first excited state of the nucleon, the $\Delta(1232)$ with $(J^\pi, T) = (3/2, 3/2)$, was first successfully described as a resonance arising from pion–nucleon dynamics. As a further example, one of the significant achievements in the field of electromagnetic nuclear physics in recent years has been the unambiguous identification of exchange currents, additional currents present in the nuclear system arising from the flow of charged mesons between the nucleons in the nucleus.

In any extrapolation away from the traditional nuclear physics approach, it is important to incorporate general principles of physics such as quantum mechanics, special relativity, and microscopic causality. The only consistent theoretical framework we have for describing such a relativistic, interacting, many-body system is relativistic quantum field theory based on a local lagrangian density. It is convenient to refer to relativistic quantum field theories of the nuclear system based on hadronic degrees of freedom as *quantum hadrodynamics* (QHD).

At a still finer level, we now know that the hadrons are themselves composite objects made up of quarks held together by the exchange of gluons. We now have a theory of the strong interactions binding quarks and gluons into the observed hadrons. This theory is based on an internal color symmetry and is known as *quantum chromodynamics* (QCD). The theory of QCD has two absolutely remarkable properties. The first is *asymptotic freedom*, which roughly states that at very high momenta, or very short distances, the renormalized coupling constant for the basic processes in the theory goes to zero; as a consequence, one can do perturbation theory in this regime. The second property is *confinement*. The basic underlying degrees of freedom in the theory, quarks and gluons, do not exist as asymptotic, free, scattering states in the laboratory. They exist and interact only inside hadrons. You cannot hold a single quark, or single gluon in your hand. There are strong indications from lattice gauge theory, where QCD is solved at a finite number of space-time points, that confinement is indeed a dynamic property of QCD arising from the nonlinear gluon couplings. Ultimately, nucleon and nuclear physics are the study of strong-coupling QCD.

As for the other basic forces in nature, surely one of the great intellectual achievements of our era is the unification of the theories of electromagnetism and of the weak interactions. It is essential to continue to put this theory of the electroweak interactions to rigorous tests and fully explore its consequences. Nuclei and nucleons provide unique laboratories in which to conduct such tests and explorations.

The current picture of the nucleus in the *standard model* is that of a bound system of baryons and mesons, which are in turn confined triplets of quarks and of quark–antiquark pairs, respectively. The electroweak interactions of leptons (electrons and neutrinos) with the nucleus are mediated by the photon and the heavy weak vector bosons, the Z^0 and $W^{\pm}$. The electroweak interactions couple directly to the quarks; the gluons are absolutely neutral to the electroweak interactions. Thus every time one studies a nuclear gamma decay, for example, one is directly probing the quark structure of the nucleus. Once the quark is struck, it is not a quark that is emitted from the target, but a hadron. Nuclei are the ideal laboratories for studying this process of *hadronization*.

Another truly remarkable property of QCD is that the effective degrees of freedom at low energy and long wavelengths *are* the hadrons, the baryons and mesons.

In this book, the motivation for electron scattering is examined in some detail. The theoretical analysis of the process is developed, as is our current theoretical understanding of the underlying structure of nuclei and nucleons at appropriate levels of resolution and sophistication. Selected examples are given, present experimental capabilities are summarized, and future directions are previewed.

In part 1 of this book modern pictures of the nucleus and nucleon are surveyed. As an introduction to electron scattering, the optical analogy is developed. The virtues of electron scattering are described and a qualitative overview of the nuclear response surfaces in inclusive electron scattering presented. The arguments for coincidence experiments are then given.

In part 2, a general theoretical analysis of electron scattering is developed, starting from a discussion of the electromagnetic interaction with an arbitrary localized quantum mechanical system. This includes a multipole decomposition. The relativistic electrons of interest here are described by the Dirac equation, and the necessary tools are developed. A covariant analysis of the scattering of an electron by nuclear and nucleon targets is then carried out. Both the excitation of discrete target states and one-particle emission coincidence experiments are analyzed. An analysis of deep-inelastic scattering (DIS) experiments, where the four-momentum transfer squared and energy transfer both grow large, but with a fixed ratio, is presented. This section ends with a general analysis of parity violation in inclusive polarized electron scattering.

Since electrons are charged and light, they by necessity radiate during the scattering process. This is one of the technical complications of electron scattering. This radiation as well as the accompanying virtual electromagnetic effects are described by *quantum electrodynamics* (QED); part 3 presents a brief review of the essentials of QED.

Part 4 presents experimental and theoretical results for selected examples. These examples are chosen to illustrate the wide variety of incisive information that can be obtained about the structure of nuclei and nucleons, the influence electron scattering has had on the development of our pictures of these systems, the role various laboratories throughout the world have played in these developments, and, quite frankly, the beauty of this branch of physics. Theoretical background in traditional nuclear physics, relativistic mean field theory, the quark model, QCD, and the standard model is developed in sufficient depth that the reader can indeed work through the examples in detail.

In part 5, future directions for the field are discussed, building on the evolving TJNAF program, but including other world-wide developments at both intermediate and very high energy.

Nine appendixes are included which explore some of the more interesting and important technical aspects of this subject.

The book assumes only a working knowledge of quantum mechanics and special relativity and develops the theoretical analysis in a self-contained fashion up to current levels of sophistication. It is basically aimed at first-year graduate students and advanced undergraduates in physics, although it should be accessible to others in the natural sciences. Parts 1 and 5 can be read by a wider audience interested in understanding the essentials of the subject. The book should serve effectively as a text for special topics courses on this subject or as a supplemental text for nuclear or particle physics courses. It should also serve as a summary and reference for researchers already working in electron scattering as well as those in other areas.

This manuscript was typed by the author in LATEX, from which the book is printed. Figures are reproduced by permission.

Williamsburg, Virginia
April 22, 2001

John Dirk Walecka
Governor's Distinguished CEBAF
Professor of Physics
College of William and Mary

Part 1
Introduction

1

Motivation

This monograph is concerned with the study of nuclear and nucleon structure through the scattering of high energy electrons. The history of this field is well summarized in the proceedings of the *Conference on 35 Years of Electron Scattering* held at the University of Illinois in 1986 to commemorate the 1951 experiment of Lyman, Hanson, and Scott; this experiment provided the first observation of the finite size of the nucleus by electron scattering [Ly51, Il87]. Hofstadter and his colleagues, working in the High Energy Physics Laboratory (HEPL) at Stanford University in the late 1950's, beautifully and systematically exhibited the shape of the charge distributions of nuclei and nucleons through experiments at higher momentum transfer [Ho56, Ho63]. Subsequent experimental work at HEPL, the Bates Laboratory at M.I.T., Saclay in France, NIKHEF in Holland, and both Darmstadt and Mainz in Germany (as well as other laboratories), utilizing parallel theoretical analysis [Gu34, Sc54, Al56, de66, Ub71], clearly exhibited more detailed aspects of nuclear structure. Experiments at higher electron energies and momentum transfers at the Stanford Linear Accelerator Center (SLAC) by Friedman, Kendall, and Taylor, together with theoretical developments by Bjorken, for the first time demonstrated the pointlike quark–parton substructure of nucleons and nuclei [Bj69, Fr72]. This work played a key role in the development of modern theories of the strong interaction. Major efforts today at CEBAF, the Continuous Electron Beam Accelerator Facility (now known as TJNAF, the Thomas Jefferson National Accelerator Facility) in the U.S., Bates, Mainz, SLAC, DESY in Germany, and CERN in Geneva (using muons) contribute to the development of our understanding of nuclei and nucleons.

In part 1 we discuss modern pictures of the nucleus and nucleon, starting with non-relativistic nucleons interacting through static potentials and proceeding to quarks and gluons with interactions described

3

by strong-coupling *quantum chromodynamics* (QCD). As an introduction to electron scattering, the optical analogy is developed. The virtues of electron scattering are described and a qualitative overview of the nuclear response surfaces in inclusive electron scattering presented. The arguments for coincidence experiments are then given.

In part 2, a general theoretical analysis of electron scattering is developed, starting from a discussion of the electromagnetic interaction with an arbitrary localized quantum mechanical system. This includes a multipole decomposition. Since electrons are relativistic here, they are described by the Dirac equation and the necessary tools are developed. A covariant analysis of the scattering of an electron by a nuclear target is then carried out. Both the excitation of discrete target states and one-particle emission coincidence experiments are analyzed. An analysis of deep-inelastic scattering (DIS) experiments, where the momentum transfer squared and energy transfer both grow large, but with a fixed ratio, is presented. This section ends with a general analysis of parity violation in inclusive polarized electron scattering.

Since electrons are charged and light, they by necessity radiate during the scattering process. This is one of the technical complications of electron scattering. This radiation as well as the accompanying virtual electromagnetic effects are described by *quantum electrodynamics* (QED); part 3 presents a brief review of the essentials of QED.

Part 4 presents experimental and theoretical results for selected examples. These examples are chosen to illustrate the wide variety of incisive information that can be obtained about the structure of nuclei and nucleons, the influence electron scattering has had on the development of our pictures of these systems, and the role various laboratories throughout the world have played in these developments.

In part 5, future directions for the field are discussed, building on the evolving TJNAF program [Wa93, Wa94], but including other world-wide developments at both intermediate and very high energy.

One of the most attractive and powerful aspects of the field of electron scattering for the structure of nuclei and nucleons is that experimental and theoretical developments have always progressed hand in hand, with each reinforcing the other.

We start this monograph with a more detailed discussion of the motivation for studying the structure of nuclei and nucleons through the scattering of high energy electrons.

Let us go back to the beginning. Why do we do nuclear physics? Why is nuclear physics interesting? First of all, the nucleus is a unique form of matter consisting of many baryons in close proximity. All the forces of nature are present in the nucleus — strong, electromagnetic, weak, and even gravity if one includes condensed stellar objects which are nothing more than enormous nuclei held together by the gravitational attraction. The nucleus provides a microscopic laboratory to test the structure of the fundamental interactions. Furthermore, the nucleus manifests remarkable properties as a strongly interacting, quantum mechanical, relativistic, many-body system. In addition, most of the mass and energy in the visible universe comes from nuclei and nuclear reactions. Also, we now know there are new underlying degrees of freedom in the nucleus, quarks and gluons, interacting through remarkable new forces described by quantum chromodynamics (QCD). The single nucleon itself is now a complicated nuclear many-body system. The electromagnetic properties of nucleons and nuclei provide benchmarks with which to test our understanding of strong-coupling QCD and the quark substructure of matter. Moreover, nuclear physics is crucial to the understanding of the universe, for example: the early universe, formation of the elements, supernovae, and neutron stars. In sum, nuclear physics is really the study of the *structure of matter*.

Where is nuclear physics going? The nuclear science community in the U.S. recently underwent one of its periodic long-range planning exercises under the leadership of the Nuclear Science Advisory Committee (NSAC) and the Division of Nuclear Physics (DNP) of the American Physical Society (APS). In the report entitled *Nuclear Science:A Long-Range Plan* [NS96] the headings in part II on *The Scientific Frontiers* capture the present frontiers:

1. Nuclear Structure and Dynamics: Exploring the Limits

2. To the Quark Structure of Matter

3. The Phases of Nuclear Matter

4. Fundamental Symmetries and Nuclear Astrophysics

2

Pictures of the nucleus

We currently possess three levels of understanding of the nucleus within the following frameworks [Wa95]:

(I) *Traditional, Non-Relativistic, Many-Body Systems* [Fe71]. This approach uses static two-body potentials fit to two-nucleon scattering and bound-state data. These potentials are then inserted in the non-relativistic Schrödinger equation, and that equation is then solved in some approximation; with few-nucleon systems and large-scale computing capabilities, the equations can now be solved exactly. Electroweak currents constructed from the properties of free nucleons are then used to probe the nuclear system. Although this approach has had a great many successes [Bl52, Ma55, Bo69, Fe71, de74, Pr75, Fe91, Wa95], it is clearly inadequate for a more detailed understanding of the nuclear system.

(II) *Relativistic Many-Body Systems.* A more appropriate set of degrees of freedom for nuclear physics consists of the *hadrons*, the strongly interacting mesons and baryons. There are many arguments one can give for this. For example, the long-range part of the best modern two-nucleon potentials is given by meson exchange, predominantly π with $(J^\pi, T) = (0^-, 1)$, $\sigma(0^+, 0)$, $\rho(1^-, 1)$ and $\omega(1^-, 0)$ [La80, Ma89]. Furthermore, one of the successes of electromagnetic nuclear physics is the unambiguous demonstration of the existence of exchange currents, additional electromagnetic currents in the nucleus arising from the flow of charged mesons between nucleons. In addition, one daily sees copious production of mesons from nuclei in high-energy accelerators.

The only consistent theoretical framework we have for describing such a strongly-coupled, relativistic, interacting, many-body system is relativistic quantum field theory based on a local lagrangian density. It is convenient to refer to relativistic quantum field theory models of the nuclear system based on hadronic degrees of freedom as *quantum hadrodynamics* (QHD).

6

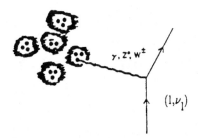

Fig. 2.1. Nucleus as a strongly-coupled system of colored quarks and gluons; electroweak interaction with a lepton.

More generally, one can view such field theories as *effective* field theories for the underlying theory of QCD [Se86, Se97].

(III) *Strongly-Coupled Colored Quarks and Gluons.* Our deepest level of understanding of nucleons, and the nucleus from which they are made, is as a strongly-coupled system of quarks and gluons (Fig. 2.1). Their interactions are described by a Yang–Mills theory [Ya54] based on an internal color symmetry (QCD). This theory has two remarkable properties: it is *asymptotically free*, which means that at very high momenta, or very short distances, the renormalized coupling constant becomes small. This has several consequences. For example, it implies that when in the appropriate kinematic regime, one scatters from essentially free point-like objects. In fact, it was the experimental observation of this phenomenon in deep inelastic scattering (DIS) that drove theorists to hunt for asymptotically free theories [Gr73a, Gr73b, Po73, Po74]. Furthermore, when the coupling is small, one can do perturbation theory. The many high-energy successes of *perturbative QCD* now provide convincing evidence that QCD is truly the underlying theory of the strong interactions.

When one scatters a lepton from a nuclear system, the electroweak interaction takes place through the exchange of one of the electroweak bosons ($\gamma, W^\pm, Z^0$), as illustrated in Fig. 2.1. These bosons couple directly to the quarks; the gluons are *absolutely neutral to the electroweak interactions*. Thus every time one observes a gamma decay or beta decay of a nucleus or nucleon, one is directly observing the quark structure of these systems!

The second remarkable property of QCD is *confinement*, which means that the underlying degrees of freedom, quarks and gluons, never appear as asymptotic, free scattering states in the laboratory. You cannot hold a free quark or gluon in your hand. Quarks and gluons, and their strong color interactions, are confined to the interior of the hadrons. At low momenta, or the large distances appropriate for nuclear physics, the renormalized coupling grows large. QCD becomes a strong-coupling theory in this limit.

There are convincing indications from lattice gauge theory (LGT), where strong-coupling QCD is solved on a finite space-time lattice [Wi74], that confinement is indeed a dynamical property of QCD arising from the nonlinear gluon couplings dictated by local color gauge invariance in this non-abelian Yang–Mills theory.

3

Some optics

To obtain insight into the electron scattering process, we appeal to some elementary optics, with which the reader is certainly familiar from an introductory physics course. If one looks through a telescope at a star, or shines a laser through a pinhole, one does not really observe a point of light, but actually a diffraction pattern with a bright disc at the center and a series of concentric rings with diminishing intensity. If the radius of the aperture through which the light passes is a, and the wavelength of the incident light is λ_1, then the angle θ to the first diffraction minimum of the central *Airy disc* is given by

$$a\theta \approx 0.61\lambda_1 \tag{3.1}$$

Here θ is measured from the central ray, starting at the aperture. Now introduce the incident wave number k_1 and "momentum transfer" κ

$$k_1 \equiv \frac{2\pi}{\lambda_1} \qquad ; \text{wave number}$$

$$\kappa \approx k_1\theta \qquad ; \text{momentum transfer} \tag{3.2}$$

Equation (3.1) can then be rewritten as

$$\kappa a \approx 1.22\,\pi \tag{3.3}$$

This relation has a marvelous consequence. Suppose one shines light from a laser of given wavelength on a pinhole, and projects the resulting diffraction pattern on a screen behind the pinhole. The angle to the first minimum can be determined by making *macroscopic measurements* of the distance of the screen from the aperture and the transverse distance on the screen out to the first minimum. Equation (3.3) then allows one to determine the radius a of the pinhole. *One can measure a radius of arbitrarily small size if only the momentum transfer is large enough!* The

9

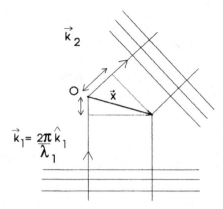

Fig. 3.1.   Optical pathlength with respect to central ray in Fraunhofer diffraction.

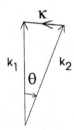

Fig. 3.2.   The "momentum transfer" $\kappa = \mathbf{k}_1 - \mathbf{k}_2$.

momentum transfer is inversely proportional to the wavelength. Thus to obtain large momentum transfer, one has to go to short wavelength. One evidently needs a wavelength comparable to the size of the aperture to make this measurement.

Let us extend these simple considerations. In Fraunhofer diffraction one has an incident plane wave and an outgoing plane wave in the direction of observation as illustrated in Fig. 3.1. The optical pathlength of an arbitrary ray with respect to the central ray is evidently given from this figure as

$$\Delta_{\mathrm{opt}} = \frac{2\pi}{\lambda_1}(\hat{\mathbf{k}}_1 \cdot \mathbf{x} - \hat{\mathbf{k}}_2 \cdot \mathbf{x}) = \kappa \cdot \mathbf{x} \qquad (3.4)$$

where $\hat{\mathbf{k}}_1$ and $\hat{\mathbf{k}}_2$ are unit vectors in the incident and outgoing directions respectively. Here the momentum transfer κ is defined by (Fig. 3.2)

$$\kappa = \mathbf{k}_1 - \mathbf{k}_2 \qquad (3.5)$$

Since the lengths of the incoming and outgoing wave numbers are identical

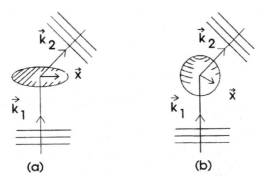

Fig. 3.3. (a) Fraunhofer diffraction of light from a circular aperture; and (b) Electron scattering through the Coulomb interaction from a spherical charge distribution.

$|\mathbf{k}_2| = |\mathbf{k}_1|$, the square of the momentum transfer is given by

$$\kappa^2 = 2k_1^2(1 - \cos\theta)$$
$$= 4k_1^2 \sin^2 \frac{\theta}{2} \tag{3.6}$$

Here θ is the angle between the incident and outgoing wave number vectors (Fig. 3.2). *Huygens Principle* says that each point on a wavefront serves as a new source of outgoing waves. The outgoing waves interfere. To determine the net outgoing wave from a circular aperture one must add the contributions from each little element of the disc weighted by $\exp\{i\Delta_{\text{opt}}\}$ as illustrated in Fig. 3.3 (a). The resulting amplitude of the light wave far from the scatterer is thus given by

$$\mathscr{A}_\gamma = \int_{\text{Aperture}} d^2x \, e^{i\boldsymbol{\kappa}\cdot\mathbf{x}} \tag{3.7}$$

The diffraction pattern evidently measures the two-dimensional Fourier transform of the aperture.

Now consider the scattering of an electron from a spherical charge distribution through the Coulomb interaction. de Broglie and quantum mechanics tell us that there is a wave associated with the electron of wavelength

$$\lambda_1 = \frac{h}{p_1} \qquad ; \text{electron} \tag{3.8}$$

Here $h \equiv 2\pi\hbar$ is Planck's constant and p_1 the incident electron momentum. The scattering amplitude from each little element will be proportional to the amount of charge there, or to the charge density $\rho_{\text{ch}}(x)$. The resulting amplitude of the electron wave far from the target is thus given in direct analogy with the above by (see Fig. 3.3 b)

$$\mathscr{A}_{\mathrm{el}} = \int_{\mathrm{Nucleus}} d^3x \, \rho_{\mathrm{ch}}(x) \, e^{i\boldsymbol{\kappa} \cdot \mathbf{x}} \tag{3.9}$$

The diffraction pattern of the scattered electron measures the three-dimensional Fourier transform of the target charge distribution.

In electron scattering, the incident electron momentum is evidently $\mathbf{p}_1 = \hbar \mathbf{k}_1$ and the momentum transferred from the electron is $\hbar \boldsymbol{\kappa}$; this is the reason for the terminology.

One can see a macroscopic diffraction pattern from arbitrarily small charge distributions if only the momentum transfer is large enough, or equivalently, if the wavelength is small enough. It follows from Eq. (3.8) that to achieve very small wavelengths, one must go to very high electron energies. It is an irony (and an expensive one!) that to look in detail with accuracy and precision at very small objects such as nuclei and nucleons one needs accurate and precise high-energy electron accelerators to produce the incident electron beams and correspondingly large, accurate and precise spectrometers to detect the scattered electrons.

One can put in some numbers. To have an electron with wavelength 1 fm = 10^{-13} cm, a typical nuclear dimension, one needs a relativistic electron of energy[1]

$$\lambda = 1 \text{ fm}$$
$$\Rightarrow E = pc = \hbar k c = 1240 \text{ MeV} \tag{3.10}$$

To obtain some insight, it is useful to evaluate the above amplitudes for the simple cases of a circular disc and unit spherical charge distribution, both of radius a. With the introduction of polar coordinates in the first case, and the use of spherical coordinates in the second, one finds [Fe80]

$$\mathscr{A}_\gamma^{\mathrm{disc}} = \int_0^a \rho \, d\rho \int_0^{2\pi} d\phi \, \exp\{i\kappa_\perp \rho \cos\phi\}$$
$$= \pi a^2 \left[\frac{2J_1(\kappa_\perp a)}{\kappa_\perp a} \right]$$
$$\mathscr{A}_{\mathrm{el}}^{\mathrm{sphere}} = \int_0^a r^2 \, dr \int d\Omega_r \, \exp\{i\kappa r \cos\theta_r\}$$
$$= \left(\frac{4\pi a^3}{3} \right) \left[\frac{3j_1(\kappa a)}{\kappa a} \right] \tag{3.11}$$

---

[1] Recall $\hbar c = 197.3 \text{ MeV fm}$. Here c is the speed of light.

Here $J_n(\alpha)$ and $j_n(\alpha)$ are cylindrical and spherical Bessel functions respectively. The quantities in square brackets in the above expressions are known as *form factors*. It is instructive to make some log plots of the square of these quantities on your PC. The first zero of $J_1(\alpha)$ occurs at $\alpha_{1,1} = 1.22\,\pi$; this is the origin of Eq. (3.3).

4

Why electron scattering?

In this section we present a brief overview of the virtues of electron scattering. We revisit most of this material in detail in the remainder of the book.

There are many reasons why inclusive electron scattering (e, e') provides a powerful tool for studying the structure of nuclei and nucleons. First, the interaction is known — it is given by *quantum electrodynamics* (QED), the most accurate physical theory we have. Second, the interaction is relatively weak — of order $\alpha = 1/137.0$, the fine-structure constant, and thus one can make measurements without greatly disturbing the structure of the target. Furthermore, the interaction is with the local electromagnetic current density in the target $\hat{J}_\mu(x)$. Hence one *knows* what is measured.

The process is governed by the S-matrix, which with one photon exchange (Fig. 4.1) takes the form[1]

$$S_{fi}^{(\gamma)} = \frac{-ee_p}{\hbar c \, \Omega} \bar{u}(k_2)\gamma_\mu u(k_1)\frac{1}{k^2} \int e^{ik \cdot x}\langle f|\hat{J}_\mu(x)|i\rangle \, d^4x \qquad (4.1)$$

What is measured is the Fourier transform with respect to the four-momentum transfer $\hbar k$ with $k \equiv k_1 - k_2$ of the transition matrix element of the current density.

In electron scattering, one can vary the three-momentum transfer and energy transfer *independently* in

$$k = (\boldsymbol{\kappa}, i\omega/c)$$
$$\kappa^2 = (\mathbf{k}_1 - \mathbf{k}_2)^2$$
$$\hbar\omega = \varepsilon_1 - \varepsilon_2 \qquad (4.2)$$

For a given energy transfer, one can map out the three-dimensional Fourier transform with respect to $\boldsymbol{\kappa}$ of the transition densities. The inversion of

---

[1] We quantize with periodic boundary conditions in a big box of volume Ω and in the end let $\Omega \to \infty$.

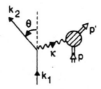

Fig. 4.1. Kinematics for electron scattering (e, e') with one photon exchange.

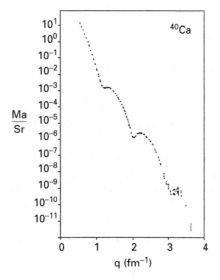

Fig. 4.2. Cross section for elastic electron scattering ^{40}Ca(e, e) vs. momentum transfer (here $q \equiv \kappa$) [Fr79].

this Fourier transform than provides the *microscopic spatial distribution of the densities.*

We give an example in Fig. 4.2. This is the diffraction pattern observed when electrons are scattered elastically from ^{40}Ca. The data are from Saclay [Fr79]. Notice the central diffraction maximum and the series of concentric rings with decreasing intensity as the scattering angle is increased. Notice also the scale on the ordinate; it runs over 13 decades. Figure 4.3 [Ho81, Se86] shows the charge distribution obtained upon inversion of the Fourier transform. The abscissa is in fermis.[2] The band in the experimental data is an estimate of the uncertainty introduced by the fact that one, by necessity, only measures a partial Fourier transform. The

---

[2] The situation is actually somewhat more complicated than this. As Z gets large, the distortion of the incident and outgoing electron wave functions by the Coulomb field of the nucleus must be taken into account, and one must perform a partial wave analysis of Coulomb scattering from the nuclear charge distribution.

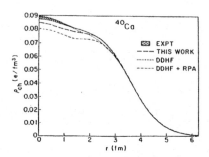

Fig. 4.3. Charge distribution of ^{40}Ca obtained from Fig. 4.2 with estimate of measurement error. Units are 1 fm = 10^{-13} cm. Heavy dashed curve shows calculation in relativistic mean field theory (RMFT) in QHD (other curves show similar results in traditional approach) [Ho81, Se86].

theoretical curves give an indication of the present level of understanding of these charge densities in nuclear physics.

Recall that there is an *inverse relationship* between the three-momentum transfer and the distance scale at which one probes the system

$$|\kappa| = \frac{2\pi}{\lambda} \qquad (4.3)$$

In electromagnetic studies in nuclear physics one focuses on how matter is put together from its constituents and on distance scales ~ 10 fm to ~ 0.1 fm. Particle physics concentrates on finer and finer details of the substructure of matter with experiments at high energy which in turn explore much shorter distances.

In electron scattering, one can moreover vary the polarization of the virtual photon in Fig. 4.1 by changing the electron kinematics; through this, the charge and current interaction can be separated. In sum, electron scattering gives rise to a precisely defined virtual quantum of electromagnetic radiation, and hence electrons provide a *precision tool* for examining the structure of nuclei and nucleons. Of course, an additional great advantage of electrons is that they can be copiously produced in the laboratory, and since they are charged, they can readily be accelerated and detected.[3]

Electron scattering is furthermore a versatile tool. One knows from the theory of electromagnetism that two currents will interact with each other. The moving electron produces such a current. Thus not only is there a Coulomb interaction between the charged electron and the charges in the target, but there is also a magnetic interaction between the moving electron and the current in the target. The nuclear current is produced

---

[3] Neutrino scattering for example, which has similar virtues for the weak interaction, lacks these properties.

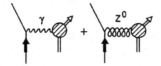

Fig. 4.4. (e, e′) amplitude as sum of γ and Z^0 exchange.

both by the convection current of the moving protons and also by the curl of the intrinsic magnetization, arising from the fact that nucleons are themselves little magnets; electron scattering measures the full transition matrix element of the target current

$$J_\lambda(x) = [\mathbf{J}_c(x) + \nabla \times \boldsymbol{\mu}(x),\ i\rho(x)] \tag{4.4}$$

In addition, with electron scattering one has the possibility of bringing out high multipoles of the current at large values of κR.

The interference between γ and Z^0 exchange (Fig. 4.4), where Z^0 is the heavy boson mediating the weak neutral current interaction, gives rise to *parity violation*. One measure of parity violation is the asymmetry arising from the difference in cross section of right- and left-handed electrons in inclusive electron scattering $(\vec{e}, e')$

$$\mathscr{A} \equiv \frac{d\sigma_\uparrow - d\sigma_\downarrow}{d\sigma_\uparrow + d\sigma_\downarrow} \tag{4.5}$$

The S-matrix for the amplitude in Fig. 4.4 takes the form[4]

$$S_{fi} = S_{fi}^{(\gamma)} - \left(\frac{\hbar}{c}\right)^2 \frac{G}{\sqrt{2}\,\Omega} \bar{u}\gamma_\mu(a + b\gamma_5)u \int e^{ik\cdot x}\langle f|\hat{\mathscr{J}}_\mu^{(0)}(x)|i\rangle\, d^4x \tag{4.6}$$

where

$$\langle f|\hat{\mathscr{J}}_\mu^{(0)}(x)|i\rangle = \langle f|\hat{J}_\mu^{(0)}(x) + \hat{J}_{\mu 5}^{(0)}(x)|i\rangle \tag{4.7}$$

Here $\hat{\mathscr{J}}_\mu^{(0)}(x)$ is the weak neutral current operator for the target and $G = 1.027 \times 10^{-5}/m_p^2$ is Fermi's weak coupling constant. Parity violation arises from the interference of the first term in Eq. (4.6) with the two contributions linear in the axial vector current in the second. If the first term has been measured and is assumed known, then the parity-violation asymmetry measures the second. Hence parity violation in $(\vec{e}, e')$ *doubles* the information content in electron scattering as it provides a means of

---

[4] In the standard model of the electroweak interactions $a = -(1 - 4\sin^2\theta_W)$ and $b = -1$ [Wa95].

measuring the spatial distribution of weak neutral current in nuclei and nucleons.

The cross section for inclusive electron scattering (e, e′) with one photon exchange is characterized by two response surfaces (see below) which are each functions of two Lorentz invariants. These invariants can be taken to be the four-momentum transfer squared $\hbar^2 k^2$ and the scalar product $v \equiv -k \cdot p / M_T$ where $\hbar p$ is the initial four-momentum of the target, m_T its mass, and M_T is its inverse Compton wavelength.

$$M_T \equiv \frac{m_T c}{\hbar} \qquad\qquad M \equiv \frac{m_p c}{\hbar} \qquad\qquad (4.8)$$

The second invariant v, when evaluated in the laboratory frame where the target is initially at rest, reduces to the energy loss of the electron $v = \hbar \omega_{\text{lab}} / \hbar c$. The deep-inelastic region (DIS) for electron scattering from the nucleon is defined by letting $k^2 \to \infty$ and $v \to \infty$ while keeping their *ratio* $x \equiv k^2 / 2Mv$ fixed. In deep-inelastic scattering the two response surfaces are observed to satisfy *Bjorken scaling*. They become independent of k^2 and are finite functions of the single variable x [Bj69, Fr72]. There is no form factor for the constituents from which one is scattering in this region. DIS provided the first dynamical evidence for the point-like quark substructure of hadrons. It also provides a measurement of the quark momentum distribution. Furthermore, QCD predictions for the $\ln k^2$ corrections in the approach to scaling can also be tested in DIS [Ro90].

The initial experiments at SLAC on parity violation in DIS [Pr78, Pr79] gave the first clear evidence that the weak neutral current has the structure predicted by the standard model of the electroweak interactions.

Further experiments, originated at SLAC, on the scattering of polarized electrons by polarized nucleons [Hu83] allow one to examine the strong-interaction spin structure functions of the nucleon.

5

Target response surfaces

As we shall see, the target response in inclusive electron scattering (e, e′) is summarized in the following Lorentz tensor

$$
\begin{aligned}
W_{\mu\nu} &= \frac{(2\pi)^3}{\hbar c} \overline{\sum_i \sum_f} \langle i|\hat{J}_\nu(0)|f\rangle \langle f|\hat{J}_\mu(0)|i\rangle (\Omega E)\delta^{(4)}(p'-p-k) \\
&= W_1(k^2, k\cdot p)\left(\delta_{\mu\nu} - \frac{k_\mu k_\nu}{k^2}\right) \\
&\quad + W_2(k^2, k\cdot p)\frac{1}{M_T^2}\left(p_\mu - \frac{p\cdot k}{k^2}k_\mu\right)\left(p_\nu - \frac{p\cdot k}{k^2}k_\nu\right)
\end{aligned}
\tag{5.1}
$$

The cross section is expressed in terms of the two, two-dimensional response surfaces as

$$
\begin{aligned}
\frac{d^2\sigma}{d\Omega_2 dk_2} &= \sigma_M \frac{1}{M_T}\left[W_2(k^2, v) + 2W_1(k^2, v)\tan^2\frac{\theta}{2}\right] \\
\sigma_M &= \frac{\alpha^2 \cos^2\theta/2}{4k_1^2 \sin^4\theta/2}
\end{aligned}
\tag{5.2}
$$

Here $k_1 = |\mathbf{k}_1|$. The square of the four-momentum transfer is given for a relativistic electron by

$$
\begin{aligned}
k^2 &= (\mathbf{k}_1 - \mathbf{k}_2)^2 - (k_1 - k_2)^2 \\
&= 4k_1 k_2 \sin^2\theta/2 \qquad \text{; in laboratory}
\end{aligned}
\tag{5.3}
$$

The second Lorentz scalar is written as the kinematic variable

$$
\begin{aligned}
v &\equiv -k\cdot p/M_T \\
&= k_1 - k_2 \qquad \text{; in laboratory}
\end{aligned}
\tag{5.4}
$$

There are three lepton variables in electron scattering, the initial and final electron energies ε_1 and ε_2 and the scattering angle θ, or equivalently

19

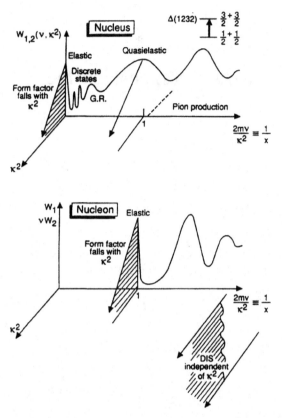

Fig. 5.1. Qualitative sketch of response surfaces $W_{1,2}(v,k^2)$ for nuclei and nucleons. One axis is the square of the four-momentum transfer k^2 (denoted in this figure by κ^2), the other is $2Mv/k^2 = 1/x$.

(k^2, v, θ). The two response surfaces can be separated by varying the electron scattering angle θ at fixed v and k^2. Alternatively, one can work at back angles $\theta = \pi$ where only the term in W_1 contributes.

For orientation, a qualitative sketch of the response surfaces $W_{1,2}(v,k^2)$ for electron scattering (e, e′) from both nuclei and nucleons is given in Fig. 5.1.[1] For a nucleus, one has the following features. First there is elastic scattering with a form factor that falls in the k^2 direction indicating the extended charge distribution in the target. One then sees inelastic scattering leading to excitation of discrete nuclear levels. The form factors for these inelastic transitions characterize the spatial distribution of the *transition* charge and current densities. At higher energy loss, above particle

---

[1] Electrons are light and radiate as they scatter; these *radiative corrections* must always be unfolded from the data before one gets at the underlying nuclear physics. We go into this in some detail in the section on QED.

emission threshold, one observes nuclear giant resonances (GR) with broader widths. The subsequent quasielastic peak is essentially scattering from a free nucleon, Doppler broadened by the Fermi motion of the nucleons in the nucleus. At energy losses higher than the pion mass, pion production occurs. At still higher energy loss, one observes production of the internal excitations of the nucleon itself, the first and most prominent being the $\Delta(1232)$ with $(J^\pi, T) = (3/2^+, 3/2)$. The k^2 dependence of the form factors for the excitation of nucleon resonances characterizes the spatial distribution of the transition charge and current densities in the nucleon.

For a single nucleon target, one sees first elastic scattering with a form factor which falls with k^2, again indicating a spatially extended structure in the nucleon — Robert Hofstadter won the Nobel prize for the measurement of the charge distributions of nuclei and the charge and magnetization distributions of the nucleon. At sufficiently high energy loss there is production of the nucleon resonances with the characteristic k^2 dependence of the inelastic form factors. Since all the nucleon resonances lie above the pion production threshold, they have strong-interaction widths. While the $\Delta(1232)$ appears as a distinct isolated peak, the higher nucleon resonances, as with giant resonances in nuclei, present multiple broad overlapping structures.

With higher energy accelerators, one can push into the region of deep-inelastic scattering (DIS) where the electron energy loss gets very large $v \to \infty$ and the four-momentum transfer also grows very large $k^2 \to \infty$ but where the ratio of these quantities $x \equiv k^2/2Mv$ is fixed at a finite value

$$
\begin{aligned}
v &\equiv -k \cdot p/M \;\to\; \infty \\
k^2 &\to\; \infty \\
x &\equiv\; k^2/2Mv \qquad \text{; fixed in DIS}
\end{aligned}
\tag{5.5}
$$

In DIS something quite remarkable happens. The two response surfaces are *independent of k^2* and satisfy *Bjorken scaling*, becoming finite functions of the single variable x [Bj69, Fr72]

$$
\begin{aligned}
\frac{v}{M} W_2(k^2, v) &\;\to\; F_2(x) \\
2W_1(k^2, v) &\;\to\; F_1(x) \qquad \text{; Bjorken scaling in DIS}
\end{aligned}
\tag{5.6}
$$

The fact that the structure functions become independent of k^2 indicates that the objects inside the nucleon from which one is scattering have no spatially extended structure, that is, one is scattering from *point-like constituents*. Friedman, Kendall, and Taylor won the Nobel prize for their discovery of this dynamic evidence for a point-like quark substructure of the nucleon.

As we shall see, scattering of polarized electrons on polarized targets allows one to access additional spin structure functions.

6

Why coincidence experiments?

There are many reasons why the ability to perform coincident electron scattering measurements, provided by a "continuous wave" (c.w.) accelerator greatly increases the power of electron scattering. Let us first review some of the essentials.

The kinematics for the coincident electron scattering process (e, e′ X) are defined in Fig. 6.1. Here the incident and scattered electron determine a scattering plane and an orthonormal system of unit vectors $\mathbf{e}_i$ with $\mathbf{e}_3$ along $\boldsymbol{\kappa} \equiv \mathbf{k}_1 - \mathbf{k}_2$ and $\mathbf{e}_2$ in-plane. Note that this frame is invariant under a Lorentz transformation along $\boldsymbol{\kappa}$ to the C-M system of the target and virtual photon. We use $\mathbf{q}$ to denote the momentum of the produced particle X. The reaction plane is then defined by the two vectors $(\boldsymbol{\kappa}, \mathbf{q})$. The orientation of $\mathbf{q}$ and the reaction plane are specified by polar and azimuthal angles (θ_q, ϕ_q) in the orthonormal system (Fig. 6.1). The angles $\phi_q = \pi/2, 3\pi/2$ produce an *in-plane* configuration.

The S-matrix for the process (e, e′ X) is given by

$$S_{fi} = -\frac{ee_p}{\hbar c\,\Omega}\bar{u}\gamma_\mu u\,\frac{1}{k^2}\int e^{ik\cdot x}\langle\Psi_{p'}; q^{(-)}|\hat{J}_\mu(x)|\Psi_p\rangle\,d^4x \qquad (6.1)$$

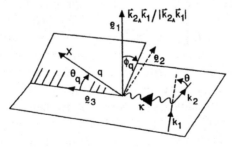

Fig. 6.1.   Kinematics for basic coincident electron scattering process (e, e′ X).

23

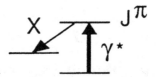

Fig. 6.2. Intermediate state J^π characterizes angular distribution of emitted particle X.

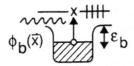

Fig. 6.3. Basic nuclear coincidence process (e, e′ N).

What one measures is again the Fourier transform of the transition matrix element of the electromagnetic current density between exact Heisenberg states of the target. The final state now consists asymptotically of a target state $|\Psi_{p'}\rangle$ and an emitted particle X with four-momentum q; it is constructed with incoming wave boundary conditions.

What can one learn about the structure of nuclei and nucleons from such experiments? First, if the reaction (e, e′ X) proceeds through an intermediate state of the target with given J^π (Fig. 6.2), then that J^π characterizes the angular distribution of X. The virtual photon orients the target along $\boldsymbol{\kappa}$. Angular correlation measurements of the emitted particle with respect to the virtual photon determine the contributing multipolarities. Furthermore, all values of J^π at any ω can again be accessed by increasing (κR).

Moreover, in contrast to inclusive scattering (e, e′) where the cross section is given by the sums of squares of the transition multipoles (see part 2), (e, e′ X) involves interference between amplitudes. One then has the ability to determine small, but important, amplitudes through interference effects.

Consider the basic nuclear coincidence process (e, e′ N) where N is a single nucleon, as illustrated in Fig. 6.3. This process creates a *hole* in the final nucleus [Ja66, Ja73]. Let the initial nucleon binding energy be ε_b and wave function be $\phi_b(\mathbf{x})$. Consider for illustration only the Coulomb interaction and assume the final nucleon can be described by a plane wave: $\exp\{i\mathbf{q} \cdot \mathbf{x}\}$. A measurement of all energies in (e, e′ N) determines the binding energy of the final hole state ε_b. A measurement of all momenta measures the Fourier transform of the hole-state wave function $\tilde{\phi}_b(\boldsymbol{\kappa} - \mathbf{q})$; by basic quantum mechanics, this is the amplitude of the momentum distribution in the state ϕ_b.

In addition, coincidence capability implies that multiple scattering experiments can be performed. The polarization $\mathbf{P}_X$ of the produced particle X, for example, can be measured through a second scattering. Polarization transfer experiments $(\vec{e}, e'\,\vec{X})$ that provide precision measurements of the charge form factor of the nucleon [Ar81] now form an important part of electromagnetic nuclear physics.

Furthermore, strangeness is conserved in the strong and electromagnetic interactions; one then only has *associated production* of strange particles, for example through the reaction $^{1}H(e, e'\,K^{+})\Lambda$.[1] With high enough incident electron energy, the reaction $(e, e'\,K^{+})$ can be accessed. This reaction produces a tagged hypernucleus. By varying the momentum transferred to the nucleus, the Fourier transform of the wave function of the deposited hyperon can be determined.

Moreover, with *multiple* coincidence experiments such as $(e, e'\,2N)$ and extreme kinematics, one can investigate the short-range behavior of two nucleons in the nuclear medium.

At the quark level, when quarks are struck in an electroweak interaction, it is not the quarks that emerge from the nucleus, rather it is a hadron. The *hadronization* of quarks is studied in the coincidence reaction $(e, e'\,X)$.

---

[1] The reaction notation $A(b, c\,d\ldots)E$ used in this book is a very convenient one. The first and last symbols denote the initial and final target states and the symbols in parenthesis indicate the incident and final detected particles; in the generic case, the last and first symbols may be suppressed. We denote elastic electron scattering by (e, e), inelastic scattering by (e, e'), and coincidence reactions by $(e, e'\,X)$.

7

Units and conventions

To define the units and conventions used in this book, and to set the stage for the subsequent analysis, we conclude this introduction by writing Maxwell's equations for the electromagnetic field in vacuum with sources. With the use of Heaviside–Lorentz (rationalized c.g.s.) units these equations are[1]

$$
\begin{aligned}
\nabla \cdot \mathbf{E} &= \rho \\
\nabla \cdot \mathbf{H} &= 0 \\
\nabla \times \mathbf{H} - \frac{1}{c}\frac{\partial \mathbf{E}}{\partial t} &= \mathbf{j} \\
\nabla \times \mathbf{E} + \frac{1}{c}\frac{\partial \mathbf{H}}{\partial t} &= 0
\end{aligned}
\tag{7.1}
$$

Here ρ and $\mathbf{j}$ are the local charge and current density; the former is measured in e.s.u. and the latter in e.m.u. where 1 e.m.u = 1 e.s.u./c. The Lorentz force equation and fine structure constant are given respectively by

$$
\mathbf{F} = e\left(\mathbf{E} + \frac{\mathbf{v}}{c} \times \mathbf{H}\right)
$$

$$
\frac{e^2}{4\pi\hbar c} = \alpha = \frac{1}{137.04}
\tag{7.2}
$$

Introduce the antisymmetric electromagnetic field tensor

$$
F_{\mu\nu} = \begin{pmatrix}
0 & H_3 & -H_2 & -iE_1 \\
-H_3 & 0 & H_1 & -iE_2 \\
H_2 & -H_1 & 0 & -iE_3 \\
iE_1 & iE_2 & iE_3 & 0
\end{pmatrix}
\tag{7.3}
$$

---

[1] In this case the magnetic field is $\mathbf{H} \equiv \mathbf{B}$.

Straightforward algebra then shows that Maxwell's equations can be written in covariant form as

$$\frac{\partial}{\partial x_\nu} F_{\mu\nu} = j_\mu$$

$$\varepsilon_{\mu\nu\rho\sigma} \frac{\partial}{\partial x_\sigma} F_{\nu\rho} = 0 \qquad (7.4)$$

Here $\varepsilon_{\mu\nu\rho\sigma}$ is the completely antisymmetric tensor and repeated Greek indices are summed from one to four. Also

$$x_\mu = (\mathbf{x}, ict)$$

$$j_\mu = (\mathbf{j}, i\rho) \qquad (7.5)$$

The second set of Maxwell's Equations can be satisfied identically with the introduction of a vector potential

$$F_{\mu\nu} = \frac{\partial}{\partial x_\mu} A_\nu - \frac{\partial}{\partial x_\nu} A_\mu$$

$$A_\mu = (\mathbf{A}, i\Phi) \qquad (7.6)$$

Comparison with Eq. (7.3) then allows the identification

$$\mathbf{H} = \nabla \times \mathbf{A}$$

$$\mathbf{E} = -\nabla\Phi - \frac{1}{c}\frac{\partial}{\partial t}\mathbf{A} \qquad (7.7)$$

While k and q are used interchangeably in the following for the four-momentum transfer in inclusive electron scattering (e, e'), with a direction defined in context,[2] when the coincidence process (e, e' X) is discussed, k is reserved for the four-momentum transfer of the electron to the target and q for the four-momentum of the produced particle X.

---

[2] Unfortunately, this is common usage. The four-momentum transfer will be denoted $(\kappa, i\omega/c)$, with a direction again defined in context.

Part 2
General analysis

8

Electromagnetic interactions

In this section we discuss the interaction of nuclei, nucleons, or any finite quantum mechanical system with the electromagnetic field. Much of what we know about nuclei and nucleons comes from such interactions. We start with the general multipole analysis of the interaction of a nucleus with the quantized radiation field [Bl52, Sc54, de66, Wa95]. In the following $e_p = |e|$ is the proton charge.

The starting point in this analysis is the total hamiltonian for the nuclear system, the free photon field, and the electromagnetic interaction

$$
\begin{aligned}
H_{\text{total}} \quad = \quad & H_{\text{nuclear}} + \sum_{\mathbf{k}} \sum_{\rho=1,2} \hbar \omega_k a_{\mathbf{k}\rho}^\dagger a_{\mathbf{k}\rho} \\
& -\frac{e_p}{c} \int \mathbf{J}_N(\mathbf{x}) \cdot \mathbf{A}(\mathbf{x}) \, d^3x + \frac{e_p^2}{8\pi} \int \int \frac{\rho_N(\mathbf{x})\rho_N(\mathbf{x}')}{|\mathbf{x} - \mathbf{x}'|} \, d^3x \, d^3x'
\end{aligned} \tag{8.1}
$$

This is the hamiltonian of *quantum electrodynamics* (QED); it is written in the Coulomb gauge. $\mathbf{A}$ is the vector potential for the quantized radiation field, which in the Schrödinger picture takes the form

$$
\mathbf{A}(\mathbf{x}) = \sum_{\mathbf{k}} \sum_{\rho=1,2} \left(\frac{\hbar c^2}{2\omega_k \Omega} \right)^{1/2} [\mathbf{e}_{\mathbf{k}\rho} a_{\mathbf{k}\rho} e^{i\mathbf{k}\cdot\mathbf{x}} + \text{h.c.}] \tag{8.2}
$$

Here $\mathbf{e}_{\mathbf{k}\rho}$ with $\rho = (1, 2)$ represent two unit vectors transverse to $\mathbf{k}$ (see Fig. 8.1). The hermitian conjugate is denoted by h.c. We quantize with periodic boundary conditions (p.b.c.) in a large box of volume Ω, and in the end let $\Omega \to \infty$. With this choice

$$
\frac{1}{\Omega} \int_{\text{Box}} d^3x \, e^{i(\mathbf{k}_1 - \mathbf{k}_2)\cdot\mathbf{x}} = \delta_{\mathbf{k}_1, \mathbf{k}_2} \tag{8.3}
$$

where the expression on the right is a Kronecker delta.

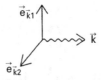

Fig. 8.1.   Transverse unit vectors.

The only assumption made about the target is the existence of local current and charge density operators $\mathbf{J}_N(\mathbf{x})$ and $\rho_N(\mathbf{x})$. These quantities must exist for any true quantum mechanical system. H_{nuclear} could be given in terms of potentials, or it could be for a coupled baryon and meson system, or it could be for a system of quarks and gluons; it does not matter at this point.[1]

It is convenient to henceforth incorporate the explicit factor of $1/c$ in Eq. (8.1) into the definition of the current $\mathbf{J}_N(\mathbf{x})$ itself.

First go from plane polarization to circular polarization with the transformation (cf. Fig. 8.1).

$$\mathbf{e}_{\pm 1} \equiv \mp \frac{1}{\sqrt{2}}(\mathbf{e}_1 \pm i\mathbf{e}_2) \qquad\qquad \mathbf{e}_0 \equiv \mathbf{e}_z \equiv \frac{\mathbf{k}}{|\mathbf{k}|} \tag{8.4}$$

These circular polarization vectors satisfy the relations

$$\mathbf{e}_{\mathbf{k}\lambda}^{\dagger} = (-1)^{\lambda}\mathbf{e}_{\mathbf{k}-\lambda} \qquad\qquad \mathbf{e}_{\lambda}^{\dagger}\cdot\mathbf{e}_{\lambda'} = \delta_{\lambda\lambda'} \tag{8.5}$$

If, at the same time, one defines

$$a_{\mathbf{k}\pm 1} \equiv \mp \frac{1}{\sqrt{2}}(a_{\mathbf{k}1} \mp ia_{\mathbf{k}2}) \tag{8.6}$$

then the transformation is *canonical*

$$[a_{\mathbf{k}\lambda}, a_{\mathbf{k}'\lambda'}^{\dagger}] = \delta_{\mathbf{k}\mathbf{k}'}\delta_{\lambda\lambda'} \tag{8.7}$$

Since $\mathbf{e}_1 a_1 + \mathbf{e}_2 a_2 = \mathbf{e}_{+1} a_{+1} + \mathbf{e}_{-1} a_{-1}$ the vector potential takes the form

$$\mathbf{A}(\mathbf{x}) = \sum_{\mathbf{k}}\sum_{\lambda=\pm 1}\left(\frac{\hbar c^2}{2\omega_k \Omega}\right)^{1/2}[\mathbf{e}_{\mathbf{k}\lambda}a_{\mathbf{k}\lambda}e^{i\mathbf{k}\cdot\mathbf{x}} + \text{h.c.}] \tag{8.8}$$

The index $\lambda = \pm 1$ is the circular polarization, as we shall see, and only $\lambda = \pm 1$ appears in the expansion so $\nabla \cdot \mathbf{A}(\mathbf{x}) = 0$, characterizing the Coulomb gauge.

Now proceed to calculate the transition probability for the nucleus or

---

[1] Although Eq. (8.1) is correct in QCD, some models may have an additional term of $O(e^2\mathbf{A}^2)$ in the hamiltonian; the arguments in this section are unaffected by such a term.

nucleon to make a transition between two states and emit (or absorb) a photon. Work to lowest order in the electric charge e, use the *Golden Rule*, and compute the nuclear matrix element $\langle J_f M_f; \mathbf{k}\lambda | H' | J_i M_i \rangle$ where H' is here the term linear in the vector potential in Eq. (8.1); it is this interaction term that can create (or destroy) a photon. All that will be specified about the nuclear state at this point is that it is an eigenstate of angular momentum. It will be assumed that the target is massive and its position will be taken to define the origin; transition current densities occur over the nuclear volume and hence *all transition current densities will be localized in space*. Since the photon matrix element is $\langle \mathbf{k}\lambda | a^\dagger_{\mathbf{k}'\lambda'} | 0 \rangle = \delta_{\mathbf{k}\mathbf{k}'}\delta_{\lambda\lambda'}$, the required transition matrix element takes the form[2]

$$\langle J_f M_f; \mathbf{k}\lambda | \hat{H}' | J_i M_i \rangle = -e_p \left(\frac{\hbar c^2}{2\omega_k \Omega}\right)^{1/2} \langle J_f M_f | \int e^{-i\mathbf{k}\cdot\mathbf{x}} \mathbf{e}^\dagger_{\mathbf{k}\lambda} \cdot \hat{\mathbf{J}}(\mathbf{x})\, d^3x | J_i M_i \rangle$$

(8.9)

This expression now contains all of the physics of the target. We proceed to make a multipole analysis of it. With the aid of the Wigner–Eckart theorem we will then be able to extract two invaluable types of information:

- The angular momentum selection rules

- The explicit dependence on the orientation of the target as expressed in (M_i, M_f)

The goal of the multipole analysis is to reduce the transition operator to a sum of *irreducible tensor operators* (ITO) to which the Wigner–Eckart theorem applies [Ed74].

We recall the definition of an ITO. It is a set of $2J+1$ operators $\hat{T}(J, M)$ with $-J \leq M \leq J$ that satisfy the following commutation relations with the three components $\hat{J}_i$ of the angular momentum operator

$$[\hat{J}_i, \hat{T}(J, M)] = \sum_{M'} \langle JM' | \hat{J}_i | JM \rangle \, \hat{T}(J, M')$$

(8.10)

The above is the infinitesimal form of the integral definition of an ITO (they are fully equivalent)

$$\hat{R}_{\alpha\beta\gamma} \hat{T}(J, M) \hat{R}^{-1}_{\alpha\beta\gamma} = \sum_{M'} \mathscr{D}^J_{M'M}(\alpha\beta\gamma) \, \hat{T}(J, M')$$

(8.11)

Here $\hat{R}_{\alpha\beta\gamma}$ is the rotation operator, and $\mathscr{D}^J_{M'M}(\alpha\beta\gamma)$ are the rotation matrices, defined by [Ed74]

$$\hat{R}_{\alpha\beta\gamma} \equiv e^{i\alpha\hat{J}_3} e^{i\beta\hat{J}_2} e^{i\gamma\hat{J}_3}$$
$$\mathscr{D}^J_{M'M}(\alpha\beta\gamma) = \langle JM' | e^{i\alpha\hat{J}_3} e^{i\beta\hat{J}_2} e^{i\gamma\hat{J}_3} | JM \rangle$$

(8.12)

We proceed to the multipole analysis.

---

[2] For clarity we now use a notation where a caret over a symbol denotes an operator in the target Hilbert space.

9

Multipole analysis

Start by taking the photon momentum to define the z-axis (Fig. 9.1); the generalization follows below. In this case the plane wave can be expanded as [Fe80]

$$e^{i\mathbf{k}\cdot\mathbf{x}} = \sum_l i^l \sqrt{4\pi(2l+1)} j_l(kx) Y_{l0}(\Omega_x) \tag{9.1}$$

The vector spherical harmonics are defined by the relations [Ed74]

$$\mathscr{Y}_{Jl1}^M \equiv \sum_{m\lambda} \langle lm1\lambda|l1JM\rangle Y_{lm}(\Omega_x)\mathbf{e}_\lambda \tag{9.2}$$

Note this sum goes over all three spherical unit vectors, $\lambda = \pm 1, 0$. The definition in Eq. (9.2) can be inverted with the aid of the orthogonality properties of the Clebsch–Gordan (C–G) coefficients

$$Y_{lm}\mathbf{e}_\lambda = \sum_{JM} \langle lm1\lambda|l1JM\rangle \mathscr{Y}_{Jl1}^M \tag{9.3}$$

The $\mathbf{e}_\lambda$ are now just fixed vectors; they form a complete orthonormal set.

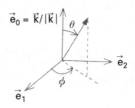

Fig. 9.1. Coordinate system with z-axis defined by photon momentum.

34

Therefore any vector can be expanded in spherical components as

$$\mathbf{v} = \sum_\lambda (\mathbf{v} \cdot \mathbf{e}_\lambda) \mathbf{e}_\lambda^\dagger = \sum_\lambda v_\lambda \mathbf{e}_\lambda^\dagger$$

$$v_{\pm 1} = \mp \frac{1}{\sqrt{2}} (v_x \pm i v_y) \qquad\qquad v_0 = v_z \qquad (9.4)$$

As we shall see, the vector spherical harmonics project an irreducible tensor operator (ITO) of rank J from any vector density operator in the nuclear Hilbert space. A combination of Eqs. (9.1) and (9.3) and use of the properties of the C–G coefficients yields[1]

$$\mathbf{e}_{k\lambda} e^{i\mathbf{k}\cdot\mathbf{x}} = \sum_l \sum_J i^l \sqrt{4\pi(2l+1)} j_l(kx) \langle l01\lambda | l1J\lambda \rangle \mathscr{Y}_{Jl1}^\lambda(\Omega_x) \qquad (9.5)$$

The C–G coefficient limits the sum on l to three terms $l = J, l = J \pm 1$, and these C–G coefficients can be explicitly evaluated to give for $\lambda = \pm 1$ [Ed74]

$$\mathbf{e}_{k\lambda} e^{i\mathbf{k}\cdot\mathbf{x}} = \sum_{J \geq 1} i^J \sqrt{\frac{4\pi(2J+1)}{2}} \left\{ - \lambda j_J(kx) \mathscr{Y}_{JJ1}^\lambda \right.$$

$$\left. -i \left[\sqrt{\frac{J+1}{2J+1}} j_{J-1}(kx) \mathscr{Y}_{J,J-1,1}^\lambda - \sqrt{\frac{J}{2J+1}} j_{J+1}(kx) \mathscr{Y}_{J,J+1,1}^\lambda \right] \right\} \qquad (9.6)$$

From [Ed74] one has

$$\nabla \times j_J(kx) \mathscr{Y}_{JJ1}^\lambda = i \left[\left(\frac{d}{dx} - \frac{J}{x} \right) j_J(kx) \sqrt{\frac{J}{2J+1}} \mathscr{Y}_{J,J+1,1}^\lambda \right.$$

$$\left. + \left(\frac{d}{dx} + \frac{J+1}{x} \right) j_J(kx) \sqrt{\frac{J+1}{2J+1}} \mathscr{Y}_{J,J-1,1}^\lambda \right] \qquad (9.7)$$

The differential operators just raise and lower the indices on the spherical Bessel functions, giving $-k j_{J+1}(kx)$ and $k j_{J-1}(kx)$, respectively. A combination of these results gives for $\lambda = \pm 1$

$$\mathbf{e}_{k\lambda} e^{i\mathbf{k}\cdot\mathbf{x}} = \sum_{J \geq 1} \sqrt{2\pi(2J+1)} \, i^J \left\{ - \lambda j_J(kx) \mathscr{Y}_{JJ1}^\lambda(\Omega_x) \right.$$

$$\left. - \frac{1}{k} \nabla \times [j_J(kx) \mathscr{Y}_{JJ1}^\lambda(\Omega_x)] \right\} \qquad ; \lambda = \pm 1 \qquad (9.8)$$

---

[1] Note this is the amplitude for photon *absorption*.

Note the divergence of both sides of this equation vanishes [Ed74].[2] Now use

$$\mathscr{Y}_{JJ1}^{\lambda\dagger} = -(-1)^{\lambda}\mathscr{Y}_{JJ1}^{-\lambda} \tag{9.9}$$

to arrive at the basic result for photon emission with $\lambda = \pm 1$

$$-e_{\mathrm{p}}\left(\frac{\hbar c^2}{2\omega_k\Omega}\right)^{1/2}\int e^{-i\mathbf{k}\cdot\mathbf{x}}\,\mathbf{e}_{\mathbf{k}\lambda}^{\dagger}\cdot\hat{\mathbf{J}}(\mathbf{x})\,d^3x \tag{9.10}$$

$$= e_{\mathrm{p}}\left(\frac{\hbar c^2}{2\omega_k\Omega}\right)^{1/2}\sum_{J\geq 1}(-i)^J\sqrt{2\pi(2J+1)}\,[\hat{T}_{J,-\lambda}^{\mathrm{el}}(k) + \lambda\hat{T}_{J,-\lambda}^{\mathrm{mag}}(k)]$$

The *transverse electric* and *magnetic multipole operators* are defined by

$$\hat{T}_{JM}^{\mathrm{el}}(k) \equiv \frac{1}{k}\int d^3x \left[\nabla \times j_J(kx)\mathscr{Y}_{JJ1}^{M}(\Omega_x)\right]\cdot\hat{\mathbf{J}}(\mathbf{x})$$

$$\hat{T}_{JM}^{\mathrm{mag}}(k) \equiv \int d^3x \left[j_J(kx)\mathscr{Y}_{JJ1}^{M}(\Omega_x)\right]\cdot\hat{\mathbf{J}}(\mathbf{x}) \tag{9.11}$$

This important result merits several observations.

In a nucleus both the convection current density arising from the motion of charged particles (e.g. protons) and the intrinsic magnetization density coming from the intrinsic magnetic moments of the nucleons contribute to the electromagnetic interaction. The appropriate interaction hamiltonian should actually be written as

$$H' = -e_{\mathrm{p}}\int \hat{\mathbf{J}}_c(\mathbf{x})\cdot\mathbf{A}(\mathbf{x})\,d^3x - e_{\mathrm{p}}\int \hat{\boldsymbol{\mu}}(\mathbf{x})\cdot[\nabla \times \mathbf{A}(\mathbf{x})]\,d^3x$$

$$= -e_{\mathrm{p}}\int \left[\hat{\mathbf{J}}_c(\mathbf{x}) + \nabla \times \hat{\boldsymbol{\mu}}(\mathbf{x})\right]\cdot\mathbf{A}(\mathbf{x})\,d^3x \tag{9.12}$$

To obtain the second line, a vector identity has been employed

$$\nabla\cdot(\mathbf{a} \times \mathbf{b}) = \mathbf{b}\cdot(\nabla \times \mathbf{a}) - \mathbf{a}\cdot(\nabla \times \mathbf{b}) \tag{9.13}$$

The total divergence has been converted to a surface integral far away from the nucleus using Gauss' theorem

$$\int_V \nabla \cdot \mathbf{v}\,d^3x = \int_S \mathbf{v}\cdot d\mathbf{S} \tag{9.14}$$

Finally, the integral over the far-away surface can be discarded for a *localized source*. A second application of this procedure yields the relation

$$\int d^3x \left[\nabla \times j_J(kx)\mathscr{Y}_{JJ1}^{M}\right]\cdot\nabla \times \hat{\boldsymbol{\mu}}(\mathbf{x}) \tag{9.15}$$

$$= \int d^3x\,\hat{\boldsymbol{\mu}}(\mathbf{x})\cdot\nabla \times [\nabla \times j_J(kx)\mathscr{Y}_{JJ1}^{M}] = k^2\int d^3x\,\hat{\boldsymbol{\mu}}(\mathbf{x})\cdot[j_J(kx)\mathscr{Y}_{JJ1}^{M}]$$

---

[2] The relation to be used is $\vec{\nabla}\cdot[j_J(kx)\vec{\mathscr{Y}}_{JJ1}^{M}] = 0$.

In arriving at the second equality the relation $\nabla \times (\nabla \times \mathbf{v}) = \nabla(\nabla \cdot \mathbf{v}) - \nabla^2 \mathbf{v}$ has been employed; the term $\nabla \cdot \mathbf{v}$ vanishes here, and in this application the remaining term satisfies the Helmholtz equation $(\nabla^2 + k^2)\mathbf{v} = 0$, as the reader can readily verify. Thus the multipole operators can be rewritten to explicitly exhibit the individual contributions of the convection current and the intrinsic magnetization densities

$$
\begin{aligned}
\hat{T}^{\text{el}}_{JM}(k) &= \frac{1}{k} \int d^3x \left\{ [\nabla \times j_J(kx)\mathscr{Y}^M_{JJ1}] \cdot \hat{\mathbf{J}}_c(\mathbf{x}) + k^2 j_J(kx)\mathscr{Y}^M_{JJ1} \cdot \hat{\boldsymbol{\mu}}(\mathbf{x}) \right\} \\
\hat{T}^{\text{mag}}_{JM}(k) &= \int d^3x \left\{ j_J(kx)\mathscr{Y}^M_{JJ1} \cdot \hat{\mathbf{J}}_c(\mathbf{x}) + [\nabla \times j_J(kx)\mathscr{Y}^M_{JJ1}] \cdot \hat{\boldsymbol{\mu}}(\mathbf{x}) \right\}
\end{aligned}
\tag{9.16}
$$

The $\hat{T}_{JM}$ are now *irreducible tensor operators of rank J in the nuclear Hilbert space*. This can be proven in general by utilizing the properties of the vector density operator $\hat{\mathbf{J}}(\mathbf{x})$ under rotations. It is easier to prove this property explicitly in any particular application. For example, consider the case where the nucleus is pictured as a collection of non-relativistic nucleons, and the intrinsic magnetization density at the point $\mathbf{x}$ is constructed in first quantization by summing over the contribution of the individual nucleons

$$
e_p \hat{\boldsymbol{\mu}}(\mathbf{x}) = \mu_N \sum_{i=1}^{A} \lambda_i \boldsymbol{\sigma}(i) \delta^{(3)}(\mathbf{x} - \mathbf{x}_i)
\tag{9.17}
$$

Here λ_i is the intrinsic magnetic moment of the ith nucleon in nuclear magnetons (see below).[3] The contribution to $\hat{T}^{\text{el}}_{JM}$, for example, then takes the form

$$
e_p \int j_J(kx)\mathscr{Y}^M_{JJ1} \cdot \hat{\boldsymbol{\mu}}(\mathbf{x}) \, d^3x =
$$

$$
\mu_N \sum_{i=1}^{A} \lambda_i j_J(kx_i) \sum_{mq} \langle Jm1q | J1JM \rangle Y_{Jm}(\Omega_i) \sigma_{1q}(i)
\tag{9.18}
$$

Here the definition of the vector spherical harmonics in Eq. (9.2) has been introduced. Each term in this sum is now recognized, with the aid of [Ed74], to be a tensor product of rank J formed from two ITO of rank J and 1, respectively.[4] Thus $\hat{T}^{\text{el}}_{JM}$ is evidently an ITO of rank J under commutation with the total angular momentum operator, which in this

---

[3] One could be dealing with a density operator in second quantization, or expressed in collective coordinates, etc; to test for an ITO, one first constructs the appropriate total angular momentum operator $\hat{\mathbf{J}}$, and then examines the commutation relations (see [Ed74]).

[4] Any spherically symmetric factor does not affect the behavior under rotations.

case takes the form

$$\hat{\mathbf{J}} = \sum_{i=1}^{A} \mathbf{J}(i) = \sum_{i=1}^{A} [\mathbf{L}(i) + \mathbf{S}(i)] \qquad ; \text{ angular momentum} \qquad (9.19)$$

As another example, the convection current in this same picture of the nucleus is

$$\hat{\mathbf{J}}_c(\mathbf{x}) = \sum_{i=1}^{Z} \frac{1}{m_p c} \{\delta^{(3)}(\mathbf{x} - \mathbf{x}_i), \mathbf{p}(i)\}_{\text{sym}} \doteq \sum_{i=1}^{Z} \delta^{(3)}(\mathbf{x} - \mathbf{x}_i) \frac{\mathbf{p}(i)}{m_p c} \qquad (9.20)$$

The need for symmetrization[5] arises from the fact that $\mathbf{p}(i)$ and $\mathbf{x}_i$ do not commute; the current density arising from the matrix element of this expression takes the appropriate quantum mechanical form $(\hbar/2im_p c)[\psi^* \nabla \psi - (\nabla \psi)^* \psi]$. The last equality in Eq. (9.20) follows since one of the symmetrized terms can be partially integrated in the required matrix elements of the current, using the hermiticity of $\mathbf{p}(i)$ and the observation that $\nabla \cdot \mathbf{A} = 0$ in the Coulomb gauge. Multipoles constructed from the convection current density in Eq. (9.20) are now shown to be ITO by arguments similar to the above.

The *parity* of the multipole operators is [Bl52]

$$\hat{\Pi} \, \hat{T}_{JM}^{\text{el}} \, \hat{\Pi}^{-1} \;=\; (-1)^J \, \hat{T}_{JM}^{\text{el}}$$

$$\hat{\Pi} \, \hat{T}_{JM}^{\text{mag}} \, \hat{\Pi}^{-1} \;=\; (-1)^{J+1} \hat{T}_{JM}^{\text{mag}} \qquad (9.21)$$

Again the general proof follows from the behavior of the current density $\hat{\mathbf{J}}(\mathbf{x})$ as a polar vector under spatial reflections. It is easy to see this behavior in any particular application. For example, it follows from Eqs. (9.17) and (9.20) if one uses the properties of the individual quantities under spatial reflection: $\sigma_{1q} \to \sigma_{1q}$; $p_{1q} \to -p_{1q}$; and $Y_{lm}(-\mathbf{x}/|\mathbf{x}|) = (-1)^l Y_{lm}(\mathbf{x}/|\mathbf{x}|)$. Parity selection rules on the matrix elements of the transverse multipole operators now follow directly.

There is no $J = 0$ term in the sum in Eq. (9.10). This arises from the fact that the vector potential is transverse, and hence there are only transverse unit vectors, or equivalently unit helicities $\lambda = \pm 1$, arising in its expansion into normal modes [see Eqs. (8.2) and (8.4)]. This has the consequence, for example, that there can be no $J = 0 \to J = 0$ real photon transitions in nuclei.

The *Wigner–Eckart theorem* [Ed74] can now be employed to exhibit the angular momentum selection rules and M-dependence of the matrix

---

[5] $\{A, B\}_{\text{sym}} \equiv (AB + BA)/2$.

element of an ITO between eigenstates of angular momentum

$$\langle J_f M_f | \hat{T}_{JM} | J_i M_i \rangle = \frac{(-1)^{J_i - M_i}}{(2J+1)^{1/2}} \langle J_f M_f J_i - M_i | J_f J_i J M \rangle \langle J_f \| \hat{T}_J \| J_i \rangle \quad (9.22)$$

The Clebsch–Gordan (C–G) coefficients provide all the relevant information. They contain the entire M-dependence, and they vanish unless the angular momentum quantum numbers satisfy the triangle inequality, e.g. $|J_i - J_f| \le J \le J_i + J_f$. We adopt the convention that this selection rule is built into the reduced matrix elements themselves, and that they are defined to be zero unless the triangle inequality is satisfied.

Note that the required matrix elements of Eq. (9.10) imply $M_f = M_i - \lambda$. This means that the photon carries away the angular momentum λ along the z-axis, which is the direction of emission of the photon in the preceding analysis (Fig. 9.1); thus the *helicity* of the photon (its angular momentum along $\mathbf{k}$) is $\lambda = \pm 1$.

If the target is unpolarized and unobserved, one can simply pick a convenient z-axis along which to quantize, and the photon momentum $\mathbf{k}$ provides such a choice. In that case, the average over initial target orientations $\overline{\sum_i} = (2J_i + 1)^{-1} \sum_{M_i}$ and sum over final target orientations $\sum_f = \sum_{M_f}$ can be immediately evaluated with the aid of the Wigner–Eckart theorem and the orthonormality properties of the C–G coefficients to give

$$\frac{1}{2J_i + 1} \sum_{M_i} \sum_{M_f} \left| \sum_J (-i)^J \sqrt{2J+1} \langle J_f M_f | \hat{T}_{JM} | J_i M_i \rangle \right|^2$$

$$= \frac{1}{2J_i + 1} \sum_J |\langle J_f \| \hat{T}_J \| J_i \rangle|^2 \quad (9.23)$$

One then proceeds directly to the transition rate given below in Eq. (9.41).

It is useful for the subsequent discussion of angular correlations to first digress and consider the more general situation where the photon is *emitted in an arbitrary direction* relative to the coordinate axes picked to describe the quantization of the nuclear system. The situation is illustrated in Fig. 9.2. The unit vectors describing the photon are assumed to have Euler angles $\{\alpha, \beta, \gamma\}$ with respect to the nuclear quantization axes. The difficulty in achieving this configuration is that the photon axes here are the axes that are assumed to be *fixed in space*, having been determined, for example, by the detection of the photon, and the rotations are to be carried out with respect to these axes.

Now one knows how to carry out a rotation of the nuclear state vector relative to a fixed set of axes. For example, consider the rotation operator that rotates a physical state vector through the angle β relative

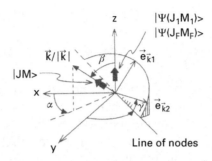

Fig. 9.2.   Photon emitted in arbitrary direction relative to quantization axes for nuclear system. Note $\{\alpha, \beta, \gamma\}$ are Euler angles.

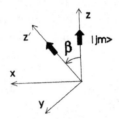

Fig. 9.3.   Rotate physical state vector by angle β about y-axis.

to a laboratory-fixed y-axis as indicated in Fig. 9.3. It follows entirely from the defining commutation relations for the angular momentum, that the operator which accomplishes this task is $\hat{R}_{-\beta} \equiv e^{-i\beta \hat{J}_y}$. This is demonstrated as follows. Introduce a new unit vector along the z' direction and dot this into the angular momentum operator

$$\mathbf{e}_{z'} = \mathbf{e}_z \cos\beta + \mathbf{e}_x \sin\beta$$
$$\mathbf{e}_{z'} \cdot \hat{\mathbf{J}} = \hat{J}_z \cos\beta + \hat{J}_x \sin\beta \tag{9.24}$$

Now make use of the following identity and basic commutation relations

$$e^{-i\beta \hat{J}_y} \hat{J}_z e^{i\beta \hat{J}_y} = \hat{J}_z + (-i\beta)[\hat{J}_y, \hat{J}_z] + \frac{(-i\beta)^2}{2!}[\hat{J}_y, [\hat{J}_y, \hat{J}_z]]$$
$$+ \frac{(-i\beta)^3}{3!}[\hat{J}_y, [\hat{J}_y, [\hat{J}_y, \hat{J}_z]]] + \cdots$$
$$[\hat{J}_i, \hat{J}_j] = i\varepsilon_{ijk}\hat{J}_k \tag{9.25}$$

One finds

$$e^{-i\beta \hat{J}_y} \hat{J}_z e^{i\beta \hat{J}_y} = \hat{J}_z \left(1 - \frac{\beta^2}{2!} + \cdots\right) + \hat{J}_x \left(\beta - \frac{\beta^3}{3!} + \cdots\right)$$
$$= \hat{J}_z \cos\beta + \hat{J}_x \sin\beta$$
$$= \mathbf{e}_{z'} \cdot \hat{\mathbf{J}} \tag{9.26}$$

Thus, from general principles,

$$(\mathbf{e}_{z'} \cdot \hat{\mathbf{J}})e^{-i\beta\hat{J}_y} = e^{-i\beta\hat{J}_y}\hat{J}_z \tag{9.27}$$

Now apply this relation to the state vector $|jm\rangle$ representing a particle with angular momentum j and z-component m, and let $\hat{J}_z$ act on this eigenstate.

$$(\mathbf{e}_{z'} \cdot \hat{\mathbf{J}})[e^{-i\beta\hat{J}_y}|jm\rangle] = m[e^{-i\beta\hat{J}_y}|jm\rangle] \tag{9.28}$$

This is the desired result. The quantity $e^{-i\beta\hat{J}_y}|jm\rangle$ is a rotated eigenstate with angular momentum m along the new z'-axis.

The goal now is to rotate the nuclear state vector $|J_iM_i\rangle$ quantized with respect to the photon axes into a nuclear state vector $|\Psi_i(J_iM_i)\rangle$ correctly quantized with respect to the indicated $\{x, y, z\}$ coordinates. A concentrated effort, after staring at Fig. 9.2, will convince the reader that the following rotations, carried out with respect to the laboratory-fixed photon coordinate system in the indicated sequence, will achieve this end

1. $-\alpha$ about $\mathbf{k}/|\mathbf{k}|$
2. $-\beta$ about $\mathbf{e}_{k2}$
3. $-\gamma$ about $\mathbf{k}/|\mathbf{k}|$

The rotation operator that accomplishes this rotation is

$$\hat{R}_{+\gamma, +\beta, +\alpha} = \exp\{i\gamma\hat{J}_3\}\exp\{i\beta\hat{J}_2\}\exp\{i\alpha\hat{J}_3\} \tag{9.29}$$

The $\{2, 3\}$ axes are now the laboratory-fixed $\{\mathbf{e}_{k2}, \mathbf{k}/|\mathbf{k}|\}$ axes. Thus

$$|\Psi_i(J_iM_i)\rangle = \hat{R}_{\gamma,\beta,\alpha}|J_iM_i\rangle = \sum_{M_k} \mathscr{D}^{J_i}_{M_kM_i}(\gamma, \beta, \alpha)|J_iM_k\rangle \tag{9.30}$$

Here the rotation matrices have been introduced that characterize the behavior of the eigenstates of angular momentum under rotation [Ed74]. It is clear from Fig. 9.2 that one can identify the usual polar and azimuthal angles that the photon makes with respect to the nuclear coordinate system according to $\beta \leftrightarrow \theta$ and $\alpha \leftrightarrow \phi$; the angle $\gamma \leftrightarrow -\phi$ of the orientation of the photon polarization vector around the photon momentum is a definition of the overall phase of the state vector, and, as such, merely involves a phase convention; the choice here is that of Jacob and Wick [Ja59]. It will be apparent in the final result that this phase is irrelevant. Equation (9.30) expresses the required nuclear state vector as a linear combination of state vectors quantized along the photon axes. Since now only matrix elements between states quantized along $\mathbf{k}$ are required, *all the previous results can be utilized*. The required photon transition matrix element takes the form

$$\langle\Psi_f(J_fM_f)|\hat{H}_{J,-\lambda}|\Psi_i(J_iM_i)\rangle = \langle J_fM_f|\hat{R}^{-1}_{\gamma,\beta,\alpha}\hat{H}_{J,-\lambda}\hat{R}_{\gamma,\beta,\alpha}|J_iM_i\rangle \tag{9.31}$$

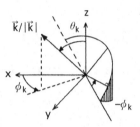

Fig. 9.4. Configuration for transition matrix element describing photon emission and nuclear process $J_i M_i \rightarrow J_f M_f$ with nuclear quantization axis along the z-axis.

Here λ is the photon helicity, and $\hat{H}_{J,-\lambda}$ indicates one of the contributions to the operator in Eq. (9.10). Evidently

$$\hat{R}^{-1}_{\gamma,\beta,\alpha} = \hat{R}_{-\alpha,-\beta,-\gamma} \qquad (9.32)$$

The definition of an ITO can now be used to simplify the calculation [Ed74]

$$\hat{R}_{-\alpha,-\beta,-\gamma} \hat{H}_{J,-\lambda} \hat{R}^{-1}_{-\alpha,-\beta,-\gamma} = \sum_{M'} \mathscr{D}^J_{M',-\lambda}(-\alpha,-\beta,-\gamma) \hat{H}_{JM'} \qquad (9.33)$$

The previous identification of the angles, and a combination of these results, permits one to write the transition matrix element describing the nuclear process $J_i M_i \rightarrow J_f M_f$ with the nuclear quantization axis along z and emission of a photon with helicity λ (Fig. 9.4) as

$$\langle \Psi_f(J_f M_f) | \hat{H}'(\mathbf{k}\lambda) | \Psi_i(J_i M_i) \rangle = \langle J_f M_f | \hat{H}^{\text{em}}_1(\mathbf{k}\lambda) | J_i M_i \rangle \qquad (9.34)$$

where the appropriate transition operator is given by

$$\hat{H}^{\text{em}}_1(\mathbf{k}\lambda) = e_{\text{p}} \left(\frac{\hbar c^2}{2\omega_k \Omega} \right)^{1/2} \sum_{JM} (-i)^J \sqrt{2\pi(2J+1)}$$
$$\times [\hat{T}^{\text{el}}_{JM}(k) + \lambda \hat{T}^{\text{mag}}_{JM}(k)] \, \mathscr{D}^J_{M,-\lambda}(-\phi_k,-\theta_k,\phi_k) \qquad (9.35)$$

The Wigner–Eckart theorem in Eq. (9.22) now permits one to extract all the angular momentum selection rules and M-dependence of the matrix element in Eq. (9.34). All M's now refer to a common set of coordinate axes.[6]

---

[6] These axes were originally the photon axes with the z-axis along $\mathbf{k}$, but they can now just as well be the nuclear $\{x,y,z\}$ axes in Fig. 9.4; the equivalence of these two interpretations is readily demonstrated by taking out the M-dependence in a C–G coefficient with the aid of the Wigner–Eckart theorem — it is the same in both cases. The two interpretations differ only by an *overall* rotation (with $\hat{R}^{-1}\hat{R}$ inserted everywhere), which leaves the physics unchanged.

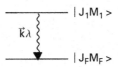

Fig. 9.5.   Nuclear transition with real photon emission.

The final $\mathscr{D}^J_{M,-\lambda}$ in Eq. (9.35) plays the role of a "photon wave function," since the square of this quantity gives the intensity distribution in (θ_k, ϕ_k) of electromagnetic radiation carrying off $\{J, -M, \lambda\}$ from the target.

We proceed to calculate the *transition rate* for the process indicated in Fig. 9.5. The total transition rate for an unoriented nucleus is given by the Golden Rule

$$\omega = \frac{2\pi}{\hbar} \sum_f \overline{\sum_i} |\langle J_f M_f; \mathbf{k}\lambda|H'|J_i M_i\rangle|^2 \delta(E_f + \omega_k - E_i) \tag{9.36}$$

The appropriate sum over final states is given by

$$\sum_f = \frac{\Omega}{(2\pi)^3} \sum_\lambda \sum_{M_f} \int d^3k \tag{9.37}$$

The $\int dk$ allows one to integrate over the energy-conserving delta function $\int dk\, \delta(E_f + \omega_k - E_i) = 1/\hbar c$. The integral over final solid angles of the photon $\int d\Omega_k$ can be performed with the aid of the orthogonality properties of the rotation matrices [Ed74]

$$\int_0^\pi \sin\theta d\theta \int_0^{2\pi} d\phi\, \mathscr{D}^J_{M,-\lambda}(-\phi, -\theta, \phi)^* \mathscr{D}^{J'}_{M',-\lambda}(-\phi, -\theta, \phi)$$

$$= \frac{4\pi}{2J+1} \delta_{JJ'} \delta_{MM'} \tag{9.38}$$

Note that since λ is the same in both functions, the dependence on the last ϕ (which was the phase convention adopted for the third Euler angle $-\gamma$ in Fig. 9.2) drops out of this expression, as advertised.

The average over initial nuclear states is performed according to $\overline{\sum_i} = (2J_i + 1)^{-1} \sum_{M_i}$. The use of the Wigner–Eckart theorem in Eq. (9.22) and the orthonormality of the C–G coefficients then permits one to perform the required sums over M_f and M_i

$$\sum_{M_f} \sum_{M_i} |\langle J_f M_f J_i - M_i|J_f J_i JM\rangle|^2 = 1 \tag{9.39}$$

The final sum on M gives $\sum_M = 2J + 1$.

Since the matrix element of one or the other multipoles must vanish by conservation of parity, assumed to hold for the strong and electromagnetic interactions, it follows that

$$|\langle J_f||\hat{T}_J^{\mathrm{el}} + \lambda\hat{T}_J^{\mathrm{mag}}||J_i\rangle|^2 = |\langle J_f||\hat{T}_J^{\mathrm{el}}||J_i\rangle|^2 + |\langle J_f||\hat{T}_J^{\mathrm{mag}}||J_i\rangle|^2 \quad (9.40)$$

This expression is now independent of λ, and the sum over final photon polarizations gives $\sum_\lambda = 2$.

A combination of these results yields the total photon transition rate for the process illustrated in Fig. 9.5

$$\omega_{fi} = 8\pi\alpha kc \,\frac{1}{2J_i+1}\sum_{J\geq 1}\left\{|\langle J_f||\hat{T}_J^{\mathrm{el}}(k)||J_i\rangle|^2 + |\langle J_f||\hat{T}_J^{\mathrm{mag}}(k)||J_i\rangle|^2\right\} \quad (9.41)$$

The multipole operators are now dimensionless. Equation (9.41) is a very general result. It holds for any localized quantum mechanical system. All that has been assumed about the target is that there is a local electromagnetic current operator. For most nuclear transitions of interest involving real photons, the wavelength is large compared to the size of the nucleus. It is thus important to consider the *long-wavelength reduction* of the multipole operators. This informative analysis is somewhat technical, and in order to not break the thread of the present development, we relegate the details to appendix A.[7]

---

[7] We leave it as an exercise for the reader to demonstrate that the inclusion of target recoil in the density of final states leads to an additional factor of r on the right side of Eq. (9.41) where $r^{-1} = 1 + \hbar k/M_T c$.

10

Dirac equation

Electrons are light, with a rest mass of

$$m_e c^2 = 0.5110 \, \text{MeV} \tag{10.1}$$

For the energies of interest here, electrons must be treated relativistically. Fortunately, for leptons, one knows how to do this with the Dirac equation [Bj65, Sc68]

$$(c\,\boldsymbol{\alpha} \cdot \mathbf{p} + \beta m_0 c^2)\psi = i\hbar \frac{\partial \psi}{\partial t}$$

$$\mathbf{p} = \frac{\hbar}{i}\nabla \tag{10.2}$$

Here ψ is a 4-component column vector and $\boldsymbol{\alpha}$ and β are 4×4 hermitian matrices satisfying the relations

$$\begin{aligned}
\beta\alpha_k + \alpha_k\beta &= 0 &&; k = 1, 2, 3 \\
\alpha_k\alpha_l + \alpha_l\alpha_k &= 2\delta_{kl} \\
\beta^2 &= 1
\end{aligned} \tag{10.3}$$

A specific (*standard*) representation of the Dirac matrices is given in 2×2 form by

$$\boldsymbol{\alpha} = \begin{pmatrix} 0 & \boldsymbol{\sigma} \\ \boldsymbol{\sigma} & 0 \end{pmatrix} \qquad \beta = \begin{pmatrix} 1 & 0 \\ 0 & -1 \end{pmatrix} \tag{10.4}$$

Introduce

$$\begin{aligned}
\gamma &\equiv i\boldsymbol{\alpha}\beta \\
\gamma_4 &\equiv \beta
\end{aligned} \tag{10.5}$$

45

It follows that these new matrices are also hermitian and satisfy the following algebra

$$\gamma_\mu\gamma_\nu + \gamma_\nu\gamma_\mu = 2\delta_{\mu\nu}$$
$$\gamma_\mu^\dagger = \gamma_\mu \qquad ; \mu = 1,\dots,4 \qquad (10.6)$$

In the standard representation, the gamma matrices are given by

$$\gamma = \begin{pmatrix} 0 & -i\boldsymbol{\sigma} \\ i\boldsymbol{\sigma} & 0 \end{pmatrix} \qquad \gamma_4 = \begin{pmatrix} 1 & 0 \\ 0 & -1 \end{pmatrix} \qquad (10.7)$$

Multiplication on the left by γ_4 and division by $\hbar c$ leads to the covariant form of the Dirac equation

$$\left(\gamma_\mu \frac{\partial}{\partial x_\mu} + \frac{m_0 c}{\hbar} \right) \psi = 0$$
$$\gamma_\mu = (\gamma, \gamma_4)$$
$$x_\mu = (\mathbf{x}, ict) \qquad (10.8)$$

Repeated Greek indices are summed from 1 to 4.

To include an *electromagnetic field* one makes the gauge invariant replacement $p_\mu \to p_\mu - (e/c)A_\mu$ or

$$\frac{\partial}{\partial x_\mu} \rightarrow \frac{\partial}{\partial x_\mu} - \frac{ie}{\hbar c}A_\mu$$
$$A_\mu = (\mathbf{A}, i\Phi) \qquad (10.9)$$

This yields the Dirac equation

$$\left[\gamma_\mu \left(\frac{\partial}{\partial x_\mu} - \frac{ie}{\hbar c}A_\mu \right) + \frac{m_0 c}{\hbar} \right] \psi = 0 \qquad (10.10)$$

The equivalent Dirac hamiltonian is obtained by working backwards

$$H = c\boldsymbol{\alpha} \cdot \left(\mathbf{p} - \frac{e}{c}\mathbf{A} \right) + \beta m_0 c^2 + e\Phi$$
$$= H_0 + H_1$$
$$H_1 = -e\boldsymbol{\alpha} \cdot \mathbf{A} + e\Phi \qquad (10.11)$$

Here $e = -|e| = -e_p$ is the charge on the electron.

The Dirac equation for the adjoint field is obtained from Eq. (10.10) by taking the adjoint and multiplying on the right with γ_4

$$\bar{\psi} \left[\gamma_\mu \left(\frac{\overleftarrow{\partial}}{\partial x_\mu} + \frac{ie}{\hbar c}A_\mu \right) - \frac{m_0 c}{\hbar} \right] = 0$$
$$\bar{\psi}(x) \equiv \psi^\dagger(x)\gamma_4 \qquad (10.12)$$

The Dirac electromagnetic current is given by

$$ej_\mu = e\left(\frac{1}{c}\psi^\dagger \boldsymbol{\alpha} c\,\psi,\ i\psi^\dagger\psi\right)$$
$$= ie\,\bar{\psi}(x)\gamma_\mu\psi(x) \tag{10.13}$$

It then follows by direct differentiation and use of the equations of motion that the Dirac current is conserved

$$\frac{\partial j_\mu}{\partial x_\mu} = 0 \tag{10.14}$$

Note that this relation holds in the presence of the electromagnetic field, as it must.

One obtains stationary state, plane wave solutions to the free Dirac equation upon substitution of the form

$$\psi = e^{-iEt/\hbar}e^{i\mathbf{p}\cdot\mathbf{x}/\hbar}\,u(\mathbf{p}) \tag{10.15}$$

The resulting equations only have solutions for eigenvalues of the energy. We denote these eigenvalues and the corresponding eigenfunctions by

$$\begin{aligned}
E &= +E_p &&; \text{ solution } u(\mathbf{p})\\
E &= -E_p &&; \text{ solution } v(\mathbf{p})\\
E_p &\equiv \sqrt{\mathbf{p}^2c^2 + m_0^2c^4} && \tag{10.16}
\end{aligned}$$

That it yield the correct relativistic energy–momentum relation is one of the requirements used to derive the Dirac equation. A little algebra shows that the four eigenfunctions (u_1, u_2, v_1, v_2) can be exhibited explicitly as the columns of the following modal matrix, again expressed in 2×2 form,

$$\mathcal{M} = \left(\frac{E_p + m_0c^2}{2E_p}\right)^{1/2}\begin{bmatrix} 1 & -\dfrac{c\boldsymbol{\sigma}\cdot\mathbf{p}}{E_p + m_0c^2} \\[3mm] \dfrac{c\boldsymbol{\sigma}\cdot\mathbf{p}}{E_p + m_0c^2} & 1 \end{bmatrix} \tag{10.17}$$

They satisfy the orthonormality conditions

$$\begin{aligned}
u_i^\dagger u_j &= v_i^\dagger v_j = \delta_{ij}\\
u_i^\dagger v_j &= v_i^\dagger u_j = 0 \tag{10.18}
\end{aligned}$$

Evidently, from the Dirac equation, these solutions satisfy

$$\begin{aligned}
(i\gamma_\mu p_\mu + m_0c)u(\mathbf{p}) &= 0\\
(i\gamma_\mu p_\mu - m_0c)v(-\mathbf{p}) &= 0\\
p_\mu &\equiv (\mathbf{p}, iE_p/c) \tag{10.19}
\end{aligned}$$

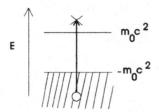

Fig. 10.1. Promotion of particle from a negative energy to a positive energy state in Dirac's hole theory.

Now **p** is the momentum eigenvalue. Note that the second equation is written for $v(-\mathbf{p})$. This solution can be interpreted with the aid of Dirac's *hole theory*.

A heuristic understanding of the role of the negative energy states was given by Dirac. Since particles in the positive energy states could just keep cascading down endlessly, he invoked the Pauli Exclusion Principle and assumed that in the vacuum the negative energy states are all filled. One always measures quantities with respect to the vacuum and the constant contribution of the filled states has no consequence for this theory.

This picture does have implications. A particle in one of the filled negative energy states can be promoted by some mechanism to one of the positive energy states as illustrated in Fig. 10.1. Since if one fills the negative energy state one recovers the vacuum, the hole (absence of a particle) must have the opposite properties of the particle. Dirac called these *antiparticles*. The antiparticle of the electron is the positron. If $v(-\mathbf{p}, \lambda)$ is the negative energy solution of a particle with charge $e = -|e|$, momentum $-\mathbf{p}$ and helicity λ with respect to $-\mathbf{p}$, then it represents a positron with charge $+|e|$, momentum $+\mathbf{p}$ and helicity λ with respect to $+\mathbf{p}$. (Since the spin also reverses, the helicity, or component of spin along the momentum, is unchanged.) Another immediate consequence of Dirac hole theory is that the vacuum becomes a dynamical quantity; it is polarizable for example, and relativistic quantum mechanics immediately confronts one with the relativistic quantum many-body problem.

The solutions in Eq. (10.17) can be combined to yield the projection operators

$$\sum_{\text{spins, } E>0} u(\mathbf{p}, s)_\alpha \bar{u}(\mathbf{p}, s)_\beta = \left(\frac{m_0 c^2 - i\gamma_\mu p_\mu c}{2E_p} \right)_{\alpha\beta}$$

$$\sum_{\text{spins, } E<0} v(-\mathbf{p}, s)_\alpha \bar{v}(-\mathbf{p}, s)_\beta = \left(\frac{-m_0 c^2 - i\gamma_\mu p_\mu c}{2E_p} \right)_{\alpha\beta} \qquad (10.20)$$

The relativistic quantum field for a free electron can be expanded in terms of the normal-model solutions to the Dirac equation obtained above. The coefficients in the expansion become creation and destruction operators satisfying canonical anti-commutation relations (for fermions). After a canonical transformation to particles and holes, the field in the Schrödinger representation takes the form [Bj65, Fe71]

$$\psi(\mathbf{x}) = \frac{1}{\sqrt{\Omega}} \sum_{\mathbf{k},\lambda} \left[a_{\mathbf{k},\lambda} u(\mathbf{k}\lambda) e^{i\mathbf{k}\cdot\mathbf{x}} + b_{\mathbf{k},\lambda}^{\dagger} v(-\mathbf{k}\lambda) e^{-i\mathbf{k}\cdot\mathbf{x}} \right] \tag{10.21}$$

We again quantize in a big box of volume Ω and use periodic boundary conditions. The Dirac current is given in terms of the field by

$$j_\mu(\mathbf{x}) = i\bar{\psi}(\mathbf{x})\gamma_\mu\psi(\mathbf{x}) \tag{10.22}$$

The hamiltonian in first quantization for a Dirac particle in an external, time-dependent field $A_\mu^{\text{ext}}(\mathbf{x}, t)$ is given by Eq. (10.11). In second quantization this hamiltonian takes the form [Bj65, Fe71]

$$\hat{H} = \int \hat{\psi}^\dagger(\mathbf{x}) \left\{ c\,\boldsymbol{\alpha} \cdot \left[\mathbf{p} - \frac{e}{c}\mathbf{A}^{\text{ext}}(\mathbf{x}, t) \right] + \beta m_0 c^2 + e\Phi^{\text{ext}}(\mathbf{x}, t) \right\} \hat{\psi}(\mathbf{x})\, d^3x \tag{10.23}$$

Here $\hat{\psi}(\mathbf{x})$ and $\hat{\psi}^\dagger(\mathbf{x})$ are field operators in the Schrödinger picture satisfying canonical anti-commutation relations [Eq. (10.21) provides a convenient representation]. The interaction hamiltonian in the external electromagnetic field is evidently

$$\hat{H}^{(1)} = -e\,\hat{j}_\mu(\mathbf{x})A_\mu^{\text{ext}}(\mathbf{x}, t) \tag{10.24}$$

This hamiltonian can be used to determine the relativistic, quantum behavior of an electron in an arbitrary, time-dependent external electromagnetic field. It also governs pair production processes.

There are several readily established relations on the traces of the gamma matrices which are useful in the calculation of rates and cross sections [Bj65].

$$\begin{aligned}
\text{trace}\,\gamma_\mu &= 0 \\
\text{trace}\,\gamma_\mu\gamma_\nu &= 4\delta_{\mu\nu} \\
\text{trace}\,\gamma_\mu\gamma_\nu\gamma_\rho &= 0 \\
\text{trace}\,\gamma_\mu\gamma_\nu\gamma_\rho\gamma_\sigma &= 4(\delta_{\mu\nu}\delta_{\rho\sigma} - \delta_{\mu\rho}\delta_{\nu\sigma} + \delta_{\mu\sigma}\delta_{\nu\rho})
\end{aligned} \tag{10.25}$$

Other relations are given in appendix D.

Since $\hbar$ and c have now served their purpose, and we know where all the factors are, it is convenient to go over to a more common set of units used in nuclear and particle physics where

$$\hbar = c = 1 \tag{10.26}$$

This simplifies the algebra considerably, and we shall henceforth assume this to be the case. All momenta and energies now become inverse lengths with the conversion factor

$$\hbar c = 197.3 \text{ MeV fm} \tag{10.27}$$

We shall take care to ensure that all final results are written in transparent and dimensionally correct form.

11

Covariant analysis

Consider the scattering of a Dirac electron in an external field created by an electromagnetic transition current density in a hadronic target. We assume to start with that the target makes a transition from the ground state to some discrete excited state. The external vector potential $A_\mu^{\text{ext}}(\mathbf{x}, t)$ can be related to that current density through the use of Maxwell's equations[1]

$$\left(\frac{\partial}{\partial x_\nu}\right)^2 A_\mu^{\text{ext}}(\mathbf{x}, t) = -e_p \langle f | \hat{J}_\mu(\mathbf{x}, t) | i \rangle \tag{11.1}$$

Here $|i\rangle$ and $|f\rangle$ are exact Heisenberg eigenstates of the target with energies E, E' respectively. It follows that the time dependence of the target matrix element can be extracted as

$$\left(\frac{\partial}{\partial x_\nu}\right)^2 A_\mu^{\text{ext}}(\mathbf{x}, t) = -e_p \langle f | \hat{J}_\mu(\mathbf{x}) | i \rangle e^{-i(E-E')t} \tag{11.2}$$

The states can similarly be taken as eigenstates of four-momentum $p_\mu = (\mathbf{p}, iE)$, so the entire space-time dependence can be extracted as

$$\left(\frac{\partial}{\partial x_\nu}\right)^2 A_\mu^{\text{ext}}(\mathbf{x}, t) = -e_p \langle f | \hat{J}_\mu(0) | i \rangle e^{i(p-p')\cdot x} \tag{11.3}$$

First-order time-dependent perturbation theory and the interaction of Eq. (10.24) lead to the scattering operator

$$\hat{S} \doteq -i \int \hat{\mathscr{H}}_I^{(1)}(\mathbf{x}, t) \, d^4x \tag{11.4}$$

---

[1] These are Maxwell's equations in the Lorentz gauge for the external field where, by current conservation, $\partial A_\mu^{\text{ext}}/\partial x_\mu = 0$.

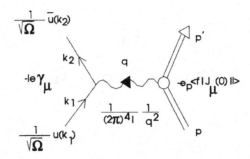

Fig. 11.1.   Feynman diagram for electron scattering from a hadronic target.

Here the interaction representation for the electrons has been introduced and they carry the free field time dependence $\exp(\pm ik \cdot x)$. Now take matrix elements of the scattering operator between appropriate initial and final electron states

$$\langle \mathbf{k}_2, s_2 | \hat{S} | \mathbf{k}_1, s_1 \rangle = -\frac{e}{\sqrt{\Omega^2}} \bar{u}(\mathbf{k}_2, s_2) \gamma_\mu u(\mathbf{k}_1, s_1) \int e^{-iq \cdot x} A_\mu^{\text{ext}}(x) \, d^4 x$$
$$q \equiv k_2 - k_1 \tag{11.5}$$

If one proceeds directly to the continuum limit, what is required is the four-dimensional Fourier transform of the external field

$$\tilde{A}_\mu^{\text{ext}}(q) = \int e^{-iq \cdot x} A_\mu^{\text{ext}}(x) \, d^4 x \tag{11.6}$$

The Fourier transform is inverted with the relation

$$A_\mu^{\text{ext}}(x) = \int e^{iq \cdot x} \tilde{A}_\mu^{\text{ext}}(q) \frac{d^4 q}{(2\pi)^4} \tag{11.7}$$

Substitute Eq. (11.7) on the left hand side of Eq. (11.3). The right hand side is then reproduced if one chooses

$$-q^2 \tilde{A}_\mu^{\text{ext}}(q) = -e_p \langle f | \hat{J}_\mu(0) | i \rangle (2\pi)^4 \delta^{(4)}(p - p' - q) \tag{11.8}$$

The required S-matrix thus takes the form

$$\langle f | \hat{S} | i \rangle = -\frac{ee_p}{\sqrt{\Omega^2}} \bar{u}(k_2) \gamma_\mu u(k_1) \frac{1}{q^2} \langle f | \hat{J}_\mu(0) | i \rangle (2\pi)^4 \delta^{(4)}(p - p' - q)$$
$$q = k_2 - k_1 = p - p' \tag{11.9}$$

The spin quantum numbers for the electron have been suppressed.

The amplitude in Eq. (11.9) can be represented as a *Feynman diagram* as shown in Fig. 11.1. There is a corresponding set of *Feynman rules* for the S-matrix:

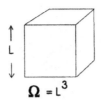

$\Omega = L^3$

Fig. 11.2. Quantization volume.

1. Include a factor of $(-i)$ for each order of perturbation theory; here second order;

2. Include a factor of $(-eJ_\mu)$ for each vertex; here

   - $-ie\gamma_\mu$; for electron vertex
   - $-e_p\langle f|\hat{J}_v(0)|i\rangle$; for hadronic vertex (lowest order)

3. Include factors of $u(k_1)/\sqrt{\Omega}$ and $\bar{u}(k_2)/\sqrt{\Omega}$ for the initial and final electron legs;

4. Include the following factor for the virtual photon propagator

$$\frac{1}{(2\pi)^4 i} \frac{1}{q^2} \delta_{\mu\nu} \tag{11.10}$$

Since both the electron and target currents are conserved, one could just as well use the following expression for the photon propagator

$$\frac{1}{(2\pi)^4 i} \frac{1}{q^2} \left[\delta_{\mu\nu} - \frac{q_\mu q_\nu}{q^2}(1 - \bar{\alpha}) \right] \tag{11.11}$$

The extra term in $q_\mu q_\nu$ vanishes in the S-matrix element (see below). Here different choices of $\bar{\alpha}$ correspond to different gauges for the internal vector potential;

5. Include a factor $(2\pi)^4\delta^{(4)}(\sum_i p_i)$ at each vertex;

6. Integrate $\int d^4p$ over internal lines.

The factors in the above diagram can be checked according to $(-i)^2(-ie)$ $\times(-e_p)(2\pi)^8/(2\pi)^4 i\sqrt{\Omega^2} = -ee_p(2\pi)^4/\sqrt{\Omega^2}$.

As indicated previously, we choose to quantize in a big box of volume Ω (Fig. 11.2) and in the end let $\Omega \to \infty$. This fictitious volume must

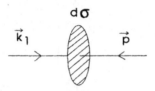

Fig. 11.3. Flux and cross section in any frame obtained by a Lorentz transformation along the incident electron direction.

disappear from any physical result. The end term in Eqn. (11.9) should really be written in the form

$$(2\pi)^3 \delta^{(3)}(\mathbf{p} - \mathbf{p}' - \mathbf{q}) = \int_{\text{box}} e^{i(\mathbf{p}-\mathbf{p}'-\mathbf{q})\cdot\mathbf{x}} d^3x = \Omega\, \delta_{\mathbf{p}',\mathbf{p}-\mathbf{q}} \qquad (11.12)$$

Thus the S-matrix for the problem at hand actually takes the form

$$\langle f|\hat{S}|i\rangle \equiv -2\pi i \delta(W_f - W_i) T_{fi}$$

$$T_{fi} = -iee_p \bar{u}(k_2)\gamma_\mu u(k_1)\frac{1}{q^2}\langle f|\hat{J}_\mu(0)|i\rangle\, \delta_{\mathbf{p}',\mathbf{p}-\mathbf{q}} \qquad (11.13)$$

Here W_f, W_i are the final and initial total energies. The cross section can now be evaluated immediately with the aid of the *Golden Rule*

$$d\sigma = \frac{R_{fi}}{\text{Flux}} = 2\pi |T_{fi}|^2 \delta(W_f - W_i) d\rho_f \frac{1}{\text{Flux}} \qquad (11.14)$$

It follows immediately that

$$[\delta_{\mathbf{p}',\mathbf{p}-\mathbf{q}}]^2 = \delta_{\mathbf{p}',\mathbf{p}-\mathbf{q}} \qquad (11.15)$$

Only one of the momenta in the final state is independent. The number of states for the final electron in a big box with periodic boundary conditions is

$$d\rho_f = \frac{\Omega}{(2\pi)^3} d^3k_2 \qquad (11.16)$$

Consider the incident flux in any frame obtained by a Lorentz transformation along the incident electron direction as shown in Fig. 11.3. The incident flux is defined by v_{rel}/Ω and is given by

$$\begin{aligned}
\text{Flux} &= \frac{1}{\Omega} v_{\text{rel}} = \frac{1}{\Omega}\left(\frac{k_1}{\varepsilon_1} + \frac{p}{E}\right) \\
&= \frac{1}{\Omega\varepsilon_1 E}(\varepsilon_1 p + k_1 E) \\
&= \frac{1}{\Omega\varepsilon_1 E}\sqrt{(k_1 \cdot p)^2} \qquad (11.17)
\end{aligned}$$

The last relation follows for $\mathbf{k}_1$ antiparallel to $\mathbf{p}$ and massless electrons since then $k_1 \cdot p = -k_1 p - \varepsilon_1 E = -(k_1 E + \varepsilon_1 p)$. The reader can verify that the same result holds if $\mathbf{k}_1 \| \mathbf{p}$, which we now use to generically identify this case. A combination of the above results then leads to

$$
\begin{aligned}
d\sigma = {} & 2\pi \frac{e^2 e_p^2}{q^4} |\bar{u}(k_2)\gamma_\mu u(k_1)\langle f|\hat{J}_\mu(0)|i\rangle|^2 \delta(W_f - W_i) \frac{\Omega\, d^3 k_2}{(2\pi)^3} \\
& \times \delta_{\mathbf{p}',\mathbf{p}-\mathbf{q}} \frac{\Omega \varepsilon_1 E}{\sqrt{(k_1 \cdot p)^2}}
\end{aligned}
\tag{11.18}
$$

If the electron beam is unpolarized and its final polarization unmeasured, one must average over initial electron spins and sum over final spins.[2] If the target is unpolarized and unobserved, one must average over initial target states and sum over final states. The cross section thus takes the form (recall $e^2/4\pi = \alpha$, the fine-structure constant)

$$
\begin{aligned}
d\sigma = {} & \frac{4\alpha^2}{q^4} \left(\frac{d^3 k_2}{2\varepsilon_2} \right) \frac{1}{\sqrt{(k_1 \cdot p)^2}} (2\varepsilon_1\varepsilon_2) \frac{1}{2} \sum_{s_1} \sum_{s_2} \overline{\sum_i} \sum_f \\
& \times |\bar{u}(k_2)\gamma_\mu u(k_1)\langle f|\hat{J}_\mu(0)|i\rangle|^2 (\Omega E)(2\pi)^3 \delta^{(4)}(p - p' - q)
\end{aligned}
\tag{11.19}
$$

Here Eqn. (11.12) has again been employed. The product of the matrix element and its complex conjugate can now be written out. In taking the complex conjugate, one must use $\{\gamma_4, \gamma_i\} = 0$ to restore the gamma matrices to the proper order and also remember that $\hat{J}_\mu = (\hat{\mathbf{J}}, i\hat{\rho})$ has an imaginary fourth component. We therefore arrive at the final, important result

$$
d\sigma = \frac{4\alpha^2}{q^4} \left(\frac{d^3 k_2}{2\varepsilon_2} \right) \frac{1}{\sqrt{(k_1 \cdot p)^2}} \eta_{\mu\nu} W_{\mu\nu}
\tag{11.20}
$$

$$
\eta_{\mu\nu} \equiv -2\varepsilon_1\varepsilon_2 \frac{1}{2} \sum_{s_1} \sum_{s_2} \bar{u}(k_1)\gamma_\nu u(k_2)\bar{u}(k_2)\gamma_\mu u(k_1)
$$

$$
W_{\mu\nu} = (2\pi)^3 \overline{\sum_i} \sum_f \langle i|\hat{J}_\nu(0)|f\rangle\langle f|\hat{J}_\mu(0)|i\rangle (\Omega E)\delta^{(4)}(p' - p + q)
$$

This expression represents the cross section in any frame where $\mathbf{k}_1 \| \mathbf{p}$. It is evident from Fig. 11.3 that $d\sigma$ is a small element of transverse area and, as such, is invariant under Lorentz transformations along the incident electron direction. The initial factors in this result are all Lorentz invariant. The quantity $\eta_{\mu\nu}$ transforms as a second rank tensor (see below). Hence

---

[2] We shall later relax these conditions.

one concludes that $W_{\mu\nu}$ must also be a second rank tensor.[3] The right hand side of Eq. (11.20) is explicitly Lorentz invariant and can now be evaluated in any Lorentz frame; it represents the physical cross section in any frame where $\mathbf{k}_1 \| \mathbf{p}$.

If one were doing elastic scattering from a point Dirac particle, the matrix elements of the current would each be proportional to $1/\Omega$ and the final momentum would be determined by the use of Eq. (11.12). The quantity $W_{\mu\nu}$ would thus be independent of Ω and the quantization volume would then cancel from Eq. (11.20), as it must. This is in fact a general result, as we shall see in all our applications.

Although Eq. (11.20) has been derived under the assumption of a discrete final state of the target, the generalization to an arbitrary final state of the target, which might include the production of many particles, is now immediate. One simply calculates the appropriate inelastic matrix element of the current and then sums over the correct number of final states at given (p, q) in $W_{\mu\nu}$.

With the aid of the positive-energy projection operators for the Dirac equation in Eq. (10.20), the lepton response tensor can be evaluated for massless electrons as follows

$$
\begin{aligned}
\eta_{\mu\nu} &\equiv -2\varepsilon_1\varepsilon_2 \frac{1}{2} \sum_{s_1}\sum_{s_2} \bar{u}(k_1)\gamma_\nu u(k_2)\bar{u}(k_2)\gamma_\mu u(k_1) \\
&= -\varepsilon_1\varepsilon_2 \, \text{trace}\left[\gamma_\nu \left(\frac{-i\gamma_\lambda k_{2\lambda}}{2\varepsilon_2}\right)\gamma_\mu \left(\frac{-i\gamma_\rho k_{1\rho}}{2\varepsilon_1}\right)\right] \\
&= \frac{1}{4}4\left[k_{2\nu}k_{1\mu} + k_{1\nu}k_{2\mu} - (k_1 \cdot k_2)\delta_{\mu\nu}\right] \\
\eta_{\mu\nu} &= k_{2\nu}k_{1\mu} + k_{1\nu}k_{2\mu} - (k_1 \cdot k_2)\delta_{\mu\nu}
\end{aligned}
\tag{11.21}
$$

This is evidently a second rank Lorentz tensor, as advertised.

The target response tensor $W_{\mu\nu}$ is a second rank Lorentz tensor built out of the two remaining independent four-vectors p and q; everything else has been summed over. The electromagnetic current is conserved. With the aid of the Heisenberg equations of motion, one concludes that

$$
\frac{\partial}{\partial x_\mu}\langle f|\hat{J}_\mu(x)|i\rangle = e^{i(p-p')\cdot x}i(p-p')_\mu\langle f|\hat{J}_\mu(0)|i\rangle = 0
$$

$$
q_\mu\langle f|\hat{J}_\mu(0)|i\rangle = 0
\tag{11.22}
$$

Hence current conservation for the target implies

$$
q_\mu W_{\mu\nu} = W_{\mu\nu}q_\nu = 0
\tag{11.23}
$$

---

[3] This can be proven directly from the Lorentz transformation properties of the states, but the argument is more involved.

The Dirac equation for the massless electrons implies that

$$\bar{u}(k_2)\gamma_\lambda q_\lambda u(k_1) = \bar{u}(k_2)(\gamma_\lambda k_{2\lambda} - \gamma_\lambda k_{1\lambda})u(k_1) = 0 \tag{11.24}$$

It follows that the lepton response tensor in Eq. (11.21) obeys the same conditions

$$q_\mu \eta_{\mu\nu} = \eta_{\mu\nu} q_\nu = 0 \tag{11.25}$$

The two independent Lorentz scalars that can be constructed from p and q are q^2 and $q \cdot p$. Recall $p^2 = -M_T^2$ is fixed by the target mass. In the laboratory frame, for massless electrons, $q^2 = (\mathbf{k}_2 - \mathbf{k}_1)^2 - (k_2 - k_1)^2 = 2k_1 k_2(1 - \cos\theta)$ where θ is the scattering angle. Furthermore, the target is at rest in that frame so $p = (\mathbf{0}, iM_T)$. Hence one can identify these scalers in the laboratory frame according to

$$q^2 = 4k_1 k_2 \sin^2 \frac{\theta}{2} \qquad ; \text{laboratory frame}$$
$$q \cdot p = -q_0 M_T \tag{11.26}$$

The conditions in Eq. (11.23) then imply that the target response tensor must have the following form

$$W_{\mu\nu} = W_1(q^2, q \cdot p)\left(\delta_{\mu\nu} - \frac{q_\mu q_\nu}{q^2}\right)$$
$$+ W_2(q^2, q \cdot p)\frac{1}{M_T^2}\left(p_\mu - \frac{p \cdot q}{q^2}q_\mu\right)\left(p_\nu - \frac{p \cdot q}{q^2}q_\nu\right) \tag{11.27}$$

This result is due to Bjorken [Bj60], Von Gehlen [Vo60], and Gourdin [Go61]. It forms the basis for the subsequent analysis. It makes use only of Lorentz covariance and current conservation, and it holds for *any* hadronic target, independent of its internal structure. Note that it is $\overline{\sum_i}\sum_f$ that yields the simplicity of the form in Eq. (11.27). Upon substitution of this expression, the cross section in Eq. (11.20) is then exact to lowest order in α. We proceed to the proof of this important result.

Write the most general tensor[4] one can make out of the four-vectors p and q

$$W_{\mu\nu} = W_1\delta_{\mu\nu} + W_2\frac{p_\mu p_\nu}{M_T^2} + A\frac{q_\mu q_\nu}{M_T^2} + B\frac{1}{M_T^2}(p_\mu q_\nu + p_\nu q_\mu)$$
$$+ C\frac{1}{M_T^2}(p_\mu q_\nu - p_\nu q_\mu) \tag{11.28}$$

---

[4] Note that $\varepsilon_{\mu\nu\rho\sigma}q_\rho p_\sigma$ is a *pseudo*tensor. We shall return to this later.

Use the current conservation relations in Eq. (11.23)

$$W_1 q_\nu + W_2 \frac{p \cdot q \, p_\nu}{M_T^2} + A \frac{q^2 q_\nu}{M_T^2} + B \frac{1}{M_T^2} (p \cdot q \, q_\nu + q^2 \, p_\nu)$$

$$+ C \frac{1}{M_T^2} (p \cdot q \, q_\nu - q^2 \, p_\nu) = 0$$

$$W_1 q_\mu + W_2 \frac{p \cdot q \, p_\mu}{M_T^2} + A \frac{q^2 q_\mu}{M_T^2} + B \frac{1}{M_T^2} (p \cdot q \, q_\mu + q^2 \, p_\mu)$$

$$+ C \frac{1}{M_T^2} (q^2 \, p_\mu - p \cdot q \, q_\mu) = 0 \qquad (11.29)$$

Since p and q are linearly independent four-vectors, their coefficients must individually vanish

$$W_1 + \frac{q^2}{M_T^2} A + \frac{p \cdot q}{M_T^2} B + \frac{p \cdot q}{M_T^2} C = 0$$

$$\frac{p \cdot q}{M_T^2} W_2 + \frac{q^2}{M_T^2} B - \frac{q^2}{M_T^2} C = 0$$

$$W_1 + \frac{q^2}{M_T^2} A + \frac{p \cdot q}{M_T^2} B - \frac{p \cdot q}{M_T^2} C = 0$$

$$\frac{p \cdot q}{M_T^2} W_2 + \frac{q^2}{M_T^2} B + \frac{q^2}{M_T^2} C = 0 \qquad (11.30)$$

The solution to these linear equations is

$$C = 0$$

$$B = -\frac{p \cdot q}{q^2} W_2$$

$$A = -\frac{M_T^2}{q^2} W_1 + \left(\frac{p \cdot q}{q^2} \right)^2 W_2 \qquad (11.31)$$

This is the desired result.

The next task is to combine the expressions in Eqs. (11.21, 11.27) to get the cross section in Eq. (11.20). With the aid of Eq. (11.25), the required expression reduces to

$$\eta_{\mu\nu} W_{\mu\nu} = (k_{1\mu} k_{2\nu} + k_{1\nu} k_{2\mu} - k_1 \cdot k_2 \, \delta_{\mu\nu}) \left(W_1 \delta_{\mu\nu} + W_2 \frac{p_\mu p_\nu}{M_T^2} \right)$$

$$= W_1(-2 k_1 \cdot k_2) + W_2 \frac{1}{M_T^2} (2 p \cdot k_1 \, p \cdot k_2 - p^2 \, k_1 \cdot k_2) \qquad (11.32)$$

Now employ some kinematics in the laboratory frame. Since the electrons are massless here

$$
\begin{aligned}
q &= k_2 - k_1 \\
q^2 &= -2k_1 \cdot k_2 = -2\mathbf{k_1} \cdot \mathbf{k_2} + 2\varepsilon_1\varepsilon_2 \\
&= 2\varepsilon_1\varepsilon_2(1 - \cos\theta) = 4\varepsilon_1\varepsilon_2 \sin^2\frac{\theta}{2}
\end{aligned}
\tag{11.33}
$$

Also, since $p = (\mathbf{0}, iM_T)$ in the laboratory frame,

$$
\begin{aligned}
p^2 &= -M_T^2 \\
(p \cdot k_1)(p \cdot k_2) &= M_T^2 \varepsilon_1\varepsilon_2
\end{aligned}
\tag{11.34}
$$

Hence

$$
\begin{aligned}
\eta_{\mu\nu} W_{\mu\nu} &= 4\varepsilon_1\varepsilon_2 \sin^2\frac{\theta}{2} W_1 + 2\varepsilon_1\varepsilon_2\left(1 - \sin^2\frac{\theta}{2}\right)W_2 \\
&= 2\varepsilon_1\varepsilon_2\left(W_2\cos^2\frac{\theta}{2} + 2W_1\sin^2\frac{\theta}{2}\right)
\end{aligned}
\tag{11.35}
$$

The double differential cross section in the laboratory frame in Eq. (11.20) can therefore be written

$$
\frac{d^2\sigma}{d\varepsilon_2 d\Omega_2} = \frac{\alpha^2}{4\varepsilon_1^2\varepsilon_2^2\sin^4\theta/2}\left(\frac{\varepsilon_2^2}{2\varepsilon_2}\right)\left(\frac{1}{M_T\varepsilon_1}\right) 2\varepsilon_1\varepsilon_2\left(W_2\cos^2\frac{\theta}{2} + 2W_1\sin^2\frac{\theta}{2}\right)
\tag{11.36}
$$

Introduce the *Mott cross section* for the scattering of a relativistic (massless) Dirac electron from a point charge[5]

$$
\sigma_M \equiv \frac{\alpha^2\cos^2(\theta/2)}{4\varepsilon_1^2\sin^4(\theta/2)}
\tag{11.37}
$$

The double differential cross section in the laboratory frame for the scattering of a relativistic Dirac electron from an arbitrary hadronic target to order α^2 then takes the form

$$
\frac{d^2\sigma}{d\Omega_2 d\varepsilon_2} = \sigma_M\frac{1}{M_T}[W_2(q^2, q\cdot p) + 2W_1(q^2, q\cdot p)\tan^2\frac{\theta}{2}]
\tag{11.38}
$$

This is a central result.

It is useful at this point to demonstrate the relation to the *photoabsorption* cross section. The process is illustrated in Fig. 11.4. This cross section

---

[5] Two factors of Eq. (10.27) restore the correct dimensions (recall $\varepsilon = \hbar kc$).

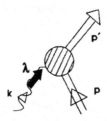

Fig. 11.4.   The process of photoabsorption by a hadronic target.

measures one slice of the two-dimensional response surface $W_1(q^2, q \cdot p)$. In fact

$$\sigma_\gamma = \frac{(2\pi)^2 \alpha}{\sqrt{(k \cdot p)^2}} W_1(k^2, -k \cdot p) \qquad ; k^2 = 0 \qquad (11.39)$$

Here $k = (\mathbf{k}, i\omega_k)$ is the four-momentum of the incoming photon. This result is derived as follows.

Start from the interaction with the transverse quantized radiation field

$$H' = -e_p \int \mathbf{J}(\mathbf{x}) \cdot \mathbf{A}(\mathbf{x}) \, d^3x \qquad (11.40)$$

Everything is now in the Schrödinger representation. The quantized radiation field is expanded in normal modes according to Eq. (8.8) with helicity unit vectors $\lambda = \pm 1$ defined in Eq. (8.4) and Fig. 8.1. The scattering operator is again given in lowest order by Eq. (11.4) where now everything is in the interaction representation. The appropriate S-matrix element of this scattering operator is

$$\langle f|\hat{S}|i \rangle = ie_p \left(\frac{1}{2\omega_k \Omega} \right)^{1/2} \mathbf{e}_{\mathbf{k}\lambda} \cdot \langle f|\hat{\mathbf{J}}(0)|i \rangle \, (2\pi)^4 \delta^{(4)}(p + k - p') \quad (11.41)$$

The system is again quantized in a big box with periodic boundary conditions so that Eq. (11.12) should actually be employed. The T-matrix is identified as in Eq. (11.13)

$$S_{fi} = -2\pi i \delta(W_f - W_i) T_{fi}$$

$$T_{fi} = -e_p \left(\frac{1}{2\omega_k \Omega} \right)^{1/2} \mathbf{e}_{\mathbf{k}\lambda} \cdot \langle f|\hat{\mathbf{J}}(0)|i \rangle \, \Omega \, \delta_{\mathbf{p}',\mathbf{p}-\mathbf{k}} \qquad (11.42)$$

The cross section is again given by

$$\sigma_\gamma = \frac{\text{Rate}}{\text{Flux}} = 2\pi |T_{fi}|^2 \delta(W_f - W_i) \frac{1}{\text{Flux}}$$

$$\text{Flux} = \frac{1}{\Omega} \frac{\sqrt{(k \cdot p)^2}}{\omega_k E} \qquad (11.43)$$

With unpolarized and unobserved targets, one must again average over initial states and sum over final states and with an unpolarized beam, one must average over photon polarizations. With the use of Eq. (11.15) and the identification of the target response tensor in Eq. (11.20), one finds

$$\sigma_\gamma = \frac{2\pi^2\alpha}{\sqrt{(k\cdot p)^2}} \sum_{\lambda=\pm 1} (\mathbf{e}_{\mathbf{k},\lambda}^\dagger)_i W_{ij}(\mathbf{e}_{\mathbf{k},\lambda})_j \tag{11.44}$$

It is now necessary to carry out the polarization sums, and with the insertion of the expressions for the helicity unit vectors one has

$$\sum_{\lambda\pm 1} (\mathbf{e}_{\mathbf{k}\lambda}^\dagger)_i (\mathbf{e}_{\mathbf{k}\lambda})_j$$
$$= \frac{1}{2}[(\mathbf{e}_{\mathbf{k}1} - i\mathbf{e}_{\mathbf{k}2})_i(\mathbf{e}_{\mathbf{k}1} + i\mathbf{e}_{\mathbf{k}2})_j + (\mathbf{e}_{\mathbf{k}1} + i\mathbf{e}_{\mathbf{k}2})_i(\mathbf{e}_{\mathbf{k}1} - i\mathbf{e}_{\mathbf{k}2})_j]$$
$$= (\mathbf{e}_{\mathbf{k}1})_i(\mathbf{e}_{\mathbf{k}1})_j + (\mathbf{e}_{\mathbf{k}2})_i(\mathbf{e}_{\mathbf{k}2})_j$$
$$= \delta_{ij} - \frac{k_i k_j}{\mathbf{k}^2} \tag{11.45}$$

The last relation follows since the set of unit vectors in Fig. 8.1 is complete. Current conservation can now be employed on the last term

$$\begin{aligned} k_\mu \langle f|\hat{J}_\mu(0)|i\rangle &= 0 \\ \mathbf{k}\cdot \langle f|\hat{\mathbf{J}}(0)|i\rangle &= |\mathbf{k}|\,\langle f|\hat{J}_0(0)|i\rangle \end{aligned} \tag{11.46}$$

The required expression in Eq. (11.44) can therefore be written as a covariant polarization sum

$$\sum_\lambda (\mathbf{e}_{\mathbf{k},\lambda}^\dagger)_i W_{ij}(\mathbf{e}_{\mathbf{k},\lambda})_j = W_{\mu\mu} \tag{11.47}$$

One has to be careful with the limit $k^2 \to 0$ of the general expression for the target response tensor in Eq. (11.27). From its definition in terms of matrix elements of the current in Eq. (11.20), $W_{\mu\nu}$ cannot be singular in this limit. Thus, by inspection

$$\begin{aligned} W_2 &\to O(q^2) &&; q^2 \to 0 \\ -W_1 + \frac{(p\cdot q)^2}{M_T^2 q^2}W_2 &\to O(q^2) \end{aligned} \tag{11.48}$$

The trace of the response tensor is given in general by

$$W_{\mu\mu} = 3W_1 + W_2 \frac{1}{M_T^2}\left[p^2 - \frac{(p\cdot q)^2}{q^2}\right] \tag{11.49}$$

With the use of Eqs. (11.48) one has

$$
\begin{aligned}
W_{\mu\mu} &\rightarrow 3W_1 - W_2 - W_1 + O(q^2) \qquad ; q^2 \to 0 \\
W_{\mu\mu} &\rightarrow 2W_1 + O(q^2) \\
W_{\mu\mu}(k^2 = 0) &= 2W_1(k^2 = 0)
\end{aligned}
\qquad (11.50)
$$

This is the desired result, and Eq. (11.39) holds as claimed.

12

Excitation of discrete states in (e, e')

Consider the simplest case of excitation of a discrete state of the target in inclusive electron scattering (e, e'). The kinematics are illustrated in Fig. 12.1. If $M_T^\star$ is the final target mass, then

$$
\begin{aligned}
p' &= p - q \\
-M_T^{\star 2} &= -M_T^2 - 2p \cdot q + q^2 \\
2p \cdot q &= M_T^{\star 2} - M_T^2 + q^2 \\
&= 2M_T(\varepsilon_1 - \varepsilon_2)
\end{aligned}
\tag{12.1}
$$

The last relation evaluates $p \cdot q = -M_T q_0$ in the laboratory frame. One observes that here only q^2 is an independent variable.

The integral over the energy-conserving delta function appearing in the response tensor can be performed according to

$$
\begin{aligned}
\int d\varepsilon_2 \delta(E' - M_T + q_0) &= \int d\varepsilon_2 \delta(W_f - W_i) \\
&= \int dW_f \, \delta(W_f - W_i) \left(\frac{\partial \varepsilon_2}{\partial W_f} \right) \\
&= \left(\frac{\partial \varepsilon_2}{\partial W_f} \right) \qquad ; W_i = M_T + \varepsilon_1
\end{aligned}
\tag{12.2}
$$

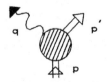

Fig. 12.1. Kinematics in inclusive electron scattering.

This jacobian is evaluated as follows

$$
\begin{aligned}
W_f &= (\mathbf{p}'^2 + M_T^{\star 2})^{1/2} + \varepsilon_2 \\
&= [(\mathbf{k}_1 - \mathbf{k}_2)^2 + M_T^{\star 2}]^{1/2} + \varepsilon_2 \\
&= [\varepsilon_1^2 + \varepsilon_2^2 - 2\varepsilon_1\varepsilon_2 \cos\theta + M_T^{\star 2}]^{1/2} + \varepsilon_2 \\
\frac{\partial W_f}{\partial \varepsilon_2} &= \frac{\varepsilon_2 - \varepsilon_1 \cos\theta}{E'} + 1 \\
&= \frac{\varepsilon_2 + E' - \varepsilon_1 \cos\theta}{E'} = \frac{M_T + \varepsilon_1 - \varepsilon_1 \cos\theta}{E'} \\
&= \frac{M_T}{E'}\left[1 + \frac{2\varepsilon_1 \sin^2(\theta/2)}{M_T}\right]
\end{aligned}
$$
(12.3)

Energy conservation has been used. The inverse of this relation gives the required result

$$
\begin{aligned}
\frac{\partial \varepsilon_2}{\partial W_f} &= \frac{E'}{M_T} r \\
r^{-1} &\equiv \left[1 + \frac{2\varepsilon_1 \sin^2(\theta/2)}{M_T}\right]
\end{aligned}
$$
(12.4)

Take out the following Lorentz invariant factors from the coefficients in the response tensor

$$
W_i(q^2, q \cdot p) \equiv w_i(q^2)\frac{M_T^2}{E'}\delta(p_0 - p_0' - q_0) \qquad ; i = 1, 2 \quad (12.5)
$$

Then, from the above,

$$
\int \frac{d\varepsilon_2}{M_T} W_i(q^2, q \cdot p) = w_i(q^2) r
$$
(12.6)

The remaining response tensor, which will be denoted by $w_{\mu\nu}$, is given by

$$
w_{\mu\nu}(q^2) = \overline{\sum_i}\sum_f \frac{E E' \Omega}{M_T^2}(2\pi)^3 \delta^{(3)}(\mathbf{p} - \mathbf{p}' - \mathbf{q})\langle i|J_\nu(0)|f\rangle\langle f|J_\mu(0)|i\rangle \quad (12.7)
$$

The sum over final states in the continuum limit takes the form

$$
\sum_f \rightarrow \sum_f{}' \frac{\Omega d^3 p'}{(2\pi)^3}
$$
(12.8)

Here $\sum_f'$ now goes over all other quantum numbers. The final result for the Lorentz invariant response tensor in this discrete case can then be written

$$w_{\mu\nu}(q^2) = \overline{\sum_i}\sum_f{}' \frac{EE'\Omega^2}{M_T^2} \langle i|J_\nu(0)|f\rangle\langle f|J_\mu(0)|i\rangle \tag{12.9}$$

$$= w_1(q^2)\left(\delta_{\mu\nu} - \frac{q_\mu q_\nu}{q^2}\right) + w_2(q^2)\frac{1}{M_T^2}\left(p_\mu - \frac{p\cdot q}{q^2}q_\mu\right)\left(p_\nu - \frac{p\cdot q}{q^2}q_\nu\right)$$

This relation allows an identification of the transition form factors $w_{1,2}(q^2)$. The variables q^2 and $q\cdot p$ are here related through Eq. (12.1). Furthermore, a combination of Eqs. (11.38, 12.6) yields the cross section

$$\frac{d\sigma}{d\Omega} = \sigma_M\left[w_2(q^2) + 2w_1(q^2)\tan^2\frac{\theta}{2}\right]r \tag{12.10}$$

This is an exact result, to order α^2, for the scattering of a relativistic electron with corresponding excitation of a discrete state in any quantum mechanical target.

As a simple example, consider elastic scattering from a $J^\pi = 0^+$ target. The kinematics in Eq. (12.1) yields for elastic scattering

$$M_T^\star = M_T \equiv m$$
$$2p\cdot q = q^2 \tag{12.11}$$

From Lorentz covariance and current conservation, one can write the general form of the matrix element of the current in this case as

$$\langle p';0^+|J_\mu(0)|p;0^+\rangle = \left(\frac{m^2}{EE'\Omega^2}\right)^{1/2}F_0(q^2)\frac{1}{m}\left(p_\mu - \frac{p\cdot q}{q^2}q_\mu\right) \tag{12.12}$$

Hence one can simply read off from Eqs. (12.9)

$$w_1 = 0$$
$$w_2 = |F_0(q^2)|^2 \tag{12.13}$$

As a second example, consider elastic scattering from a $J^\pi = 1/2^+$ target. It follows from Lorentz covariance and current conservation that the most general form of the matrix element of the current in this case is given by [Bj65]

$$\langle p';1/2^+|J_\mu(0)|p;1/2^+\rangle = \frac{i}{\Omega}\bar{u}(p')\left[F_1(q^2)\gamma_\mu + F_2(q^2)\sigma_{\mu\nu}q_\nu\right]u(p)$$
$$\sigma_{\mu\nu} \equiv \frac{1}{2i}[\gamma_\mu,\gamma_\nu] \tag{12.14}$$

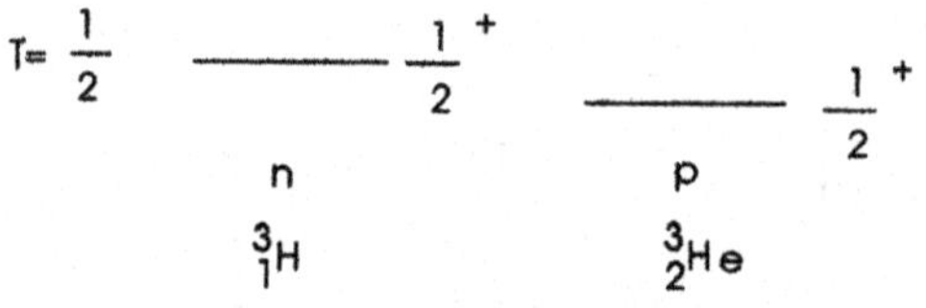

Fig. 12.2.   Elastic scattering from isodoublets.

This time one has to do a little work, but with the aid of the positive energy projection operator, and by taking the resulting traces, one identifies

$$w_1(q^2) = \frac{q^2}{4m^2}|F_1 + 2mF_2|^2$$

$$w_2(q^2) = |F_1|^2 + \frac{q^2}{4m^2}|2mF_2|^2 \qquad (12.15)$$

Substitution in Eq. (12.10) then yields the celebrated *Rosenbluth cross section*. Although very instructive, we leave the derivation of Eqs. (12.14, 12.15) to the dedicated reader.[1]

The Rosenbluth cross section is quite general. It of course applies to the nucleon, the isodoublet (p, n). It also applies to the nuclear isodoublet $(^3_2\text{He}, \,^3_1\text{H})$ as illustrated in Fig. 12.2. In both cases, one can actually do elastic scattering experiments on the higher energy state. In the case of the neutron, one uses a deuteron ^2_1H in which the neutron is bound to a proton. In the case of tritium ^3_1H, this nucleus lives long enough that one can make targets of it for external beam experiments.

It is useful to make the isospin dependence of the form factor manifest in the case of scattering from an isodoublet target. The general isospin structure of the electromagnetic current operator in any description of nuclei and nucleons (mesons and baryons, quarks and gluons, etc.) is

$$J_\mu^\gamma = J_\mu^S + J_\mu^{V_3} \qquad (12.16)$$

Here the superscript describes the behavior under isospin transformations, either scalar or third component of an isovector. It follows from the Wigner–Eckart theorem that the matrix elements of the current must reflect that behavior. Thus the form factors must have the structure

$$F_i = \frac{1}{2}(F_i^S + F_i^V \tau_3) \qquad ; i = 1, 2 \qquad (12.17)$$

Here τ are the Pauli matrices, and the target isospinors are suppressed in

---

[1] Hermiticity of the current implies that the form factors in these examples are real.

Fig. 12.3. Inelastic transition $0^+ \to 1^+$.

Eq. (12.14). For the nucleon, it follows that

$$
\begin{aligned}
F_i^p &= \frac{1}{2}(F_i^S + F_i^V) & F_i^S &= F_i^p + F_i^n \\
F_i^n &= \frac{1}{2}(F_i^S - F_i^V) & F_i^V &= F_i^p - F_i^n
\end{aligned} \qquad (12.18)
$$

It is useful to summarize the following numerical values for the nucleon $(m \equiv m_p)$

$$
\begin{aligned}
F_1^p(0) &= 1 & F_1^S(0) &= 1 \\
F_1^n(0) &= 0 & F_1^V(0) &= 1 \\
2mF_2^p(0) &= +1.793 & 2mF_2^S(0) &= -.120 \\
2mF_2^n(0) &= -1.913 & 2mF_2^V(0) &= +3.706
\end{aligned} \qquad (12.19)
$$

Consider next an inelastic transition $0^+ \to 1^+$ as illustrated in Fig. 12.3. From Lorentz covariance, the general form of the transition matrix element of the current can be written as

$$
\langle p'; 1^+\lambda | J_\mu(0) | p; 0^+ \rangle = \left(\frac{mm^\star}{EE'\Omega^2} \right)^{1/2} \frac{f(q^2)}{\sqrt{2}\, mm^\star} \varepsilon_{\mu\nu\rho\sigma}\, \varepsilon_\nu^{(\lambda)\star} p_\rho q_\sigma \qquad (12.20)
$$

Here λ is the helicity of the final 1^+ particle, and its polarization four-vector is given by

$$
\varepsilon_\nu^{(\lambda)\star} \equiv (\boldsymbol{\varepsilon}^{(\lambda)\dagger}, \, i\varepsilon_0^{(\lambda)\dagger}) \qquad (12.21)
$$

Because of the intrinsic parity of the 1^+ particle, this matrix element must transform as an axial vector. Note that current conservation is automatically maintained since $q_\mu \varepsilon_{\mu\nu\rho\sigma} q_\sigma = 0$.

To calculate the cross section one needs the sum over polarization vectors, which for a spin one particle is given by[2]

$$
\sum_\lambda \varepsilon_\mu^{(\lambda)\star} \varepsilon_\nu^{(\lambda)} = \delta_{\mu\nu} + \frac{p'_\mu p'_\nu}{m^{\star 2}} \qquad (12.22)
$$

---

[2] This must be a second rank tensor, and the polarization vectors satisfy $p' \cdot \varepsilon^{(\lambda)} = 0$.

After some algebra, one then obtains

$$w_1 = \frac{\mathbf{q}^2}{2mm^\star}|f(q^2)|^2$$

$$w_2 = \frac{q^2}{2mm^\star}|f(q^2)|^2 \tag{12.23}$$

The cross section then follows from Eq. (12.10). We shall see from the subsequent multipole analysis that this represents a "pure M1 cross section."

The reader can now write his or her own "elementary cross section." Just pick a transition and use Lorentz covariance and current conservation to write the general form of the matrix element of the current. The response functions $w_{1,2}(q^2)$ are then identified from Eqs. (12.9) and the cross section from Eq. (12.10).

Let us now make a connection to the analysis of real photon transitions in chapter 9 and make a *multipole analysis* of the electron scattering cross section. We start by going back a step and restoring the spatial dependence to the matrix elements in Eq. (11.20) through the use of the Heisenberg equations of motion

$$W_{\mu\nu} = \overline{\sum_i}\sum_f \delta(p_0 - p_0' - q_0)\langle i| \int e^{i\mathbf{q}\cdot\mathbf{x}} J_\nu(\mathbf{x})\, d^3x |f\rangle$$
$$\times \langle f| \int e^{-i\mathbf{q}\cdot\mathbf{x}} J_\mu(\mathbf{x})\, d^3x |i\rangle (E) \tag{12.24}$$

This relation is still exact since if the initial and final states are eigenstates of momentum, one has

$$\langle f| \int e^{-i\mathbf{q}\cdot\mathbf{x}} J_\mu(\mathbf{x})\, d^3x |i\rangle = \Omega\delta_{\mathbf{p},\mathbf{p}'+\mathbf{q}}\langle f|J_\mu(0)|i\rangle$$

and then; $\quad \langle i| \int e^{i\mathbf{q}\cdot\mathbf{x}} J_\nu(\mathbf{x})\, d^3x |f\rangle = \Omega\langle i|J_\nu(0)|f\rangle \tag{12.25}$

Thus in the limit $\Omega \to \infty$

$$W_{\mu\nu} = (2\pi)^3 \overline{\sum_i}\sum_f \delta^{(4)}(p - p' - q)\langle i|J_\nu(0)|f\rangle\langle f|J_\mu(0)|i\rangle (E\,\Omega) \tag{12.26}$$

This is our previous result.

Assume one goes to a discrete state with mass $M_T^\star$, then just as before

$$W_{\mu\nu} \equiv \frac{M_T^2}{E'}\delta(p_0 - p_0' - q_0)w_{\mu\nu} \tag{12.27}$$

$$w_{\mu\nu} = \sum_i \sum_f \left(\frac{EE'}{M_T^2}\right)\langle i| \int e^{i\mathbf{q}\cdot\mathbf{x}} J_\nu(\mathbf{x})\, d^3x |f\rangle\langle f| \int e^{-i\mathbf{q}\cdot\mathbf{x}} J_\mu(\mathbf{x})\, d^3x |i\rangle$$

$$= w_1(q^2)\left(\delta_{\mu\nu} - \frac{q_\mu q_\nu}{q^2}\right) + w_2(q^2)\frac{1}{M_T^2}\left(p_\mu - \frac{p\cdot q}{q^2}q_\mu\right)\left(p_\nu - \frac{p\cdot q}{q^2}q_\nu\right)$$

The cross section is again given by Eq. (12.10).

Let us now solve Eqs. (12.27) for the functions $w_{1,2}(q^2)$. First take $\mu = \nu = 4$ and make use of the fact that in the laboratory frame $p = (\mathbf{0}, iM_T)$. This yields

$$w_1 \left(1 + \frac{q_0^2}{q^2}\right) - w_2 \left(1 + \frac{q_0^2}{q^2}\right)^2 = w_1 \frac{\mathbf{q}^2}{q^2} - w_2 \frac{\mathbf{q}^4}{q^4}$$

$$= -\overline{\sum_i} \sum_f \left(\frac{EE'}{M_T^2}\right) |\langle f| \int e^{-i\mathbf{q}\cdot\mathbf{x}} \hat{\rho}(\mathbf{x}) \, d^3x |i\rangle|^2 \qquad (12.28)$$

Next dot the spatial part of the tensor $W_{\mu\nu}$ into the spherical unit vectors $\mathbf{e}_{\mathbf{q}\lambda}$ from the left and $\mathbf{e}_{\mathbf{q}\lambda}^\dagger$ from the right. Here these spherical unit vectors are defined with respect to the direction of the momentum transfer $\mathbf{q}$ [see Fig. 8.1 and Eqs. (8.4)]. For $\lambda = \pm 1$ they satisfy

$$\begin{aligned} \mathbf{e}_{\mathbf{q}\lambda} \cdot \mathbf{e}_{\mathbf{q}\lambda}^\dagger &= 1 \\ \mathbf{e}_{\mathbf{q}\lambda} \cdot \mathbf{q} &= 0 \qquad ; \lambda = \pm 1 \end{aligned} \qquad (12.29)$$

As a result of these observations, the term in w_2 no longer contributes. Finally, take $\sum_{\lambda = \pm 1}$ to simplify things. The result of these operations is

$$2w_1 = \sum_{\lambda = \pm 1} \overline{\sum_i} \sum_f \left(\frac{EE'}{M_T^2}\right) |\langle f| \int e^{-i\mathbf{q}\cdot\mathbf{x}} \mathbf{e}_{\mathbf{q}\lambda}^\dagger \cdot \hat{\mathbf{J}}(\mathbf{x}) \, d^3x |i\rangle|^2 \qquad (12.30)$$

These equations can now be solved for $w_{1,2}(q^2)$ with the result

$$2w_1(q^2) = \sum_{\lambda = \pm 1} \overline{\sum_i} \sum_f \left(\frac{EE'}{M_T^2}\right) |\langle f| \int e^{-i\mathbf{q}\cdot\mathbf{x}} \mathbf{e}_{\mathbf{q}\lambda}^\dagger \cdot \hat{\mathbf{J}}(\mathbf{x}) \, d^3x |i\rangle|^2 \qquad (12.31)$$

$$w_2(q^2) = \frac{q^2}{\mathbf{q}^2} w_1(q^2) + \frac{q^4}{\mathbf{q}^4} \overline{\sum_i} \sum_f \left(\frac{EE'}{M_T^2}\right) |\langle f| \int e^{-i\mathbf{q}\cdot\mathbf{x}} \hat{\rho}(\mathbf{x}) \, d^3x |i\rangle|^2$$

These equations are still exact.

Now assume, just as in the analysis of real photon transitions in chapters 8 and 9, that

- The target is heavy and the transition densities are well localized in space

- The initial and final states are eigenstates of angular momentum

Thus one imagines that the target is heavy and "nailed down" (at the origin, say). It makes a transition, and the localized transition density scatters the electron. Here target recoil (i.e. the C-M motion of the target) is neglected in the transition matrix elements;[3] it is included correctly where it is most important through the recoil phase space factor r.

The multipole analysis now proceeds exactly as in chapter 9. The essential difference is that the argument of the spherical Bessel functions in the multipoles, instead of being given by $|\mathbf{k}|$ the momentum of the photon (with $|\mathbf{k}| = \omega$), is now given by $\kappa = |\mathbf{q}|$ the momentum transfer in the electron scattering process.

$$\kappa \equiv |\mathbf{q}| \tag{12.32}$$

In addition to the transverse electric and magnetic multipoles of Eq. (9.11)

$$\hat{T}_{JM}^{\text{el}}(\kappa) \equiv \frac{1}{\kappa} \int d^3x \left[\nabla \times j_J(\kappa x) \mathcal{Y}_{JJ1}^M(\Omega_x) \right] \cdot \hat{\mathbf{J}}(\mathbf{x})$$

$$\hat{T}_{JM}^{\text{mag}}(\kappa) \equiv \int d^3x \left[j_J(\kappa x) \mathcal{Y}_{JJ1}^M(\Omega_x) \right] \cdot \hat{\mathbf{J}}(\mathbf{x}) \tag{12.33}$$

there is now a Coulomb multipole of the charge density defined by

$$\hat{M}_{JM}(\kappa) \equiv \int d^3x \, j_J(\kappa x) Y_{JM}(\Omega_x) \, \hat{\rho}(\mathbf{x}) \tag{12.34}$$

This is the same multipole that appears at long wavelength in the expansion of $\hat{T}_{JM}^{\text{el}}(k)$ in Eq. (A.13).

The use of the Wigner–Eckart theorem allows one to do the sum and average over nuclear states, and exactly as in chapter 9 one arrives at the relations

$$2w_1(q^2) = \frac{E'}{M_T} \frac{4\pi}{2J_i + 1} \sum_{J \geq 1} \left\{ |\langle J_f || \hat{T}_J^{\text{mag}}(\kappa) || J_i \rangle|^2 + |\langle J_f || \hat{T}_J^{\text{el}}(\kappa) || J_i \rangle|^2 \right\}$$

$$w_2(q^2) = \frac{q^2}{\mathbf{q}^2} w_1(q^2) + \frac{q^4}{\mathbf{q}^4} \frac{E'}{M_T} \frac{4\pi}{2J_i + 1} \sum_{J \geq 0} |\langle J_f || \hat{M}_J(\kappa) || J_i \rangle|^2 \tag{12.35}$$

The Wigner–Eckart theorem limits the sums on multipoles appearing in these expressions to values satisfying the triangle inequality $|J_f - J_i| \leq J \leq J_f + J_i$.

The cross section follows from Eq. (12.10) as

$$\frac{d\sigma}{d\Omega} = \sigma_M \frac{4\pi}{2J_i + 1} \left\{ \frac{q^4}{\mathbf{q}^4} \sum_{J \geq 0} |\langle J_f || \hat{M}_J(\kappa) || J_i \rangle|^2 \right. \tag{12.36}$$

$$\left. + \left(\frac{q^2}{2\mathbf{q}^2} + \tan^2 \frac{\theta}{2} \right) \sum_{J \geq 1} \left(|\langle J_f || \hat{T}_J^{\text{mag}}(\kappa) || J_i \rangle|^2 + |\langle J_f || \hat{T}_J^{\text{el}}(\kappa) || J_i \rangle|^2 \right) \right\} \bar{r}$$

---

[3] The C-M motion can, in fact, be handled correctly in the usual non-relativistic nuclear physics problem using, for example, the approach in appendix A of [Fo69]. We reproduce that analysis here in appendix B.

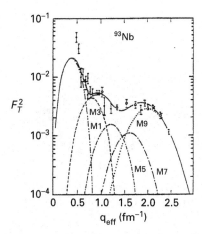

Fig. 12.4.   Elastic magnetic scattering response F_T^2 for $^{93}_{41}$Nb(e, e). Here $q_{\text{eff}} \equiv \kappa$. The single-particle shell model configuration assignment is $(1g_{9/2})_\pi$. The work is from Bates [Yo79].

Here the recoil factor $\bar{r}$ is given by

$$
\begin{aligned}
(\bar{r})^{-1} &\equiv \frac{M_T}{E'} r^{-1} \\
&= \frac{1}{E'} (M_T + \varepsilon_1 - \varepsilon_1 \cos\theta) \\
&= 1 + \frac{(\varepsilon_2 - \varepsilon_1 \cos\theta)}{E'}
\end{aligned}
\tag{12.37}
$$

Energy conservation has been used in obtaining this result (note that for most nuclear applications $M_T/E' \approx 1$).

Equation (12.36) is the general electron scattering cross section, to order α^2, from an arbitrary, localized quantum mechanical target. It forms the basis for much of our future discussion. To give the reader some feel for these results, we briefly present a few selected applications.

For real photon transitions, it is the *lowest* allowed multipole that dominates the transition (appendix A). One of the most intriguing features of electron scattering (e, e′) is that by increasing the momentum transfer κ, one can in essence dial the contributing multipolarity, even to the extent that it is the *highest* allowed multipole that dominates. We give three examples.

Figure 12.4 shows elastic magnetic scattering from $^{93}_{41}$Nb(e, e). This represents the contribution to the cross section from the transverse multipoles in the second line of Eq. (12.36). This contribution can be separated experimentally by making a straight-line *Rosenbluth plot* against $\tan^2(\theta/2)$ at fixed q^2, or by working at $\theta = 180°$ where only the transverse term

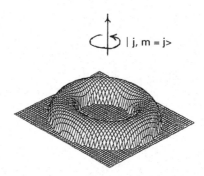

Fig. 12.5. Surface of $\mu(x)_{\mathrm{max}}/2$ for $^{51}_{23}\mathrm{V}$ with configuration assignment $(1f_{7/2})_\pi$ plotted on a 10 fm square. Here the angular momentum is aligned along the z-axis with $m_j = j$ [Do73].

contributes.[4] Parity and time-reversal invariance of the strong interactions limit one to the odd transverse magnetic multipoles in elastic scattering.[5] The work shown is from Bates [Yo79]. The single-particle shell model configuration assignment for $^{93}_{41}\mathrm{Nb}$ is $(1g_{9/2})_\pi$; recall it is predominantly the valence nucleon that gives rise to the magnetic properties of nuclei. Note how at long wavelength (low κ) the transition is all M1, while each higher multipole dominates in turn as κ is increased, until at high κ it is all M9.

What does one learn from this? Figure 12.5 shows the surface of half-maximum intrinsic magnetization density $\mu(x)_{\mathrm{max}}/2$ for $^{51}_{23}\mathrm{V}$ (chosen so that it would fit on a 10 fm square). Here the configuration assignment is $(1f_{7/2})_\pi$, and the nucleus is aligned so that its angular momentum points along the z-axis with $m_j = j$. The intrinsic magnetization tracks the location of the valence nucleon. The nucleus is a small magnet with a current loop provided by the motion of the orbiting proton. Elastic magnetic electron scattering at all κ provides a microscope to actually see the spatial structure of this small current loop [Do73].

We next recall that one of the distinguishing features of the shell model, for whose discovery Mayer and Jensen won the Nobel prize, is that levels with the highest angular momentum and opposite parity from one major shell are pushed down close to the levels of the next lower major shell. If that lower shell is filled (or partially filled), one can have low-lying, high-spin, magnetic, particle–hole transitions of the nucleus. Figure 12.6 shows

---

[4] The notation used here is $d\sigma/d\Omega \equiv \sigma_M[(q^4/\mathbf{q}^4)F_L^2 + (q^2/2\mathbf{q}^2 + \tan^2\theta/2)F_T^2]r$.

[5] An analysis similar to that for parity in chapter 9 shows that the time-reversal behavior (which includes complex conjugation) of both of the transverse multipoles is $\hat{\mathscr{T}}\hat{T}_{JM}\hat{\mathscr{T}}^{-1} = (-1)^{J+1}\hat{T}_{J-M}$. This, combined with the hermiticity of the current, leads to the quoted selection rule in the elastic case [Pr65, Do73] — see appendix E.

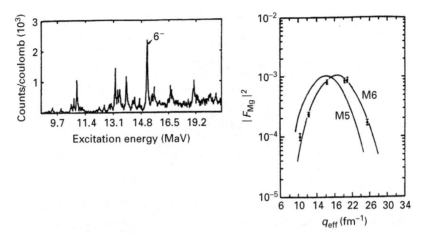

Fig. 12.6.   Transverse response for $^{24}_{12}\text{Mg}(\text{e}, \text{e}')$ at $\theta = 160°$ and $\kappa = 2.13\,\text{fm}^{-1}$ with the 6^- indicated. Also shown is the inelastic form factor F_T^2 for the 6^-, defined to be the area under the inelastic peak, vs. $\kappa \equiv q_{\text{eff}}$. This quantity is compared with the theoretical result (open-shell RPA) for a transition to the $[1f_{7/2}(1d_{5/2})^{-1}]_{6^-}$ state. The work is from Bates [Za77].

the large-angle, large κ response for electron excitation of $^{24}_{12}\text{Mg}(\text{e}, \text{e}')$. Also shown is the inelastic form factor, the area under the peak, as a function of κ for the dominant transition. This inelastic form factor manifests a characteristic M6 dependence, identifying the excited state as 6^-. The configuration assignment here is $1f_{7/2}(1d_{5/2})^{-1}$ and the 6^- is the *highest* J^π which can be formed from this configuration.[6] The work was done at Bates [Za77]. Excitations up to 14^- in $^{208}_{82}\text{Pb}$ have been similarly studied [Li79].

As a third application, there are regions of the periodic table where nuclei are deformed. Bohr, Mottelson, and Rainwater won the Nobel prize for their analysis of these systems. Suppose one measures elastic scattering, and inelastic electron scattering, to all members of the ground-state rotational band, at all κ. This requires very good energy resolution, since the ground-state band for heavy nuclei is closely spaced, and involves bringing out all the transitions of increasing multipolarity in the rotational spectrum. It is then possible to actually *see the deformed charge distribution.* This is illustrated in Fig. 12.7. The work was done at Bates [He86]. The study of the intrinsic structure of deformed nuclei is one of the most important contributions of the Bates Laboratory.

---

[6] Note that the large J^π produces a very narrow state (this state lies above particle emission threshold). Furthermore, the large isovector magnetic moment of the nucleon in Eq. (12.19), through which the state is predominantly excited from the $T = 0$ ground state, implies this excited state has isospin $T = 1$.

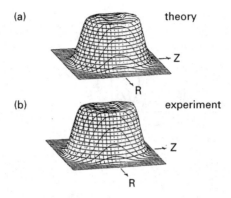

Fig. 12.7. Shape of charge distribution in the deformed nucleus $^{154}_{64}$Gd(e, e') (b) from work at Bates; (a) deformed Hartree–Fock calculation of the same [He86].

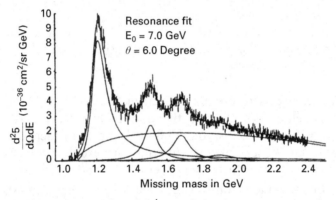

Fig. 12.8. Inelastic cross section for ^{1}H(e, e') in the resonance region measured at SLAC with $\varepsilon_1 = 7.0$ GeV and $\theta = 6.0°$. The elastic peak has been suppressed. Also shown are resonance and smooth background fits [Bl68].

As a fourth brief application, consider electron excitation of the nucleon itself. A general discussion of the process $(1/2^+, 1/2) \rightarrow (J^\pi, T)$ can be found in [Bj66].

Figure 12.8 shows the inelastic cross section for ^{1}H(e, e') in the resonance region measured at SLAC [Bl68]. These results are even more impressive when one realizes that the elastic peak has been suppressed. Also shown in this figure is a Breit–Wigner resonance fit, together with a smooth, polynomial background [Bl68, Br71]. The first resonance, the $\Delta(1232)$ with $(J^\pi, T) = (3/2^+, 3/2)$, clearly stands out. The second peak consists of at least two levels. The third has several levels, and with a good stretch of the imagination, one can even discern a fourth peak.

Figure 12.9 shows the ratio $(d\sigma_{in}/d\sigma_{el})_{6°}$ for ^{1}H(e, e')$\Delta(1232)$ as measured by the SLAC–M.I.T. collaboration [Bl68, Br71]. The inelastic cross

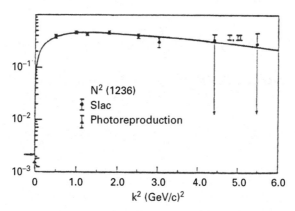

Fig. 12.9.   Ratio $(d\sigma_{in}/d\sigma_{el})_{6°}$ for ^{1}H(e, e′)Δ(1232) (see text) [Wa72]. Here $k^2 \equiv q^2$.

section for excitation of the resonance is obtained from the area under the resonance peak. The ratio to the elastic cross section is then plotted as a function of the four-momentum transfer q^2. Note that there is one point on this plot at $q^2 = 0$ obtained from photoabsorption.[7] The solid curve is a covariant, gauge-invariant calculation formulated in terms of hadronic degrees of freedom [Pr69, Wa72]. The calculation uses the multipole projections of the hadronic pole terms from (π, ω, N) exchange, with a resonant final-state enhancement factor determined from the π-N phase shift; it is discussed in some detail in chapter 28. This calculation can be viewed as a synthesis of a great deal of work on this process by Fubini Nambu and Wataghin [Fu58], Dennery [De61], Zagury [Za66], Vik [Vi67], Adler [Ad68], and others. It is remarkable that a hadronic description of the excitation of the first excited state of the nucleon can succeed down to distance scales $q^2 \sim 100\,\mathrm{fm}^{-2}$.

---

[7] In $d\sigma_{el}$ the factors $\sigma_{Mott}r$ are evaluated at the same (ε_1, θ) while $w_{1,2}(q^2)_{el}$ are evaluated at the resonance peak. The resulting ratio in Fig. 12.9 is essentially independent of θ for small θ and all the $q^2 \neq 0$ points [Wa72].

13

Coincidence experiments (e, e′ X)

With the advent of high-energy, high-intensity, high-resolution electron accelerators with continuous beams (c.w.), a whole new class of coincidence reactions becomes accessible. It is important to have a detailed understanding of such processes. In this section, a covariant analysis of the amplitude and cross section for the coincidence reaction (e, e′ X) will be developed. The results will be exact with one photon exchange, that is, to order α^2 in the cross section. The particle X can be anything. The kinematic situation is illustrated in Fig. 13.1. The four-momentum transfer from the electron is now consistently denoted by

$$
\begin{aligned}
k &\equiv k_1 - k_2 \\
k^2 &= -2k_1 \cdot k_2 \\
&= 4\varepsilon_1 \varepsilon_2 \sin^2 \frac{\theta}{2} \qquad \text{; lab frame}
\end{aligned}
\tag{13.1}
$$

The second relation holds for relativistic (massless) electrons, and the third relation holds in the laboratory frame. The four-momentum of the emitted particle X will be consistently denoted by $q = (\mathbf{q}, i\omega_q)$, and conservation

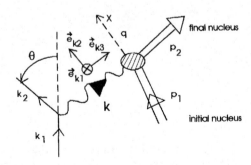

Fig. 13.1. Kinematic situation for a (e, e′ X) coincidence experiment.

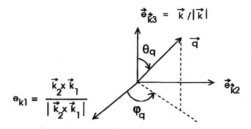

Fig. 13.2. Angles for particle X in the C-M system.

of four-momentum states that

$$k + p_1 = q + p_2 \tag{13.2}$$

Here $p_1(p_2)$ are the four-momenta of the initial (final) nucleus or nucleon.

Two distinct Lorentz frames are of primary interest. The center of momentum (C-M) frame is defined by the relation

$$q + p_2 = k + p_1 = (\mathbf{0}, iW) \qquad ; \text{C-M frame} \tag{13.3}$$

Note that W is the total energy in the C-M frame. The laboratory frame is defined by

$$p_1 = (\mathbf{0}, iM_1) \qquad ; \text{lab frame} \tag{13.4}$$

The C-M frame is reached from the laboratory frame by making a Lorentz transformation along the direction of the three-momentum transfer $\mathbf{k}$.

Introduce the orthonormal system of unit vectors in the laboratory (lab), as defined in Fig. 13.1

$$\mathbf{e}_{k3} \equiv \frac{\mathbf{k}}{|\mathbf{k}|} \qquad ; \mathbf{e}_{k1} \equiv \frac{\mathbf{k}_2 \times \mathbf{k}_1}{|\mathbf{k}_2 \times \mathbf{k}_1|} \qquad ; \mathbf{e}_{k2} = \mathbf{e}_{k3} \times \mathbf{e}_{k1} \tag{13.5}$$

It is important to note that since $\mathbf{e}_{k1}$ and $\mathbf{e}_{k2}$ are transverse to $\mathbf{k}$, they are *unchanged* under the Lorentz transformation along $\mathbf{k}$ from the lab to the C-M system. $\mathbf{e}_{k3}$, defined as the third unit vector in this orthonormal system, is thus also uniquely defined in the C-M system (it lies along the direction of $\mathbf{k}$).

In addition, we define the angles (θ_q, ϕ_q) that the particle X makes with respect to this orthonormal basis as seen in the C-M frame; this is indicated in Fig. 13.2.

From the general discussion of electron scattering in chapter 11, one has

$$d\sigma = \frac{4\alpha^2}{k^4} \frac{d^3k_2}{2\varepsilon_2} \frac{1}{\sqrt{(k_1 \cdot p_1)^2}} \eta_{\mu\nu} W_{\mu\nu} \tag{13.6}$$

$$\eta_{\mu\nu} = k_{1\mu}k_{2\nu} + k_{1\nu}k_{2\mu} - (k_1 \cdot k_2)\,\delta_{\mu\nu}$$

$$W_{\mu\nu} = (2\pi)^3 \overline{\sum_i} \sum_f \delta^{(4)}(p' - p_1 - k)\langle i|J_\nu(0)|f\rangle\langle f|J_\mu(0)|i\rangle(\Omega E_1)$$

For definiteness and clarity, specify to a two-particle final state of particle X plus a second nucleus or nucleon (denoted with subscript 2)[1]

$$\langle f|J_\mu(0)|i\rangle = \langle p_2 q^{(-)}|J_\mu(0)|p_1\rangle \tag{13.7}$$

Here $|p_2 q^{(-)}\rangle$ is an exact eigenstate of the total hamiltonian; it is a two-particle scattering state with incoming wave boundary conditions. To go to states with Lorentz invariant norm, one defines (c.f. chapter 12)

$$J_\mu \equiv \left(\frac{2\omega_q E_1 E_2 \Omega^3}{M_1 M_2}\right)^{1/2} \langle p_2 q^{(-)}|J_\mu(0)|p_1\rangle \tag{13.8}$$

Here $J_\mu = (\mathbf{J}, iJ_0)$, and this quantity now properly transforms as a four-vector under Lorentz transformations. The hadronic response tensor then takes the form

$$W_{\mu\nu} = (2\pi)^3 \overline{\sum_i} \sum_f {}' \frac{\Omega d^3 q}{(2\pi)^3} \frac{\Omega d^3 p_2}{(2\pi)^3} \delta^{(4)}(p_2 + q - p_1 - k)$$

$$\times \frac{M_1 M_2}{2\omega_q E_1 E_2 \Omega^3} (\Omega E_1) J_\nu^\star J_\mu \tag{13.9}$$

Here $\sum'$ indicates a sum over all the remaining variables. The complex four-vector $J_\nu^\star$ is defined by

$$J_\nu^\star \equiv (\mathbf{J}^\star, iJ_0^\star) \tag{13.10}$$

Thus

$$W_{\mu\nu} = \frac{2M_1 M_2}{(2\pi)^3} \overline{\sum_i} \sum_f {}' \delta^{(4)}(p_2 + q - p_1 - k) \frac{d^3 q}{2\omega_q} \frac{d^3 p_2}{2E_2} J_\nu^\star J_\mu \tag{13.11}$$

This expression is now manifestly Lorentz covariant.

---

[1] As long as one sums over everything else in $\sum_f$ the subsequent results for the general form of the coincidence cross section hold for arbitrary nuclear final states.

Consider next the Lorentz invariant combination $\eta_{\mu\nu}J_\nu^\star J_\mu$. It follows from Eq. (13.6) that

$$\eta_{\mu\nu}J_\nu^\star J_\mu = (k_1 \cdot J^\star)(k_2 \cdot J) + (k_2 \cdot J^\star)(k_1 \cdot J) - (k_1 \cdot k_2)(J^\star \cdot J) \quad (13.12)$$

Current conservation states that

$$\begin{aligned} k \cdot J &= 0 \\ k_1 \cdot J &= k_2 \cdot J \end{aligned} \quad (13.13)$$

Hence

$$\eta_{\mu\nu}J_\nu^\star J_\mu = 2(k_1 \cdot J^\star)(k_1 \cdot J) + \frac{k^2}{2}J^\star \cdot J \quad (13.14)$$

This expression is explicitly Lorentz invariant. Let us proceed to evaluate it in the C-M frame. Since $\mathbf{k}_1$ has no projection on $\mathbf{e}_{k1}$, which is perpendicular to the electron scattering plane, one can write in the C-M system (recall $k_1^2 = 0$)

$$k_{1\mu} = [(\mathbf{k}_1 \cdot \mathbf{e}_2)\mathbf{e}_2 + (\mathbf{k}_1 \cdot \mathbf{e}_3)\mathbf{e}_3, ik_1] \quad (13.15)$$

Now use current conservation

$$\mathbf{e}_3 \cdot \mathbf{J} = \frac{\mathbf{k} \cdot \mathbf{J}}{|\mathbf{k}|} = \frac{\omega_k J_0}{|\mathbf{k}|} \quad (13.16)$$

Thus

$$k_1 \cdot J = (\mathbf{k}_1 \cdot \mathbf{e}_2)(\mathbf{e}_2 \cdot \mathbf{J}) + \left[\frac{\omega_k}{\mathbf{k}^2}(\mathbf{k}_1 \cdot \mathbf{k}) - k_1\right]J_0 \quad (13.17)$$

The Coulomb amplitude is defined by $J_0 \equiv J_C$, hence

$$\begin{aligned} J_\mu &= (\mathbf{J}, iJ_C) \\ J^\star \cdot J &= |\mathbf{J}_\perp|^2 + |\mathbf{J} \cdot \mathbf{e}_3|^2 - |J_C|^2 \\ |\mathbf{J}_\perp|^2 &\equiv |\mathbf{J} \cdot \mathbf{e}_1|^2 + |\mathbf{J} \cdot \mathbf{e}_2|^2 \end{aligned} \quad (13.18)$$

Use current conservation again

$$\begin{aligned} J^\star \cdot J &= |\mathbf{J}_\perp|^2 + \left(\frac{\omega_k^2}{\mathbf{k}^2} - 1\right)|J_C|^2 \\ &= |\mathbf{J}_\perp|^2 - \frac{k^2}{\mathbf{k}^2}|J_C|^2 \end{aligned} \quad (13.19)$$

A combination of these results yields the following expression in the C-M system

$$\eta_{\mu\nu} J_\nu^\star J_\mu = 2 \left\{ \frac{1}{4} k^2 |\mathbf{J}_\perp|^2 + (\mathbf{k}_1 \cdot \mathbf{e}_2)^2 |\mathbf{e}_2 \cdot \mathbf{J}|^2 \right.$$

$$+ |J_C|^2 \left[\left(\frac{\omega_k}{\mathbf{k}^2}(\mathbf{k}_1 \cdot \mathbf{k}) - k_1 \right)^2 - \frac{k^4}{4\mathbf{k}^2} \right]$$

$$\left. + (\mathbf{k}_1 \cdot \mathbf{e}_2) \left[\frac{\omega_k}{\mathbf{k}^2}(\mathbf{k}_1 \cdot \mathbf{k}) - k_1 \right] 2\,\mathrm{Re}\,[(\mathbf{e}_2 \cdot \mathbf{J}) J_C^\star] \right\}_{CM} \quad (13.20)$$

The next step is to re-express the electron variables appearing in this expression in the laboratory frame. Start by observing that the combination $\mathbf{k}_1 \cdot \mathbf{e}_2$ is transverse and hence unaffected by the Lorentz transformation from the lab to the C-M system

$$\mathbf{k}_1 \cdot \mathbf{e}_2 = \mathbf{k}_1 \cdot (\mathbf{e}_3 \times \mathbf{e}_1) = \mathbf{k}_1 \cdot \left[\frac{\mathbf{k}}{|\mathbf{k}|} \times \frac{(\mathbf{k}_2 \times \mathbf{k}_1)}{|\mathbf{k}_2 \times \mathbf{k}_1|} \right]$$

$$= \frac{1}{|\mathbf{k}| k_1 k_2 \sin\theta} [(\mathbf{k}_1 \cdot \mathbf{k}_2)(\mathbf{k}_1 \cdot \mathbf{k}) - \mathbf{k}_1^2 (\mathbf{k}_2 \cdot \mathbf{k})]$$

$$= \frac{1}{|\mathbf{k}| k_1 k_2 \sin\theta} [(\mathbf{k}_1 \cdot \mathbf{k}_2)(\mathbf{k}_1^2 - \mathbf{k}_1 \cdot \mathbf{k}_2) - \mathbf{k}_1^2 (\mathbf{k}_1 \cdot \mathbf{k}_2 - \mathbf{k}_2^2)]$$

$$= \frac{1}{|\mathbf{k}| k_1 k_2 \sin\theta} \varepsilon_1^2 \varepsilon_2^2 \sin^2\theta$$

$$\{\mathbf{k}_1 \cdot \mathbf{e}_2\}_{CM} = \frac{\varepsilon_1 \varepsilon_2 \sin\theta}{\kappa} \qquad\qquad ; \text{lab variables} \quad (13.21)$$

Here κ is now the three-momentum transfer in the lab frame

$$\kappa \equiv |\mathbf{k}|_{lab}$$

$$= \sqrt{\varepsilon_1^2 + \varepsilon_2^2 - 2\varepsilon_1 \varepsilon_2 \cos\theta} \quad (13.22)$$

To distinguish C-M variables, the four-momentum transfer as seen in the C-M system will be written in the final expressions as

$$k_\mu \equiv (\mathbf{k}^\star, i\omega_k^\star) \qquad\qquad ; \text{C-M frame} \quad (13.23)$$

Then with the aid of Eq. (13.3), which defines the C-M frame, one can write

$$\mathbf{k}^{\star 2} = k^2 + \omega_k^{\star 2} = k^2 - \frac{[k \cdot (p_1 + k)]^2}{(p_1 + k)^2}$$

$$= \frac{1}{(p_1 + k)^2} [k^2 p_1^2 + 2k^2(p_1 \cdot k) + k^4 - (k \cdot p_1)^2 - 2k^2(p_1 \cdot k) - k^4]$$

$$= \frac{1}{(p_1 + k)^2} [k^2 p_1^2 - (k \cdot p_1)^2] \quad (13.24)$$

This expression is now in invariant form and can be evaluated in the lab frame defined by Eq. (13.4) to give

$$k^\star = \frac{M_1}{W}\kappa \qquad \text{; lab variables} \qquad (13.25)$$

Note that W is expressed in terms of lab variables by

$$\begin{aligned} W^2 &= -(p_1 + k)^2 \\ &= M_1^2 - 2p_1 \cdot k - k^2 \\ &= M_1^2 + 2M_1(\varepsilon_1 - \varepsilon_2) - 4\varepsilon_1\varepsilon_2 \sin^2\frac{\theta}{2} \quad \text{; lab variables} \quad (13.26) \end{aligned}$$

Next use (for massless electrons)

$$k_1 \cdot k = -k_1 \cdot k_2 = \frac{1}{2}k^2 \qquad (13.27)$$

to work out in the C-M system

$$\begin{aligned} \left[\frac{\omega_k}{\mathbf{k}^2}(\mathbf{k}_1 \cdot \mathbf{k}) - k_1\right]^2_{\text{CM}} &= \frac{1}{\mathbf{k}^4}[\omega_k(k_1 \cdot k + k_1\omega_k) - k_1(k^2 + \omega_k^2)]^2 \\ &= \frac{k^4}{\mathbf{k}^4}\left[\frac{1}{2}\omega_k - k_1\right]^2 \\ &= \frac{k^4}{\mathbf{k}^4}\left[\frac{-1}{(p_1 + k)^2}\right]\left\{-\frac{1}{2}k \cdot (p_1 + k) + k_1 \cdot (p_1 + k)\right\}^2 \\ &= -\frac{k^4}{4k^{\star 4}}\frac{1}{(p_1 + k)^2}[p_1 \cdot (k_1 + k_2)]^2 \qquad (13.28) \end{aligned}$$

This is also now in invariant form [note Eq. (13.24)] and can be evaluated in the lab frame to yield

$$\left\{\left[\frac{\omega_k}{\mathbf{k}^2}(\mathbf{k}_1 \cdot \mathbf{k}) - k_1\right]^2\right\}_{\text{CM}} = \frac{k^4}{k^{\star 4}}\frac{M_1^2}{4W^2}(\varepsilon_1 + \varepsilon_2)^2 \quad \text{; lab variables} \quad (13.29)$$

Now in the lab

$$k^2 = \kappa^2 - (\varepsilon_1 - \varepsilon_2)^2 \qquad (13.30)$$

Thus

$$\begin{aligned} (\varepsilon_1 + \varepsilon_2)^2 &= \kappa^2 - k^2 + 4\varepsilon_1\varepsilon_2 \\ &= \kappa^2 - 4\varepsilon_1\varepsilon_2 \sin^2\frac{\theta}{2} + 4\varepsilon_1\varepsilon_2 \\ &= \kappa^2 + 4\varepsilon_1\varepsilon_2 \cos^2\frac{\theta}{2} \end{aligned}$$

$$= 4\varepsilon_1\varepsilon_2 \cos^2 \frac{\theta}{2} \left(1 + \frac{\kappa^2}{k^2} \tan^2 \frac{\theta}{2} \right)$$

$$= 4\varepsilon_1\varepsilon_2 \cos^2 \frac{\theta}{2} \left(\frac{k^{\star 2}}{k^2} \right) \left(\frac{k^2}{k^{\star 2}} + \frac{W^2}{M_1^2} \tan^2 \frac{\theta}{2} \right) \quad (13.31)$$

Also, since $\kappa^2 = \varepsilon_1^2 + \varepsilon_2^2 - 2\varepsilon_1\varepsilon_2 \cos\theta$,

$$\left\{ \left[\frac{\omega_k}{\mathbf{k}^2}(\mathbf{k}_1 \cdot \mathbf{k}) - k_1 \right]^2 - \frac{k^4}{4\mathbf{k}^2} \right\}_{\mathrm{CM}} = \frac{k^4}{k^{\star 4}} \left[\frac{M_1^2}{4W^2}(\varepsilon_1 + \varepsilon_2)^2 \right.$$

$$\left. - \frac{1}{4} \frac{M_1^2}{W^2}(\varepsilon_1^2 + \varepsilon_2^2 - 2\varepsilon_1\varepsilon_2 \cos\theta) \right]$$

$$= \frac{k^4}{k^{\star 4}} \frac{M_1^2}{W^2} \varepsilon_1\varepsilon_2 \cos^2 \frac{\theta}{2} \; ; \text{ lab variables}$$

$$(13.32)$$

Note that since $k^2 = \mathbf{k}^2 - \omega_k^2 \geq 0$ in electron scattering, one can determine the sign of the quantity in square brackets in Eq. (13.29) as

$$k_1 - \frac{\omega_k(\mathbf{k} \cdot \mathbf{k}_1)}{\mathbf{k}^2} \geq 0 \quad (13.33)$$

In *summary* the expressions involving the electron variables in the cross section are Lorentz transformed from the C-M to the laboratory frame according to

$$\{(\mathbf{k}_1 \cdot \mathbf{e}_2)^2\}_{\mathrm{CM}} = \varepsilon_1\varepsilon_2 \cos^2 \frac{\theta}{2} \left(\frac{M_1^2}{W^2} \frac{k^2}{k^{\star 2}} \right) \quad (13.34)$$

$$\left\{ \left[\frac{\omega_k}{\mathbf{k}^2}(\mathbf{k}_1 \cdot \mathbf{k}) - k_1 \right]^2 \right\}_{\mathrm{CM}} = \varepsilon_1\varepsilon_2 \cos^2 \frac{\theta}{2} \left(\frac{M_1^2}{W^2} \frac{k^2}{k^{\star 2}} \right)$$

$$\times \left(\frac{k^2}{k^{\star 2}} + \frac{W^2}{M_1^2} \tan^2 \frac{\theta}{2} \right)$$

$$\left\{ \left[\frac{\omega_k}{\mathbf{k}^2}(\mathbf{k}_1 \cdot \mathbf{k}) - k_1 \right]^2 - \frac{k^4}{4\mathbf{k}^2} \right\}_{\mathrm{CM}} = \varepsilon_1\varepsilon_2 \cos^2 \frac{\theta}{2} \left(\frac{M_1^2}{W^2} \frac{k^2}{k^{\star 2}} \right) \frac{k^2}{k^{\star 2}}$$

Here

$$W^2 = -(p_1 + k)^2$$

$$= M_1^2 + 2M_1(\varepsilon_1 - \varepsilon_2) - 4\varepsilon_1\varepsilon_2 \sin^2 \frac{\theta}{2}$$

$$\mathbf{k}^{\star 2} = \mathbf{k}^2 - \frac{[\mathbf{k} \cdot (p_1 + k)]^2}{(p_1 + k)^2} = \frac{M_1^2}{W^2}|\mathbf{k}_{\mathrm{lab}}|^2 \quad (13.35)$$

are respectively the squares of the total energy and three-momentum transfer in the C-M system. The quantities $(\varepsilon_1, \varepsilon_2, \theta)$ with $\mathbf{k}_{\text{lab}}^2 = \varepsilon_1^2 + \varepsilon_2^2 - 2\varepsilon_1\varepsilon_2 \cos\theta$ are the electron scattering variables in the lab.

The remaining task is to work out the phase space integral. The Lorentz invariant expression is

$$\Phi \equiv \int \frac{d^3q}{2\omega_q} \int \frac{d^3p_2}{2E_2} \delta^{(4)}(k_1 + p_1 - k_2 - p_2 - q) \tag{13.36}$$

We choose to evaluate this in the C-M frame. The $\int d^3p_2$ can be immediately evaluated with the aid of the $\delta^{(3)}$ to give

$$\begin{aligned}
\Phi &= \int \frac{q^2 d\Omega_q}{4\omega_q E_2} \left(\frac{\partial q}{\partial W_f} \right) \delta(W_f - W_i) dW_f \\
&= \frac{q^2}{4\omega_q E_2} \left(\frac{\partial q}{\partial W_f} \right) d\Omega_q
\end{aligned} \tag{13.37}$$

Next use

$$\begin{aligned}
W_f &= \sqrt{\mathbf{q}^2 + m_X^2} + \sqrt{\mathbf{q}^2 + M_2^2} \qquad ; \; W_i \equiv W \\
\frac{\partial W_f}{\partial q} &= \frac{q}{\omega_q} + \frac{q}{E_2} = \frac{qW}{\omega_q E_2}
\end{aligned} \tag{13.38}$$

One has finally

$$\int \frac{d^3q}{2\omega_q} \int \frac{d^3p_2}{2E_2} \delta^{(4)}(k_1 + p_1 - k_2 - p_2 - q) = \frac{q}{4W} d\Omega_q \; ; \text{ C-M frame} \tag{13.39}$$

Note that the first of Eqs. (13.38) allows a determination of $q(W)$.

The above results are now combined to yield the laboratory cross section

$$\begin{aligned}
d\sigma &= \frac{4\alpha^2}{k^4} \frac{d^3k_2}{2\varepsilon_2} \frac{1}{\sqrt{(k_1 \cdot p_1)^2}} \eta_{\mu\nu} W_{\mu\nu} \\
&= \frac{4\alpha^2}{k^4} \frac{\varepsilon_2^2 d\varepsilon_2 d\Omega_2}{2\varepsilon_2} \frac{1}{M_1 \varepsilon_1} \frac{2M_1 M_2}{(2\pi)^3} \left(\frac{q}{4W} d\Omega_q \right) 2 \left\{ \left(\frac{M_1^2}{W^2} \right) \varepsilon_1 \varepsilon_2 \cos^2 \frac{\theta}{2} \right. \\
&\quad \times \left[\frac{W^2}{M_1^2} \tan^2 \frac{\theta}{2} |\mathbf{J}_\perp|^2 + \frac{k^2}{k^{\star 2}} |\mathbf{J} \cdot \mathbf{e}_2|^2 + \frac{k^4}{k^{\star 4}} |J_C|^2 \right. \\
&\quad \left. \left. - \frac{k^2}{k^{\star 2}} \left(\frac{k^2}{k^{\star 2}} + \frac{W^2}{M_1^2} \tan^2 \frac{\theta}{2} \right)^{1/2} 2 \operatorname{Re} J_C^\star (\mathbf{J} \cdot \mathbf{e}_2) \right] \right\}
\end{aligned} \tag{13.40}$$

Define[2]

$$\mathscr{J}_\mu = \frac{\sqrt{M_1 M_2}}{4\pi W} J_\mu = \frac{\sqrt{M_1 M_2}}{4\pi W} \left(\frac{2\omega_q E_1 E_2 \Omega^3}{M_1 M_2} \right)^{1/2} \langle q p_2^{(-)} | J_\mu(0) | p_1 \rangle \quad (13.41)$$

The differential cross section in the lab is then given by

$$\frac{d^5\sigma}{d\varepsilon_2 d\Omega_2 d\Omega_q} = \sigma_M \left(\frac{q M_1}{\pi W} \right) \left\{ \frac{k^4}{k^{\star 4}} |\mathscr{J}_C|^2 + \frac{k^2}{k^{\star 2}} |\mathscr{J} \cdot \mathbf{e}_2|^2 + \frac{W^2}{M_1^2} \tan^2 \frac{\theta}{2} |\mathscr{J}_\perp|^2 \right.$$

$$\left. - \frac{k^2}{k^{\star 2}} \left(\frac{k^2}{k^{\star 2}} + \frac{W^2}{M_1^2} \tan^2 \frac{\theta}{2} \right)^{1/2} 2 \operatorname{Re} \left[\mathscr{J}_C^\star (\mathscr{J} \cdot \mathbf{e}_2) \right] \right\} \quad (13.42)$$

Here $(\varepsilon_1, \varepsilon_2, \theta)$ are electron scattering variables in the lab, and $[W, q(W), k^\star, \theta_q, \phi_q]$ are C-M variables, the first three of which can be calculated in terms of electron lab variables by utilizing the Lorentz invariant expressions in Eqs. (13.35). The current is evaluated in the C-M system.

It is useful to rewrite this cross section in terms of helicity polarization vectors for the virtual photon.[3] Define helicity unit vectors (see Fig. 13.1) according to

$$\mathbf{e}_{k\pm 1} = \mp \frac{1}{\sqrt{2}} (\mathbf{e}_{k1} \pm i\mathbf{e}_{k2}) \quad (13.43)$$

Since these are still transverse, they are also unchanged under the Lorentz transformation from the lab to the C-M system. Inversion of the definition gives (we again suppress the $\mathbf{k}$ subscript)

$$\mathbf{e}_2 = \frac{i}{\sqrt{2}} (\mathbf{e}_{+1} + \mathbf{e}_{-1})$$

$$\mathbf{e}_1 = \frac{1}{\sqrt{2}} (\mathbf{e}_{-1} - \mathbf{e}_{+1}) \quad (13.44)$$

Define

$$\mathscr{J}^\lambda \equiv \mathbf{e}_\lambda \cdot \mathscr{J} \quad (13.45)$$

It follows that

$$|\mathscr{J}_\perp|^2 = |\mathscr{J} \cdot \mathbf{e}_1|^2 + |\mathscr{J} \cdot \mathbf{e}_2|^2 = |\mathscr{J}^{+1}|^2 + |\mathscr{J}^{-1}|^2$$

$$|\mathscr{J} \cdot \mathbf{e}_2|^2 = \frac{1}{2} |\mathscr{J}_\perp|^2 + \operatorname{Re} (\mathscr{J}^{+1})^\star (\mathscr{J}^{-1})$$

$$2 \operatorname{Re} \mathscr{J}_C^\star (\mathscr{J} \cdot \mathbf{e}_2) = -\sqrt{2} \operatorname{Im} \mathscr{J}_C^\star (\mathscr{J}^{+1} + \mathscr{J}^{-1}) \quad (13.46)$$

---

[2] By looking at a simple example for the matrix element, the reader can establish that this expression still has dimensions $[M]^{-1}$.

[3] Think of this as the *annihilation* of a photon.

Thus one arrives at the basic result for the (e, e′ X) coincidence cross section in the laboratory frame

$$\frac{d^5\sigma}{d\varepsilon_2 d\Omega_2 d\Omega_q} = \sigma_M \left(\frac{qM_1}{\pi W}\right) \left[\frac{k^4}{k^{\star 4}} |\mathscr{I}_C|^2 + \left(\frac{k^2}{2k^{\star 2}} + \frac{W^2}{M_1^2} \tan^2 \frac{\theta}{2}\right) |\mathscr{I}_\perp|^2 \right.$$

$$+\frac{k^2}{2k^{\star 2}} 2\,\mathrm{Re}\,(\mathscr{I}^{+1})^\star (\mathscr{I}^{-1})$$

$$\left. +\frac{k^2}{k^{\star 2}} \left(\frac{k^2}{k^{\star 2}} + \frac{W^2}{M_1^2} \tan^2 \frac{\theta}{2}\right)^{1/2} \sqrt{2}\,\mathrm{Im}\,\mathscr{I}_C^\star (\mathscr{I}^{+1} + \mathscr{I}^{-1}) \right] \quad (13.47)$$

In this expression $k^\star$ is the three-momentum transfer, W is the total energy, and $q = |\mathbf{q}|$ and $d\Omega_q$ refer to the momentum of particle X, all in the C-M system. The electron variables $(k^2, k^\star, W, \theta)$ appearing in the cross section are functions of $(k^2, k \cdot p_1, \theta)$ where θ is the electron scattering angle in the laboratory frame. The appropriate relations for $(k^\star, W)$ as functions of $(k^2, k \cdot p_1)$ are given in Eqs. (13.35). There are three independent electron scattering variables in the lab, $(\varepsilon_1, \varepsilon_2, \theta)$; hence it is possible to fix $(k^2, k \cdot p_1)$ and vary θ. The current is evaluated in the C-M system.

There are four target responses appearing in the cross section expressed as bilinear combinations of current matrix elements where the current is defined by Eq. (13.41) with $\mathscr{I}_\mu \equiv (\mathscr{I}, i\mathscr{I}_C)$ and $\mathscr{I}^\lambda \equiv \mathbf{e}_\lambda \cdot \mathscr{I}$. These four responses are functions of the variables $(k^2, W, \theta_q, \phi_q)$ or $(k^2, k \cdot p_1, \theta_q, \phi_q)$. The dependence on the angle variables will be made explicit in the subsequent analysis. The dependence on the "out-of-plane" angle ϕ_q, whose content must be transmitted through the virtual photon, turns out to be particularly simple. It is explicitly exhibited as

$$|\mathscr{I}_C|^2$$
$$|\mathscr{I}^{+1}|^2 + |\mathscr{I}^{-1}|^2$$
$$2\,\mathrm{Re}\,(\mathscr{I}^{+1})^\star (\mathscr{I}^{-1}) \propto \cos 2\phi_q$$
$$\sqrt{2}\,\mathrm{Im}\,\mathscr{I}_C^\star (\mathscr{I}^{+1} + \mathscr{I}^{-1}) \propto \sin \phi_q \quad (13.48)$$

The dependence on (θ, ϕ_q) in Eqs. (13.47, 13.48) now allows a complete kinematic separation of the four target response functions at fixed $(k^2, k \cdot p_1, \theta_q)$, or equivalently fixed (k^2, W, θ_q). Since the term in $\cos 2\phi_q$ takes the same value at $\phi_q = \pi/2$ and $\phi_q = 3\pi/2$, for which the reaction plane and electron scattering plane in Fig. 13.1 coincide, an out-of-plane measurement is needed to separate its contribution. Conversely, the term in $\sin \phi_q$ can be isolated with two in-plane measurements at these two values.

This derivation is from appendix C of [Pr69]. Other work on coincidence experiments is contained in [de67, Wa79, Kl83]. Work on coinci-

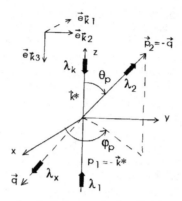

Fig. 13.3.   Configuration for helicity analysis of current matrix elements in the C-M system. Here everything is referred to the incoming and outgoing target states with momenta and helicities $(\mathbf{p}_1\lambda_1)$ and $(\mathbf{p}_2\lambda_2)$ respectively. Note how the (x, y, z) coordinate system is related to the original system defined by $\mathbf{e}_i$ with $i = 1, 2, 3$. Note in particular the relations $\theta_q = \theta_p$ and $\phi_q + \phi_p = 2\pi$.

dence experiments in pion electroproduction is contained in [Be66, Pr70]. Coincidence experiments with a polarized electron beam are discussed in [Ad68, Ra89][4] and with both a polarized electron beam and polarized target in [Ra89].

The next step is to demonstrate the angular dependence in the nuclear matrix elements. This will be done through the use of a *helicity analysis* of the current matrix elements in the C-M system. Let us go back to the form of the cross section before the $\overline{\sum}_i \sum_f$ has been carried out. The cross section is then being calculated for given initial and final helicities of all the particles in the C-M system, and of the virtual photon. The situation is illustrated in Fig. 13.3. The analysis parallels that of Jacob and Wick [Ja59]. First, recall some of the basic results from that work

For two-particles in the C-M system, the transformation from a state where the relative momentum is directed at an angle (θ, ϕ), to a state of definite angular momentum (J, M) is given by

$$\langle JM\lambda_1'\lambda_2'|\theta\phi\lambda_1\lambda_2\rangle = \delta_{\lambda_1\lambda_1'}\delta_{\lambda_2\lambda_2'}\left(\frac{2J+1}{4\pi}\right)^{1/2}\mathscr{D}_{M,\lambda}^J(-\phi,-\theta,\phi) \qquad (13.49)$$

Here $\lambda \equiv \lambda_1 - \lambda_2$ is the net helicity of the state. We have seen this expression before as the "photon wave function" in Eq. (9.35). The transformation in Eq. (13.49) is unitary.

---

[4] In a coincidence reaction with a polarized electron beam $(\vec{e}, e'\, X)$ there is an additional, *fifth response function*, sensitive to final-state interactions, which can only be accessed with out-of-plane measurements [Ra89].

The S-matrix for an arbitrary two-particle process in the C-M system can be written as

$$\langle \mathbf{p}_c \mathbf{p}_d \lambda_c \lambda_d | S | \mathbf{p}_a \mathbf{p}_b \lambda_a \lambda_b \rangle = \frac{(2\pi)^4}{\Omega^2} \delta^{(4)}(P_\mu - P'_\mu) \left[\frac{(2\pi)^2 \sqrt{v'v}}{p'p} \right]$$

$$\times \langle \theta' \phi' \lambda_c \lambda_d | S(P_\mu) | \theta \phi \lambda_a \lambda_b \rangle \qquad (13.50)$$

Here (p, v) are relative momenta and velocity in the C-M system and P is the total four-momentum in that frame. For transitions, the S-matrix is related to the T-matrix by $S = 1 + iT$.

With the aid of completeness, one then establishes the following relation for the required S-matrix in the C-M system

$$\langle \theta \phi \lambda_c \lambda_d | S(W) | 0 0 \lambda_a \lambda_b \rangle = \sum_{JM} \sum_{J'M'} \langle \theta \phi \lambda_c \lambda_d | JM \lambda_c \lambda_d \rangle$$

$$\times \langle JM \lambda_c \lambda_d | S(W) | J'M' \lambda_a \lambda_b \rangle \langle J'M' \lambda_a \lambda_b | 0 0 \lambda_a \lambda_b \rangle \qquad (13.51)$$

The scattering operator S is a scalar under rotations; it commutes with the angular momentum operator **J**. The Wigner–Eckart theorem then implies that the matrix element of S must be diagonal in J and independent of M. Use

$$\left(\frac{2J+1}{4\pi} \right)^{1/2} \mathscr{D}^J_{M,\lambda}(0,0,0) = \left(\frac{2J+1}{4\pi} \right)^{1/2} \delta_{M\lambda} \qquad (13.52)$$

Here the initial angular momentum along the z-axis is $M = \lambda = \lambda_a - \lambda_b$. A combination of the above results then yields the expression

$$\langle \theta \phi \lambda_c \lambda_d | S(W) | 0 0 \lambda_a \lambda_b \rangle =$$

$$\sum_J \left(\frac{2J+1}{4\pi} \right) \mathscr{D}^J_{\lambda_i, \lambda_f}(-\phi, -\theta, \phi)^\star \langle \lambda_c \lambda_d | S^J(W) | \lambda_a \lambda_b \rangle$$

$$; \lambda_i = \lambda_a - \lambda_b \qquad \quad ; \lambda_f = \lambda_c - \lambda_d \qquad (13.53)$$

There are various conditions on the helicity matrix elements of the scattering operator that follow from unitarity and symmetry properties of the strong interactions. Parity invariance essentially cuts the number of independent matrix elements in half. The parity operator reflects momentum and leaves particle spins unchanged; hence it reflects the helicity. It leaves the angular momentum and z-component of the angular momentum unchanged. The parity operator thus has the following effect on a two-particle state [Ja59]

$$P | JM \lambda_1 \lambda_2 \rangle = (-1)^{J - S_1 - S_2} \eta_1 \eta_2 | JM - \lambda_1 - \lambda_2 \rangle \qquad (13.54)$$

Here the η_i are intrinsic parities and the overall phase is conventional.[5] If the scattering operator is invariant under the parity transformation, i.e. if it commutes with P, then [Ja59]

$$\langle -\lambda_c - \lambda_d | S^J(W) | - \lambda_a - \lambda_b \rangle = \eta_a \eta_b \eta_c^\star \eta_d^\star \langle \lambda_c \lambda_d | S^J(W) | \lambda_a \lambda_b \rangle \quad (13.55)$$

Now the electroproduction process (e, e' X) in the C-M system presents *exactly the same problem* as discussed above.[6] The behavior under rotation of all quantities is exactly the same. The only new feature is that k^2, the mass of the virtual photon, provides an additional kinematic variable in the C-M system. To make the analogy more explicit, recall Low's first reduction of the S-matrix [Lo55]. For non-forward pion–nucleon scattering it takes the form

$$\frac{\langle p'q'|S|pq \rangle}{\langle 0|S|0 \rangle} = -(2\pi)^4 i \delta^{(4)}(p' + q' - p - q) \frac{1}{\sqrt{2\omega_q \Omega}} \langle p'q'^{(-)} | J(0) | p \rangle \quad (13.56)$$

Here $J(0)$ is the pion current (the isospin label is suppressed). This expression now has exactly the same form in terms of target matrix elements of the current as that we have been studying. The only difference is that in our case it is the matrix element of the electromagnetic current that is required. With the Low reduction, one shifts the transformation properties from the state vector (which we *do not* have for a virtual photon) to those of the current (which we *do* have).

With the electromagnetic current, one can use current conservation to relate the Coulomb and longitudinal matrix elements

$$\mathbf{e}_{k3} \cdot \mathscr{J} \equiv \mathscr{J}^{(0)} = \frac{\omega_k^\star}{k^\star} \mathscr{J}_C \quad (13.57)$$

This reduces the problem to the study of one or the other of these.

As a result of the above discussion, the helicity matrix elements of the electromagnetic current for the hadronic target in the C-M system, required for the cross section in Eq. (13.47), must have the following angular dependence

$$(\mathscr{J}_C)_{\lambda_f, \lambda_i} = \frac{k^\star}{\omega_k^\star} \frac{1}{\sqrt{4k^\star q}} \sum_J (2J + 1) \mathscr{D}_{\lambda_i, \lambda_f}^J (-\phi_p, -\theta_p, \phi_p)^\star$$
$$\times \langle \lambda_2 \lambda_X | T^J(W, k^2) | \lambda_1 \lambda_k \rangle \qquad ; \lambda_k = 0$$

$$\left(\mathscr{J}^{\lambda_k}\right)_{\lambda_f, \lambda_i} = \frac{1}{\sqrt{4k^\star q}} \sum_J (2J + 1) \mathscr{D}_{\lambda_i, \lambda_f}^J (-\phi_p, -\theta_p, \phi_p)^\star$$
$$\times \langle \lambda_2 \lambda_X | T^J(W, k^2) | \lambda_1 \lambda_k \rangle \qquad ; \lambda_k = \pm 1 \quad (13.58)$$

---

[5] For the photon $(-1)^{S_\gamma} \eta_\gamma = 1$.

[6] The particular coordinate system chosen in Fig. 13.3, which might appear somewhat perverse to the reader, was chosen to make this analogy explicit.

The normalization is conventional. Here

$$\lambda_i = \lambda_1 - \lambda_k \qquad\qquad ; \lambda_f = \lambda_2 - \lambda_X \qquad (13.59)$$

All the angular dependence is now explicit. From Fig. 13.3

$$\begin{aligned} \theta_p &= \theta_q \\ \phi_p &= 2\pi - \phi_q \end{aligned} \qquad (13.60)$$

The angular dependence with respect to the angles of Fig. 13.2 is then given by the relation

$$\mathscr{D}^J_{\lambda_i,\lambda_f}(-\phi_p,-\theta_p,\phi_p)^\star = \mathscr{D}^J_{\lambda_f,\lambda_i}(\phi_q,\theta_q,-\phi_q) \qquad (13.61)$$

The proof follows from [Ed74] and the fact that $\lambda_f - \lambda_i$ is an integer

$$\begin{aligned} \mathscr{D}^J_{\lambda_i,\lambda_f}(-\phi_p,-\theta_p,\phi_p)^\star &= e^{i\lambda_i\phi_p} d^J_{\lambda_i,\lambda_f}(-\theta_p) e^{-i\lambda_f\phi_p} \\ &= e^{-i\lambda_i\phi_q} d^J_{\lambda_f,\lambda_i}(\theta_q) e^{i\lambda_f\phi_q} \\ &= \mathscr{D}^J_{\lambda_f,\lambda_i}(\phi_q,\theta_q,-\phi_q) \qquad (13.62) \end{aligned}$$

In the expression for the (e, e′ X) cross section, for a given set of particle helicities, one needs the bilinear expression

$$\left(\mathscr{J}^{\lambda_k}\right)^\star_{\lambda_f,\lambda_i} \left(\mathscr{J}^{\lambda'_k}\right)_{\lambda_f,\lambda'_i} = \frac{1}{4k^\star q} \sum_J \sum_{J'} (2J+1)(2J'+1)\langle \lambda_2\lambda_X | T^J | \lambda_1\lambda_k\rangle^\star$$

$$\times\langle\lambda_2\lambda_X | T^{J'} | \lambda_1\lambda'_k\rangle \mathscr{D}^J_{\lambda_i,\lambda_f}(-\phi_p,-\theta_p,\phi_p)\mathscr{D}^{J'}_{\lambda'_i,\lambda_f}(-\phi_p,-\theta_p,\phi_p)^\star \qquad (13.63)$$

Here

$$\lambda_f = \lambda_2 - \lambda_X \qquad ; \lambda_i = \lambda_1 - \lambda_k \qquad ; \lambda'_i = \lambda_1 - \lambda'_k \quad (13.64)$$

This expression is required for values of λ_k and λ'_k of 0 and ± 1. With the aid of Eq. (13.62) and formulas in [Ed74], the angular functions appearing in these bilinear combinations can be written as

$$\mathscr{D}^J_{\lambda_f,\lambda_i}(\phi_q,\theta_q,-\phi_q)^\star \, \mathscr{D}^{J'}_{\lambda_f,\lambda'_i}(\phi_q,\theta_q,-\phi_q) = \qquad (13.65)$$

$$(-1)^{\lambda_f-\lambda_i}\mathscr{D}^J_{-\lambda_f,-\lambda_i}(\phi_q,\theta_q,-\phi_q)\,\mathscr{D}^{J'}_{\lambda_f,\lambda'_i}(\phi_q,\theta_q,-\phi_q)$$

Now use the composition law for rotation matrices [Ed74] to rewrite the r.h.s. of this expression as

$$\text{r.h.s.} = (-1)^{\lambda_f-\lambda_i}\sum_{lmm'}(2l+1)\begin{pmatrix} J & J' & l \\ -\lambda_f & \lambda_f & m \end{pmatrix}\mathscr{D}^l_{m,m'}(\phi_q,\theta_q,-\phi_q)^\star$$

$$\times\begin{pmatrix} J & J' & l \\ -\lambda_i & \lambda'_i & m' \end{pmatrix} \qquad (13.66)$$

Since m must vanish by the properties of the 3-j symbols, use

$$\mathscr{D}^{l}_{0,m'}(\phi_q, \theta_q, -\phi_q)^{\star} = \left(\frac{4\pi}{2l+1}\right)^{1/2} Y_{l,m'}(\theta_q, \phi_q) \qquad (13.67)$$

Since $\lambda_i - \lambda'_i = \lambda'_k - \lambda_k$, one finally has

$$\left(\mathscr{J}^{\lambda_k}\right)^{\star}_{\lambda_f,\lambda_i} \left(\mathscr{J}^{\lambda'_k}\right)_{\lambda_f,\lambda'_i} = \frac{1}{4k^{\star}q}(-1)^{\lambda_i-\lambda_f} \sum_{J}\sum_{J'}(2J+1)(2J'+1)$$

$$\times \langle \lambda_2\lambda_X | T^J | \lambda_1\lambda_k \rangle^{\star} \langle \lambda_2\lambda_X | T^{J'} | \lambda_1\lambda'_k \rangle \sum_{l} \sqrt{4\pi(2l+1)}$$

$$\times \begin{pmatrix} J & J' & l \\ \lambda_f & -\lambda_f & 0 \end{pmatrix} Y_{l,\lambda'_k-\lambda_k}(\theta_q, \phi_q) \begin{pmatrix} J & J' & l \\ \lambda_i & -\lambda'_i & \lambda_k - \lambda'_k \end{pmatrix} \qquad (13.68)$$

This formula gives the general angular dependence of the bilinear forms of the current appearing in the cross section for an arbitrary set of helicities of the reaction participants.[7] As such, it can be used to calculate the angular distributions in the C-M system for any polarization of the initial and final systems. *It is a central result.*

If the target is unpolarized, and the final particles are unobserved, one must average over initial helicities and sum over final helicities. We denote these sums with a bar over the bilinear combinations of currents

$$\overline{\mathscr{J}^{\star}\mathscr{J}} \equiv \overline{\sum_{\lambda_1}\sum_{\lambda_2}\sum_{\lambda_X}} \mathscr{J}^{\star}\mathscr{J} \qquad (13.69)$$

The transition matrix elements are functions of (W, k^2). Parity invariance of the strong and electromagnetic interactions implies

$$\langle -\lambda_2 - \lambda_X | T^J(W, k^2) | -\lambda_1 - \lambda_k \rangle =$$
$$\eta_2^{\star}\eta_X^{\star}\eta_1(-1)^{S_2+S_X-S_1}\langle \lambda_2\lambda_X | T^J(W, k^2)|\lambda_1\lambda_k \rangle \qquad (13.70)$$

A change of dummy helicity sum values to their negatives, use of the parity relation, and use of the symmetry properties of the 3-j symbols allow us to write the bilinear products of current matrix elements required in the electron scattering cross section in Eq. (13.47) in the following form[8]

$$\overline{|\mathscr{J}_{\mathscr{C}}|^2} = \frac{1}{4k^{\star}q}\sum_{l} A_l P_l(\cos\theta_q) \qquad (13.71)$$

---

[7] This includes, for example, the process of "virtual Compton scattering," now studied extensively through the coincidence reaction p(e, e'p)γ.

[8] The spherical harmonics are defined by $Y_{l,m} = (-1)^m \left[\frac{(2l+1)(l-m)!}{4\pi(l+m)!}\right]^{1/2} P_l^m(\cos\theta)e^{im\phi}$ for $m \geq 0$ while for $m < 0$ one has $Y^{\star}_{l,m} = (-1)^m Y_{l,-m}$. Here $P_l^m(\cos\theta)$ are the associated Legendre polynomials [Ed74], which for positive m are given by $P_l^m(x) = (1-x^2)^{m/2}d^m P_l(x)/dx^m$.

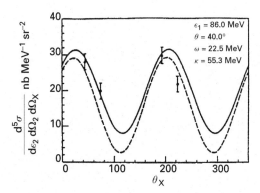

Fig. 13.4.   In-plane angular distribution of protons in $^{12}_{6}$C(e, e′ p$_0$)$^{11}_{5}$B through the giant dipole resonance measured with the SCA at HEPL [Kl83]. Data from [Ca80].

$$\overline{|\mathscr{J}^{+1}|^2} + \overline{|\mathscr{J}^{-1}|^2} = \frac{1}{4k^{\star}q} \sum_l B_l P_l(\cos\theta_q)$$

$$\mathrm{Im}\,\overline{\mathscr{J}_{\mathscr{C}}^{\star}\left(\mathscr{J}^{+1} + \mathscr{J}^{-1}\right)} = \frac{1}{4k^{\star}q} \sum_l C_l P_l^{(1)}(\cos\theta_q)\,\sin\phi_q$$

$$\mathrm{Re}\,\overline{\left(\mathscr{J}^{+1}\right)^{\star}\left(\mathscr{J}^{-1}\right)} = \frac{1}{4k^{\star}q} \sum_l \eta\,D_l P_l^{(2)}(\cos\theta_q)\,\cos 2\phi_q$$

These expressions provide the general angular distributions in the C-M system for (e, e′ X) for any target particles and any X. The coefficients (A_l, B_l, C_l, D_l) are bilinear combinations of helicity amplitudes; they are functions of (W, k^2). They are developed in detail in appendix F. The quantity $\eta = \eta_1 \eta_2^{\star} \eta_X^{\star}$ is the real combination of intrinsic parities. These expressions are further analyzed and tabulated in [Kl83].

The claim made in exhibiting the dependence on the out-of-plane angle ϕ_q in Eqs. (13.48) has now been established.

To give the reader some feel for coincident electron scattering, we present three brief examples. First, consider Fig. 13.4 which shows the coincidence cross section for $^{12}_{6}$C(e, e′ p$_0$)$^{11}_{5}$B [Ca80]. This is the first coincidence experiment done with the superconducting accelerator (SCA) at the Stanford High Energy Physics Laboratory (HEPL), a machine that proved to be the prototype for CEBAF . The energy transfer is controlled so that ^{12}C is excited to the giant dipole resonance. The in-plane angular distribution of the emitted proton leading to the ground state of ^{11}B is then measured with respect to the momentum transfer κ. This is an example of the angular correlation measurement discussed above, where the inelastic scattering of the electron first aligns the target along the direction of the momentum transfer. Notice the very nice dipole pattern

$$^{12}C\,(e,\,eP_0)''B$$

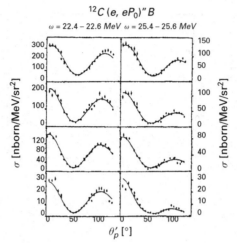

Fig. 13.5. Same reaction as in Fig. 13.4 with subsequent data from Mainz [De86, Ca94]. Here $\kappa = 0.25, 0.34, 0.41, 0.59\,\text{fm}^{-1}$.

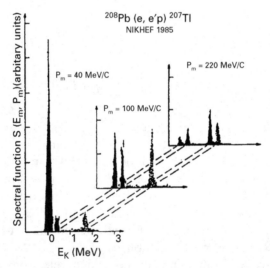

Fig. 13.6. Nuclear response for the reaction $^{208}_{82}\text{Pb}(e, e'\,p)^{207}_{81}\text{Tl}$ measured at NIKHEF [de86].

of the subsequently emitted proton.[9] The two theoretical curves in Fig. 13.4 are calculations carried out within the particle–hole model of the giant dipole resonance in ^{12}C [Kl83]. Now one may well say that the four points do not determine an angular distribution, and it is hard to disagree; however, Fig.13.5 shows the quality of the data one can now

---

[9] The initial aligned ^{12}C nucleus has $J^\pi = 1^-$ (hence the phrase "dipole pattern"). The final ground state of ^{11}B is $(3/2)^-$ and the emitted proton conserves angular momentum.

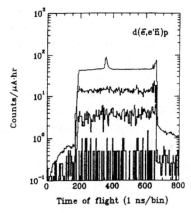

Fig. 13.7. Triple coincidence signal from $^{2}_{1}$H($\vec{e}$, e′ $\vec{n}$) experiment done at Bates [Ma92, Wa93].

obtain using the new generation of c.w. electron accelerators on the same reaction — this data is from Mainz [De86, Ca94]. The dipole pattern is now beautifully displayed.

As a second example, Fig.13.6 shows the nuclear response function for the reaction $^{208}_{82}$Pb(e, e′ p)$^{207}_{81}$Tl measured at NIKHEF [de86]. This example illustrates the discussion of (e, e′ p) in chapter 6.[10] One sees the ground state ($E_x = 0$), and then several excited hole states of $^{207}_{81}$Tl. Consider first the ground state. As $\kappa - \mathbf{q}$ is increased, the data exhibit the fall-off of the Fourier transform of the $(3s_{1/2})^{-1}_{\pi}$ wave function. The growth and fall-off of the Fourier transform of the $(2d_{3/2})^{-1}_{\pi}$ first excited state is then seen. At somewhat larger E_x, the high-multipolarity transition to the $(1h_{11/2})^{-1}_{\pi}$ appears from nowhere until it *dominates* the spectrum at the highest $\kappa - \mathbf{q}$. Note that one requires good resolution at high momenta to resolve the states.

This class of experiments represents one of the most important results coming from NIKHEF. These data are even more impressive when one realizes that they were obtained with only a few percent duty factor (d.f.) — the new generation of c.w. accelerators provides a significant advance. With this reaction, one can take the nucleus apart layer by layer and probe the limits of the single-nucleon description of nuclei.

As a third example, Fig. 13.7 shows the timing signal from the polarization transfer experiment $^{2}_{1}$H($\vec{e}$, e′ $\vec{n}$) carried out at Bates [Ma92]. This experiment provides an excellent example of how one can use interference in coincidence experiments to measure small quantities, in this case the electric form factor of the neutron which interferes with the well-known

---

[10] Here $(\varepsilon_b, \vec{\kappa} - \vec{q}) \equiv (E_m, \vec{P}_m)$.

magnetic form factor [Ar81]. This is really a triple coincidence experiment. The electron is detected, then the produced neutron, then the up or down scattering of the neutron to measure its polarization. The final signal is the small peak in the middle of the figure; the background consists of accidentals. The experiment was performed with an accelerator with $\sim 1\%$ d.f.. Now imagine that the signal forms a sea mount and the background an ocean. With a c.w. (100% d.f.) accelerator, one can lower the ocean level by over two orders of magnitude, and the small peak sticking up becomes a mountain. This is the most dramatic example, of which the author is aware, of what one gains with a c.w. machine.

14

Deep-inelastic scattering from the nucleon

We proceed to a discussion of inclusive deep-inelastic electron scattering from the nucleon $N(e, e')_{DIS}$. Here both the four-momentum transfer q^2 and energy transfer $v = q \cdot p/m$ become very large.[1] It is through these experiments, initially carried out at the Stanford Linear Accelerator Center (SLAC), that the first dynamic evidence for a point-like substructure of hadrons was obtained [Bj69, Fr72]. The structure functions exhibit this point-like substructure through Bjorken *scaling*, which implies $F_i(q^2, v) \rightarrow F_i(q^2/v)$ as $q^2 \rightarrow \infty$ and $v \rightarrow \infty$ at fixed q^2/v. To set the stage for the discussion in this section, we first review some of our general considerations on electron scattering [Qu83, Wa84] which form an essential basis for what follows. The experimental deep-inelastic results are then summarized [Fr72, Bj69, Qu83]. Finally, the quark–parton model is developed. It is through the quark–parton model that the deep-inelastic scaling was first understood [Fe69, Bj69a, Ha84, Ai89, Ma90].[2] The change of the structure functions in nuclei (EMC effect) gives direct evidence for the modification of quark properties in the nuclear medium [Au83], and this is briefly discussed.

The kinematics for electron scattering employed in this section are shown in Fig. 14.1. Here the four-momentum transfer is defined by[3]

$$
\begin{aligned}
q &= k_2 - k_1 = p - p' \\
q^2 &= 4\varepsilon_1 \varepsilon_2 \sin^2 \frac{\theta}{2} \qquad \text{; lab}
\end{aligned}
\tag{14.1}
$$

---

[1] We revert here to the previous notation where q denotes the momentum transfer in an inclusive process.

[2] QCD then allows a calculation of the *corrections* to scaling and the evolution equations for doing this [Al77] are discussed, for example, in [Wa95].

[3] Massless electrons are again assumed throughout this discussion.

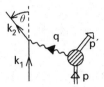

Fig. 14.1.   Kinematics in electron scattering; momenta are four-vectors.

We further define

$$
\begin{aligned}
\nu &\equiv \frac{q \cdot p}{m} \\
&= \varepsilon_1 - \varepsilon_2 \qquad\qquad ; \text{ lab} \\
x &\equiv \frac{q^2}{2m\nu}
\end{aligned}
\tag{14.2}
$$

These are the energy loss in the lab frame and Bjorken scaling variable, respectively.

The S-matrix for the process in Fig. 14.1 is given by

$$
S_{fi} = -\frac{(2\pi)^4}{\Omega}\delta^{(4)}(k_1 + p - k_2 - p')ee_p\bar{u}(k_2)\gamma_\mu u(k_1)\frac{1}{q^2}\langle p'|J_\mu(0)|p\rangle \tag{14.3}
$$

Here $J_\mu(x)$ is the local electromagnetic current operator for the target system. With box normalization,[4] momentum conservation is actually expressed through the relation

$$
\frac{(2\pi)^3}{\Omega}\delta^{(3)}(\mathbf{k}_1 + \mathbf{p} - \mathbf{k}_2 - \mathbf{p}') \doteq \delta_{\mathbf{k}_1 + \mathbf{p}, \mathbf{k}_2 + \mathbf{p}'} \tag{14.4}
$$

The incident flux in any frame where $\mathbf{k}_1\|\mathbf{p}$ is given by

$$
I_0 = \frac{1}{\Omega}\frac{\sqrt{(k_1 \cdot p)^2}}{\varepsilon_1 E_p} \tag{14.5}
$$

Then for a one-body nuclear final state

$$
\begin{aligned}
S_{fi} &\equiv -2\pi i\,\delta(\varepsilon_1 + E_p - \varepsilon_2 - E_{p'})\delta_{\mathbf{k}_1 + \mathbf{p}, \mathbf{k}_2 + \mathbf{p}'}\bar{T}_{fi} \\
d\sigma_{fi} &= 2\pi|\bar{T}_{fi}|^2\delta(W_f - W_i)\frac{\Omega d^3 k_2}{(2\pi)^3}\left[\frac{1}{\Omega}\frac{\sqrt{(k_1 \cdot p)^2}}{\varepsilon_1 E_p}\right]^{-1}
\end{aligned}
\tag{14.6}
$$

Here $W_f = \varepsilon_2 + E_{p'}$ and $W_i = \varepsilon_1 + E_p$ are the total final and initial energies, respectively. It follows that the differential cross section in any

---

[4] That is, periodic boundary conditions in a big box of volume Ω.

frame where $\mathbf{k}_1 \| \mathbf{p}$ is given in Lorentz invariant form by

$$d\sigma = \frac{4\alpha^2}{q^4} \frac{d^3 k_2}{2\varepsilon_2} \frac{1}{\sqrt{(k_1 \cdot p)^2}} \eta_{\mu\nu} W_{\mu\nu} \tag{14.7}$$

In this expression the lepton and hadron tensors for unpolarized electrons and targets, generalized to include arbitrary nuclear final states, are defined by

$$\eta_{\mu\nu} = -2\varepsilon_1 \varepsilon_2 \frac{1}{2} \sum_{s_1} \sum_{s_2} \bar{u}(k_1)\gamma_\nu u(k_2)\bar{u}(k_2)\gamma_\mu u(k_1) \tag{14.8}$$

$$W_{\mu\nu} = (2\pi)^3 \Omega \overline{\sum_i} \sum_f \delta^{(4)}(q + p' - p)\langle p|J_\nu(0)|p'\rangle\langle p'|J_\mu(0)|p\rangle E_p$$

The lepton tensor can be evaluated directly (recall the mass of the electron is neglected)

$$\eta_{\mu\nu} = -2\varepsilon_1 \varepsilon_2 \frac{1}{2} \text{trace} \frac{(-ik_{1\lambda}\gamma_\lambda)}{2\varepsilon_1} \gamma_\nu \frac{(-ik_{2\rho}\gamma_\rho)}{2\varepsilon_2} \gamma_\mu$$
$$= k_{1\mu}k_{2\nu} + k_{1\nu}k_{2\mu} - (k_1 \cdot k_2)\delta_{\mu\nu} \tag{14.9}$$

It follows from the definition in Eq. (14.8) that the lepton current is conserved

$$q_\mu \eta_{\mu\nu} = \eta_{\mu\nu} q_\nu = 0 \tag{14.10}$$

The hadron tensor depends on just the two four-vectors (q, p) and is also conserved; its general form is

$$W_{\mu\nu} = W_1(q^2, q \cdot p) \left(\delta_{\mu\nu} - \frac{q_\mu q_\nu}{q^2} \right)$$
$$+ W_2(q^2, q \cdot p) \frac{1}{m^2} \left(p_\mu - \frac{q \cdot p}{q^2} q_\mu \right) \left(p_\nu - \frac{q \cdot p}{q^2} q_\nu \right) \tag{14.11}$$

With this background, let us proceed to further analyze the hadronic response tensor. The Heisenberg equations of motion for the target are as follows:

$$\hat{O}(x) = e^{-i\hat{P}\cdot x}\hat{O}(0)e^{i\hat{P}\cdot x} \tag{14.12}$$

They can be used to exhibit the space-time dependence of a matrix element taken between eigenstates of four-momentum

$$W_{\mu\nu} = \frac{1}{2\pi}(\Omega E)\overline{\sum_i} \sum_f \int e^{iq\cdot z} d^4 z \langle p|J_\nu(z)|p'\rangle\langle p'|J_\mu(0)|p\rangle$$
$$= \frac{1}{2\pi}(\Omega E)\overline{\sum_i} \int e^{iq\cdot z} d^4 z \langle p|J_\nu(z)J_\mu(0)|p\rangle \tag{14.13}$$

Fig. 14.2.   Kinematics for crossed term.

Completeness of the final set of hadronic states has been used to obtain the second line. Consider the matrix elements of the operators in the opposite order

$$\int e^{iq\cdot z} d^4z \langle p|J_\mu(0)J_\nu(z)|p\rangle \propto \sum_f (2\pi)^4 \delta^{(4)}(p+q-p')\langle p|J_\mu(0)|p'\rangle\langle p'|J_\nu(0)|p\rangle$$

(14.14)

Here the kinematics are illustrated in Fig. 14.2

$$
\begin{aligned}
p+q &= p' \\
q_0 &= \varepsilon_2 - \varepsilon_1 < 0
\end{aligned}
$$

(14.15)

One cannot reach a physical state under these kinematic conditions since the nucleon is *stable*; thus the expression in Eq. (14.14) vanishes. One can subtract this vanishing term in Eq. (14.13) and write $W_{\mu\nu}$ as the Fourier transform of the commutator of the current density at two displaced space-time points

$$W_{\mu\nu} = \frac{1}{2\pi}(\Omega E)\overline{\sum_i} \int e^{iq\cdot z} d^4z \langle p|[J_\nu(z), J_\mu(0)]|p\rangle$$

(14.16)

Introduce states with *covariant norm*[5]

$$|p) \equiv \sqrt{2E\,\Omega}\,|p\rangle$$

(14.17)

Equation (14.13) can then be rewritten

$$-\pi W_{\mu\nu} \equiv t_{\mu\nu} = -\frac{1}{4}\overline{\sum_i} \int e^{iq\cdot z} d^4z \, (p|[J_\nu(z), J_\mu(0)]|p)$$

(14.18)

This expression is evidently covariant; it forms the absorptive part of the amplitude for forward, virtual Compton scattering. Since the currents are observables, their commutator must vanish outside the light cone. Thus the only contribution to this integral comes from inside the light cone.

---

[5] The norm of these states is $(\vec{p}\,|\vec{p}') = 2E(2\pi)^3\delta^{(3)}(\vec{p}-\vec{p}')$; this is Lorentz invariant.

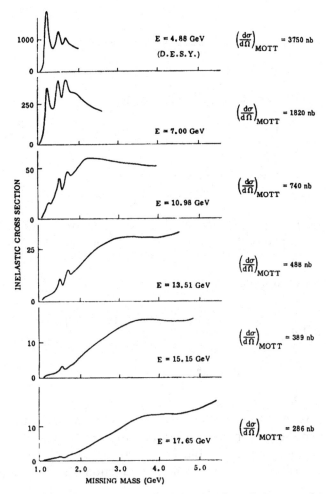

Fig. 14.3. Visual fits to spectra showing the scattering of electrons from hydrogen at $\theta = 10°$ for primary energies 4.88 to 17.65 GeV. The elastic peaks have been subtracted and radiative corrections applied. The cross sections are expressed in nanobarns/GeV/steradian [Fr72].

In the Bjorken scaling limit, the dominant contribution to this integral comes, in fact, from singularities *on* the light cone (see e.g. [De73]). This observation forms the basis for a covariant, field theory evaluation of the structure functions and systematic determination of corrections. The light-cone analysis of this expression is discussed in more detail in appendix I.

A combination of Eqs. (14.7), (14.9), and (14.11) yields the general form of the laboratory cross section for the scattering of unpolarized (massless)

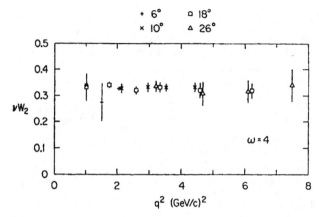

Fig. 14.4. vW_2 for the proton as a function of q^2 and total C-M energy of the proton and virtual photon $W = [-(p-q)^2]^{1/2} > 2\,\text{GeV}$ at $\omega = 1/x = 4$ [Fr72].

electrons from an unpolarized nucleon

$$\frac{d^2\sigma}{d\Omega_2 d\varepsilon_2} = \sigma_M \frac{1}{m}\left[W_2(v,q^2) + 2W_1(v,q^2)\tan^2\frac{\theta}{2}\right]$$

$$\sigma_M = \frac{\alpha^2\cos^2\theta/2}{4\varepsilon_1^2\sin^4\theta/2} \tag{14.19}$$

Here σ_M is the Mott cross section.

A qualitative overview of the SLAC data on deep-inelastic electron scattering from the proton is shown in Fig. 14.3 [Fr72]. On the basis of his analysis of various sum rules, Bjorken *predicted*, before the experiments, the following behavior of the structure functions in the deep-inelastic regime [Bj69]

$$\frac{v}{m}W_2(v,q^2) \rightarrow F_2(x) \qquad ; q^2 \rightarrow \infty, \quad v \rightarrow \infty$$

$$2W_1(v,q^2) \rightarrow F_1(x) \tag{14.20}$$

Here the scaling variable is defined by

$$x \equiv \frac{q^2}{2mv} \equiv \frac{1}{\omega} \tag{14.21}$$

These relations imply that the structure functions do not depend individually on (v,q^2) but only on their *ratio*. The scaling behavior of the SLAC data is shown in Figs. 14.4 and 14.5 [Fr72].[6] The first of these figures

---

[6] These authors use $W_{1,2} \equiv (1/m)W_{1,2}^{\text{text}}$ where $W_{1,2}^{\text{text}}$ are the structure functions used here.

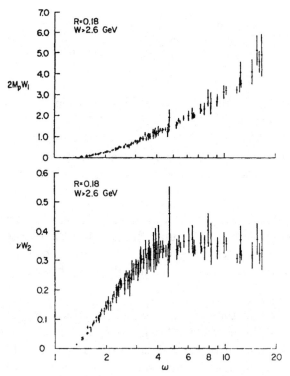

Fig. 14.5. Structure functions $2mW_1$ and vW_2 for the proton vs ω for C-M energy $W > 2.6\,\mathrm{GeV}$ and $q^2 > 1\,(\mathrm{GeV/c})^2$, and using R = 0.18 [Fr72].

illustrates the independence from q^2 at fixed $\omega = 1/x$; the second shows the extracted structure functions $F_{1,2}(x)$.[7]

Let us now turn to an interpretation of these experimental results. The empirical data on deep-inelastic electron scattering can be understood within the framework of the *quark–parton model* developed for that purpose by Feynman and Bjorken and Paschos [Fe69, Bj69a]. The basic concepts in the model are as follows:

- The calculation of the structure functions is Lorentz invariant. Go to the C-M frame of the proton and incident electron with $\mathbf{p} = -\mathbf{k}_1$. Now let the proton move very fast with $|\mathbf{p}| \to \infty$. This forms the *infinite-momentum frame*; it is illustrated in Fig. 14.6.

- Assume the nucleon is composed of a substructure of *partons*. The proper motion of the parton constituents of the hadron (here a proton) is slowed down by time dilation in the infinite-momentum

---

[7] From the SLAC data the ratio of longitudinal to transverse cross section is given by $R \equiv \sigma_1/\sigma_t = 0.18 \pm 0.10$ where $W_1/W_2 \equiv (1 + v^2/q^2)\sigma_t/(\sigma_t + \sigma_1)$.

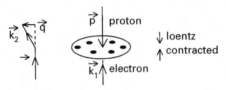

Fig. 14.6. Situation in frame where the proton is moving very rapidly with momentum $\mathbf{p} = -\mathbf{k}_1$ (the infinite-momentum frame).

frame. Thus the partons are effectively *frozen* during the scattering process. The actual interaction between the partons is then not important.

- Assume that the very-short-wavelength electrons scatter *incoherently* from the constituents. Assume further that the constituent partons have no internal electromagnetic structure and that the electrons scatter from the charged constituents as if they are *pointlike Dirac particles*.

- Assume that in the limit $q^2 \to \infty, \nu \to \infty$, the masses of the constituents can be neglected. Assume also that the transverse momentum of the parton before the collision, determined by the internal structure of the hadron and the strong-interaction dynamics, can be neglected in comparison with $\sqrt{q^2}$, the transverse momentum imparted as $|\vec{p}| \to \infty$.

We now know from subsequent developments, largely motivated by these deep-inelastic electron scattering experiments and the success of the quark–parton model, that the parton constituents of the hadron are actually quarks (charged) and gluons (neutral).

The scaling results can be understood within the framework of the *impulse approximation* applied to this model.[8] The calculation of the cross section is Lorentz invariant, and can be performed in any Lorentz frame, in particular in any frame where $\mathbf{p}\|\mathbf{k}_1$. Go to the infinite-momentum frame. The scattering situation is then illustrated in Fig. 14.7. In this frame the ith parton carries the incident four-momentum

$$p_{\text{inc}} = \eta_i\, p \qquad (14.22)$$

Here η_i is the fraction of the four-momentum p of the proton carried by the ith parton. Evidently

$$0 \le \eta_i \le 1 \qquad (14.23)$$

---

[8] This discussion is based on [Ha84, Ai89, Ma90, Wa95].

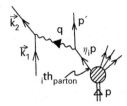

Fig. 14.7. Scattering in impulse approximation in the quark–parton model in the infinite-momentum frame.

The incident hadron is now just a collection of independent partons. The electron proceeds to scatter from one of the point-like charged partons. We do not worry here about how the parton eventually gets converted into hadrons in the final state (*hadronization*). Only the quarks are charged with charges

$$q_i \equiv Q_i e_p \tag{14.24}$$

Now

Let $f_i(\eta_i)d\eta_i$ be the number of quarks of type i with four-momentum between $\eta_i p$ and $(\eta_i + d\eta_i)p$.

The total four-momentum of the proton is then evidently given by

$$p = p_{\text{gluons}} + p_{\text{quarks}}$$
$$p = \zeta_g p + \sum_{i=1}^{N} \int_0^1 (\eta_i p) f_i(\eta_i) d\eta_i \tag{14.25}$$

Here ζ_g is the fraction of the total four-momentum of the proton carried by all the gluons, and $\sum_{i=1}^{N}$ is a sum over all types of quarks.

Cancellation of an overall factor of the four-momentum p from the last of Eqs. (14.25) gives

$$1 = \zeta_g + \sum_{i=1}^{N} \int_0^1 \eta_i f_i(\eta_i) d\eta_i \tag{14.26}$$

Introduce a dummy variable x; this *momentum sum rule* can then be written

$$1 = \zeta_g + \sum_{i=1}^{N} \zeta_i$$
$$\zeta_i \equiv \int_0^1 x f_i(x) dx \tag{14.27}$$

Now calculate the process in Fig. 14.7 using the analysis of inelastic electron scattering presented at the beginning of this chapter. With the assumption of scattering from point-like Dirac particles, the S-matrix for scattering from an isolated quark of type i is given by[9]

$$
\begin{aligned}
S^{(i)} &= \frac{-i(2\pi)^4 e e_p Q_i}{\Omega^2 q^2} \delta^{(4)}(p' + q - \eta_i p) \bar{u}(k_2) \gamma_\mu u(k_1) \bar{u}(p') \gamma_\mu u(\eta_i p) \\
&\equiv -\frac{(2\pi)^4 i}{\Omega} \delta^{(4)}(p' + q - \eta_i p) \bar{T}^{(i)}
\end{aligned}
\tag{14.28}
$$

The incident flux is given by

$$
I_0 = \frac{1}{\Omega} \frac{\sqrt{[k_1 \cdot (\eta_i p)]^2}}{\varepsilon_1(\eta_i E_p)} = \frac{1}{\Omega} \frac{\sqrt{(k_1 \cdot p)^2}}{\varepsilon_1 E_p}
\tag{14.29}
$$

The cross section for inelastic electron scattering from a point-like quark of type i, carrying four momentum $\eta_i p$ in the $|\mathbf{p}| \to \infty$ frame, in the impulse approximation follows as

$$
\begin{aligned}
d\sigma^{(i)} &= 2\pi |\bar{T}^{(i)}|^2 \delta(W_f - W_i) \frac{\Omega d^3 k_2}{(2\pi)^3} \frac{1}{I_0} \\
&= \frac{4\alpha^2}{q^4} \frac{d^3 k_2}{2\varepsilon_2} \frac{1}{\sqrt{(k_1 \cdot p)^2}} \eta_{\mu\nu} W_{\mu\nu}^{(i)}
\end{aligned}
\tag{14.30}
$$

Here the response tensor for scattering from such a quark is defined by[10]

$$
\begin{aligned}
W_{\mu\nu}^{(i)} &= -Q_i^2 E_p \sum_{\mathbf{p}'} \frac{1}{2} \sum_{s_1} \sum_{s_2} \bar{u}(p') \gamma_\mu u(\eta_i p) \bar{u}(\eta_i p) \gamma_\nu u(p') \\
&\quad \times \delta_{\mathbf{p}', \eta_i \mathbf{p} - \mathbf{q}} \, \delta(p_0' - \eta_i p_0 + q_0)
\end{aligned}
\tag{14.31}
$$

With the use of momentum conservation and the neglect of the masses of the participants, the energy-conserving delta function can be manipulated in the following manner (and this is a key step in the development)

$$
\begin{aligned}
\delta(p_0' - \eta_i p_0 + q_0) &= 2p_0' \delta[p_0'^2 - (\eta_i p_0 - q_0)^2] \\
&= 2p_0' \delta[p'^2 - (\eta_i p - q)^2] \\
&\approx 2p_0' \delta(2\eta_i p \cdot q - q^2) \\
&= \frac{2p_0'}{2p \cdot q} \delta(\eta_i - x)
\end{aligned}
\tag{14.32}
$$

Here $x \equiv q^2/2m\nu$ is the scaling variable introduced in Eq. (14.21). Hence

$$
\delta(p_0' - \eta_i p_0 + q_0) = \frac{2E_{p'}}{2m\nu} \delta(\eta_i - x)
\tag{14.33}
$$

---

[9] To avoid confusion, we here suppress the subscripts on the S-matrix $S_{fi}^{(i)}$.

[10] This assumes the target is unpolarized; polarization is discussed in the next chapter.

The required traces are the same as those evaluated in $\eta_{\mu\nu}$ at the beginning of this chapter, except that the initial momentum is $\eta_i p$. Thus

$$
\begin{aligned}
W_{\mu\nu}^{(i)} &= Q_i^2 E_p \frac{2E_{p'}}{2m\nu} \delta(\eta_i - x) \frac{4}{2E_{p'} 2(\eta_i E_p)} \frac{1}{2} \\
&\quad \times \left\{ p_\mu'(\eta_i p_\nu) + (\eta_i p_\mu) p_\nu' - (\eta_i p \cdot p') \delta_{\mu\nu} \right\} \\
&= \frac{Q_i^2}{2m\nu} \delta(\eta_i - x) \left\{ p_\mu' p_\nu + p_\nu' p_\mu - (p \cdot p') \delta_{\mu\nu} \right\}
\end{aligned}
\tag{14.34}
$$

Now use

$$
\begin{aligned}
p' &= \eta_i p - q \\
q_\mu \eta_{\mu\nu} &= \eta_{\mu\nu} q_\nu = 0
\end{aligned}
\tag{14.35}
$$

Hence, again with the neglect of masses,

$$
W_{\mu\nu}^{(i)} \doteq \frac{Q_i^2}{2} \delta(\eta_i - x) \left[\delta_{\mu\nu} + \frac{2\eta_i}{m\nu} p_\mu p_\nu \right]
\tag{14.36}
$$

The symbol $\doteq$ here indicates that the terms in q_μ and q_ν have been dropped because of Eq. (14.35).

An incoherent sum over all types of quarks with all momentum fractions now gives the response tensor for the composite nucleon

$$
W_{\mu\nu} = \sum_{i=1}^{N} \int_0^1 d\eta_i f_i(\eta_i) W_{\mu\nu}^{(i)}
\tag{14.37}
$$

Substitution of Eq. (14.36) into Eq. (14.37) demonstrates that the response functions now explicitly *exhibit Bjorken scaling* and allows one to identify [see Eqs. (14.37), (14.20), and (14.21)]

$$
\begin{aligned}
F_1(x) &= \sum_{i=1}^{N} Q_i^2 f_i(x) \\
F_2(x) &= \sum_{i=1}^{N} Q_i^2 x f_i(x) = x F_1(x)
\end{aligned}
\tag{14.38}
$$

Not only do these expressions explicitly exhibit scaling, but they also allow one to calculate the structure functions in terms of the charges of the various types of quarks and their momentum distributions as defined just below Eq. (14.24).

To proceed further, consider the nucleon to be made up of (u, d, s) quarks, with charges listed in Table 14.1, and their antiparticles. It then

Table 14.1.   Quark sector used in discussion of deep-inelastic electron scattering from the nucleon.

	u	d	s
Q_i	$2/3$	$-1/3$	$-1/3$

follows from Eq. (14.38) that

$$\frac{F_2^p(x)}{x} = \left(\frac{2}{3}\right)^2 [u^p(x) + \bar{u}^p(x)] + \left(\frac{1}{3}\right)^2 [d^p(x) + \bar{d}^p(x)]$$

$$+ \left(\frac{1}{3}\right)^2 [s^p(x) + \bar{s}^p(x)]$$

$$\frac{F_2^n(x)}{x} = \left(\frac{2}{3}\right)^2 [u^n(x) + \bar{u}^n(x)] + \left(\frac{1}{3}\right)^2 [d^n(x) + \bar{d}^n(x)]$$

$$+ \left(\frac{1}{3}\right)^2 [s^n(x) + \bar{s}^n(x)] \tag{14.39}$$

Here an obvious notation has been introduced for the momentum distributions $f_i(x)$ of the various quark types in the proton and neutron.

Strong isospin symmetry implies that the quark distributions should be invariant under the interchange $(d \rightleftharpoons u)$ and hence $(p \rightleftharpoons n)$. Thus one defines

$$u^p(x) = d^n(x) \equiv u(x)$$
$$d^p(x) = u^n(x) \equiv d(x)$$
$$s^p(x) = s^n(x) \equiv s(x) \tag{14.40}$$

The quark contributions can be divided into two types: those from *valence* quarks, from which the quantum numbers of the nucleon are constructed; and those from *sea* quarks, present, for example, from $(q\bar{q})$ pairs arising from strong vacuum polarization or mesons in the nucleon.

$$u(x) = u_V(x) + u_S(x)$$
$$d(x) = d_V(x) + d_S(x)$$
$$s(x) = s_V(x) + s_S(x) \tag{14.41}$$

Strong vacuum polarization should not distinguish greatly between the types of sea quarks; hence it will be assumed for the purposes of the present arguments that the sea quark distributions are identical

$$S(x) \equiv u_S = \bar{u}_S = d_S = \bar{d}_S = s_S = \bar{s}_S \tag{14.42}$$

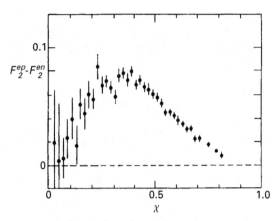

Fig. 14.8. The difference $F_2^p - F_2^n$ as a function of x, as measured in deep-inelastic scattering at the Stanford Linear Accelerator [Ha84].

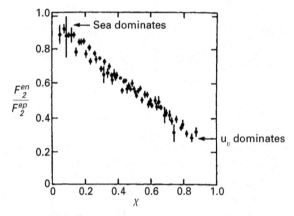

Fig. 14.9. The ratio F_2^n/F_2^p as a function of x, as measured in deep-inelastic scattering. Data are from the Stanford Linear Accelerator [Ha84].

It follows that

$$\frac{F_2^p}{x} = \frac{4}{9}u_V(x) + \frac{1}{9}d_V(x) + \frac{4}{3}S(x)$$
$$\frac{F_2^n}{x} = \frac{1}{9}u_V(x) + \frac{4}{9}d_V(x) + \frac{4}{3}S(x) \qquad (14.43)$$

The SLAC data comparing the distribution functions $F_2^{p,n}$ is shown in Figs. 14.8 and 14.9 (taken from [Ha84]). The neutron data were obtained subsequently at SLAC using a $_1^2$H target. Evidently at small x the ratio $F_2^p/F_2^n \approx 1$ and the sea quark distribution $S(x)$ dominates the structure function; at large x the ratio $F_2^n/F_2^p \approx 0.25$ and it is the valence u quark distribution $u_V(x)$ that dominates.

Consider the momentum sum rule. For simplicity, work in the *nuclear domain* where the nucleon is composed of (u, d) quarks and their anti-quarks. The contribution of these quarks to the momentum sum rule in Eq. (14.27) takes the form

$$\zeta_u \equiv \int_0^1 x \, dx (u + \bar{u})$$

$$\zeta_d \equiv \int_0^1 x \, dx (d + \bar{d}) \tag{14.44}$$

From the SLAC results [Ha84, Ma90] one finds the sum rules

$$\int_0^1 dx F_2^p(x) = \frac{4}{9}\zeta_u + \frac{1}{9}\zeta_d = 0.18$$

$$\int_0^1 dx F_2^n(x) = \frac{1}{9}\zeta_u + \frac{4}{9}\zeta_d = 0.12 \tag{14.45}$$

These results, together with Eq. (14.27), then imply

$$\zeta_u = 0.36 \qquad\qquad \zeta_d = 0.18$$
$$\zeta_g = 0.46 \tag{14.46}$$

Hence one observes that the gluons carry approximately one-half of the momentum of the proton.

We close this section with a very brief discussion of the EMC effect. This material is from [Au83, Mo86, Bi89, Dm90]. The most naive picture of the nucleus is that of a collection of free, non-interacting nucleons. In this picture the structure function one would observe from deep-inelastic electron scattering from a nucleus would be just N times the neutron structure function plus Z times that of the proton. It is an experimental fact, first established by the European Muon Collaboration (EMC), that the quark structure functions are *modified* inside the nucleus [Au83].

It is known that nucleons in the nucleus have a momentum distribution. The most elementary nuclear effect on the structure functions for the nucleus A involves a simple average over the single-nucleon momentum distribution

$$W_{\mu\nu}^{(A)}(P, q) = \sum_{i=1}^{A} \int d^3p |\phi_i(\mathbf{p})|^2 W_{\mu\nu}^{(1)}(p, q) \tag{14.47}$$

We note an immediate difficulty in the extension of the theoretical analysis to an A-body nucleus; this expression is clearly model dependent in the sense that the integration is *not covariant*. It is only with a covariant description of the nuclear many-body system that one can freely transform

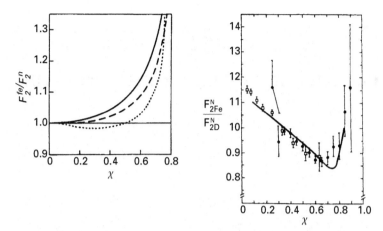

Fig. 14.10. (a) A comparison of calculations of the effect of Fermi smearing on the ratio $\mathcal{R}$ [Bi89]; (b) The ratio $\mathcal{R}$ in a relativistic version of this single-particle model compared with some early experimental data [Mo86].

between Lorentz frames, and, in particular, go to the $|\mathbf{p}| \to \infty$ frame where the parton model is developed.

It will be assumed that Eq. (14.47) holds in the laboratory frame. Define the following ratio

$$\mathcal{R} \equiv \frac{F_2^{\text{Fe}}(x)/A}{F_2^{\text{D}}(x)/2} \tag{14.48}$$

This is the ratio of the structure function for iron (per nucleon) to the structure function for deuterium (per nucleon). Calculations of $\mathcal{R}$ based on Eq. (14.47) are shown in Fig. 14.10 (a). $\mathcal{R}$ is calculated assuming the response function $W_{\mu\nu}^{(1)}(p, q)$ for a free nucleon is unmodified in the nuclear interior [Bi89]. Note that this Fermi smearing effect is sizable for large x.

The result of a relativistic version of this single-particle model is shown in Fig. 14.10 (b), along with some of the representative early experimental data [Mo86].

15

Polarization in deep-inelastic scattering

Suppose that the initial lepton beam is longitudinally polarized. If the target nucleon is unpolarized and unobserved, there is no effect on the cross section because parity is conserved in the strong and electromagnetic interactions. We consider parity violation induced by the weak interaction in the next chapter.

Suppose, however, that the target nucleon is polarized along the incident electron direction so that it is *also* longitudinally polarized in the infinite-momentum frame. There is then additional information on the strong-interaction spin structure of the nucleon in deep-inelastic scattering (DIS) experiments carried out under these conditions. Many such $\vec{N}(\vec{e}, e')_{\text{DIS}}$ experiments have now been performed, starting with the work of Vernon Hughes and collaborators at SLAC [Hu83]. A theoretical analysis of such experiments follows immediately from our discussions of the quark–parton model in chapter 14 and of the polarization of spin-1/2 fermions in appendix D.

In the extreme relativistic limit (ERL) one can simply insert the appropriate helicity projection operator for massless fermions in the lepton trace. For helicity $h = \pm 1$ one uses[1]

$$P_h = \frac{1}{2}(1 - h\gamma_5) \tag{15.1}$$

The result is that the lepton trace now takes the form

$$
\begin{aligned}
\eta^h_{\mu\nu} &= -2\varepsilon_1\varepsilon_2 \frac{1}{2} \sum_{s_1} \sum_{s_2} \bar{u}(k_1)\gamma_\nu u(k_2)\bar{u}(k_2)\gamma_\mu(1 - h\gamma_5)u(k_1) \\
&= -\varepsilon_1\varepsilon_2 \text{ trace} \left[\gamma_\nu \left(\frac{-i\gamma_\lambda k_{2\lambda}}{2\varepsilon_2} \right) \gamma_\mu(1 - h\gamma_5) \left(\frac{-i\gamma_\rho k_{1\rho}}{2\varepsilon_1} \right) \right]
\end{aligned}
$$

---

[1] Note that the h used here is twice the spin projection.

$$= \frac{1}{4}4\left[k_{2\nu}k_{1\mu} + k_{1\nu}k_{2\mu} - (k_1 \cdot k_2)\delta_{\mu\nu} + h\varepsilon_{\mu\nu\lambda\rho}k_{2\lambda}k_{1\rho}\right]$$
$$= \eta_{\mu\nu} + h\varepsilon_{\mu\nu\lambda\rho}k_{2\lambda}k_{1\rho} \tag{15.2}$$

Here, as before

$$\eta_{\mu\nu} = k_{2\nu}k_{1\mu} + k_{1\nu}k_{2\mu} - (k_1 \cdot k_2)\delta_{\mu\nu} \tag{15.3}$$

Since the helicity is a pseudoscalar under the parity transformation (witness the γ_5 in the helicity projection operator), there can now be a pseudotensor contribution $h\varepsilon_{\mu\nu\lambda\rho}k_{2\lambda}k_{1\rho}$ to the response tensor.

Assume the target is longitudinally polarized and has helicity aligned ($\uparrow$) along $\mathbf{p}$. Now carry out exactly the same impulse approximation calculation in the quark–parton model as in the previous section only now insert a projection operator for the relativistic *quarks* of helicity $h_i = \pm 1$

$$P_{h_i} = \frac{1}{2}(1 - h_i\gamma_5) \tag{15.4}$$

Evidently h_i measures the helicity of the quarks relative to the helicity of the nucleon (here $\uparrow$). An inspection of the arguments leading to Eq. (14.34), and the above analysis, indicate that one should make the following replacement in Eq. (14.34)

$$\left\{ p'_\mu p_\nu + p'_\nu p_\mu - (p \cdot p')\delta_{\mu\nu} \right\} \rightarrow$$
$$\left\{ p'_\mu p_\nu + p'_\nu p_\mu - (p \cdot p')\delta_{\mu\nu} + h_i\, \varepsilon_{\mu\nu\lambda\rho}p'_\lambda p_\rho \right\} \tag{15.5}$$

One then proceeds in exactly the same manner to Eq. (14.36) with the result

$$W^{(i)}_{\mu\nu} \doteq \frac{Q_i^2}{2}\delta(\eta_i - x)\left[\delta_{\mu\nu} + \frac{2\eta_i}{m\nu}p_\mu p_\nu + \frac{h_i}{m\nu}\varepsilon_{\mu\nu\lambda\rho}p_\lambda q_\rho\right] \tag{15.6}$$

An incoherent sum over all quarks implies that there is an additional Lorentz covariant contribution to the DIS response tensor for this nucleon with positive helicity of the form

$$\delta W^{\uparrow}_{\mu\nu} = W^{\uparrow}\frac{1}{m^2}\varepsilon_{\mu\nu\lambda\rho}p_\lambda q_\rho \tag{15.7}$$

The quark–parton model identifies

$$\frac{2\nu}{m}W^{\uparrow} = \sum_i Q_i^2 h_i f_i(x)$$
$$= \sum_i Q_i^2[f_i^{\uparrow}(x) - f_i^{\downarrow}(x)] \tag{15.8}$$

Here an obvious notation has been introduced to denote the helicity of the quark relative to the helicity of the nucleon. The quark–parton model *predicts* that the combination on the left side of Eq. (15.8) obeys Bjorken scaling in DIS, and furthermore, that it measures the helicity distribution of the quarks inside the nucleon in the infinite momentum frame

$$\frac{2v}{m} W^{\uparrow} \quad \rightarrow \quad G_1(x) \qquad\qquad ; \text{DIS}$$

$$G_1(x) \quad = \quad \sum_i Q_i^2 [f_i^{\uparrow}(x) - f_i^{\downarrow}(x)] \qquad\qquad (15.9)$$

It remains to investigate the physical consequences of the additional terms in the lepton and target response tensors in Eqs. (15.2) and (15.7) in the polarized case.

These results can be used to compute the asymmetry for scattering of the lepton by the target in the case when the helicities are both aligned or antialigned. Define this asymmetry by

$$\mathscr{A} \quad \equiv \quad \frac{d\sigma_{\uparrow\uparrow} - d\sigma_{\downarrow\uparrow}}{d\sigma_{\uparrow\uparrow} + d\sigma_{\downarrow\uparrow}} \qquad\qquad (15.10)$$

The subscripts refer to the particle *helicities*, the convention here being that the first subscript is that of the electron and the second that of the nucleon. Parity invariance of the strong and electromagnetic interactions implies that $\mathscr{A}$ will be unchanged under a reversal of *both* helicities, as the reader can readily verify explicitly from the preceeding arguments.[2]

First note that when two tensors are contracted, they must both be even or odd in the interchange of the indices μ and v to get a non-zero result. Then, since all common factors *cancel in the ratio*, the problem reduces to the evaluation of the following expression

$$\mathscr{A} \quad = \quad W^{\uparrow} \frac{1}{m^2} \frac{\varepsilon_{\mu v \lambda \rho} k_{2\lambda} k_{1\rho} \, \varepsilon_{\mu v \sigma \tau} p_{\sigma} q_{\tau}}{\eta_{\mu v} W_{\mu v}} \qquad\qquad (15.11)$$

The denominator is evaluated in Eq. (11.32)

$$
\begin{aligned}
\eta_{\mu v} W_{\mu v} \quad &= \quad (k_{1\mu} k_{2v} + k_{1v} k_{2\mu} - k_1 \cdot k_2 \, \delta_{\mu v}) \left(W_1 \delta_{\mu v} + W_2 \frac{p_{\mu} p_{v}}{m^2} \right) \\
&= \quad W_1(-2k_1 \cdot k_2) + W_2 \frac{1}{m^2} (2 p \cdot k_1 \, p \cdot k_2 - p^2 k_1 \cdot k_2) \\
&= \quad W_1 q^2 + W_2 \frac{1}{m^2} \left(2 p \cdot k_1 \, p \cdot k_2 - \frac{1}{2} m^2 q^2 \right) \qquad (15.12)
\end{aligned}
$$

---

[2] Compare with Eq. (13.55).

Recall $q = k_2 - k_1$ and $q^2 = -2k_1 \cdot k_2$ in the ERL. The numerator is evaluated using Eq. (D.18)

$$
\begin{aligned}
W^\uparrow \frac{1}{m^2} \varepsilon_{\mu\nu\lambda\rho} k_{2\lambda} k_{1\rho} \, \varepsilon_{\mu\nu\sigma\tau} p_\sigma q_\tau &= 2W^\uparrow \frac{1}{m^2} (k_2 \cdot p \, k_1 \cdot q - k_1 \cdot p \, k_2 \cdot q) \\
&= -W^\uparrow \frac{q^2}{m^2} p \cdot (k_1 + k_2)
\end{aligned}
\tag{15.13}
$$

In the quark–parton model in DIS with $x = q^2/2m\nu$ one has from before

$$
\begin{aligned}
2W_1 &= F_1(x) = \sum_i Q_i^2 [f_i^\uparrow(x) + f_i^\downarrow(x)] \\
\frac{\nu}{m} W_2 &= F_2(x) = x F_1(x) \\
\frac{2\nu}{m} W^\uparrow &= G_1(x) = \sum_i Q_i^2 [f_i^\uparrow(x) - f_i^\downarrow(x)]
\end{aligned}
\tag{15.14}
$$

Hence one can write the asymmetry $\mathscr{A}$ as

$$
\begin{aligned}
\mathscr{A} &\equiv \frac{N}{D} \\
N &= -G_1(x) \, m\nu \, [p \cdot (k_1 + k_2)] \\
D &= F_1(x) \left[m^2\nu^2 + \left(2 p \cdot k_1 \, p \cdot k_2 - \frac{1}{2} m^2 q^2 \right) \right]
\end{aligned}
\tag{15.15}
$$

An equivalent expression is

$$
\begin{aligned}
\mathscr{A} &= \frac{G_1(x)}{F_1(x)} \mathscr{D} \\
\mathscr{D} &= -\frac{m\nu \, [p \cdot (k_1 + k_2)]}{[m^2\nu^2 + (2 p \cdot k_1 \, p \cdot k_2 - m^2 q^2/2)]}
\end{aligned}
\tag{15.16}
$$

These two expressions give the quark–parton result for $\mathscr{A}$ in DIS written in Lorentz invariant form. The first factor shows that what is being measured in these experiments is the ratio $G_1(x)/F_1(x)$. The second *depolarization* factor (of the virtual photon which must transmit the spin information) is purely kinematic.

In the laboratory frame where $p = (\mathbf{0}, im)$, one has in the ERL

$$
\mathscr{D} = \frac{\varepsilon_1^2 - \varepsilon_2^2}{\varepsilon_1^2 + \varepsilon_2^2 - 2\varepsilon_1\varepsilon_2 \sin^2 \theta/2}
\tag{15.17}
$$

This reproduces the result in [Hu95].

If one retains correction terms of $O(m/\varepsilon_1)$, and correspondingly considers other directions of the polarization of the target, then the expression for the polarization asymmetry becomes more complicated, and one can, in fact, measure an additional spin structure function $G_2(x)$, whose interpretation in the quark–parton model is more ambiguous. The full response for arbitrary target polarization is given in [Vo92], where experimental results from the scattering of very-high-energy polarized muons from polarized nucleon targets is also discussed.

16

Parity violation in inclusive electron scattering

The measurement of parity violation in the scattering of longitudinally polarized electrons in inclusive deep-inelastic electron scattering from deuterium at SLAC is a classic experiment that played a pivotal role in the establishment of the weak neutral current structure of the standard model [Pr78, Pr79]. The measurement of parity violation in inclusive electron scattering from nuclear and nucleon targets $A(\vec{e}, e')_{\text{pv}}$, promises to play a central role in future developments in nuclear physics [Pa90]. In this chapter we use the previous results to develop a general description of this process.

Conservation of parity in the strong and electromagnetic interactions implies that there can be no difference in the cross section for the process $A(\vec{e}, e')$ upon reversal of the longitudinal polarization of the electron if the target is unpolarized and unobserved. This follows from general principles, for it would effectively imply a non-zero expectation value for the pseudoscalar quantity $\langle \boldsymbol{\sigma} \cdot \mathbf{k}_1 \rangle$. That the helicity-dependent lepton contribution to the cross section indeed vanishes with one photon exchange can be seen immediately from our preceeding analysis. Equation (15.2) states that a longitudinally polarized electron has an additional term in the response tensor of the form $h \, \varepsilon_{\mu\nu\lambda\rho} k_{2\lambda} k_{1\rho}$. When contracted with the response tensor for an unpolarized and unobserved hadronic target in Eq. (11.27), the result vanishes since the first expression is antisymmetric in the interchange of the indices μ and ν and the second is symmetric.

Parity violation necessitates the inclusion of the weak interaction. In addition to the exchange of a virtual photon, it is possible for an electron to exchange a $Z^{(0)}$, the heavy neutral weak vector boson with mass $M_Z = 91.19 \, \text{GeV}$. The interaction takes place through the weak neutral current, which we now know is accurately described by the *standard model* of the electroweak interactions [Sa64, We67, Gl70, We72].

To start the discussion of parity violation, consider the scattering of

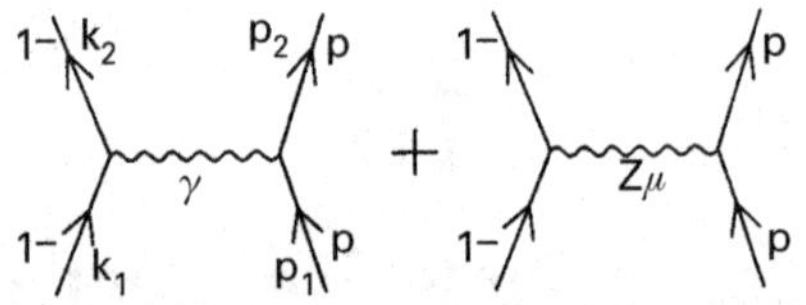

Fig. 16.1. Contributing Feynman diagrams (unitary gauge) for parity-violating asymmetry in scattering of longitudinally polarized electrons from point protons. Here $q = k_2 - k_1$.

a relativistic (massless) longitudinally polarized electron from a point proton. The contributing diagrams in the unitary gauge are shown in Fig. 16.1. The standard model is presented in detail in chapter 26 and [Wa95]. Here we simply anticipate that development and use the fact that the Feynman rules for the weak neutral current interaction of the standard model imply that the S-matrix is given by

$$S_{fi} = \frac{-(2\pi)^4 i}{\Omega^2} \delta^{(4)}(k_1 + p - k_2 - p') \left\{ \bar{u}(k_2)(e\gamma_\mu)u(k_1)\frac{\delta_{\mu\nu}}{q^2}\bar{u}(p')(-e\gamma_\nu)u(p) \right.$$

$$+ \bar{u}(k_2)\left[\frac{-g\gamma_\mu}{4\cos\theta_W}[(1 - 4\sin^2\theta_W) + \gamma_5]\right]u(k_1)\frac{(\delta_{\mu\nu} + q_\mu q_\nu/m_Z^2)}{q^2 + m_Z^2}$$

$$\left. \times\bar{u}(p')\left[\frac{g\gamma_\nu}{4\cos\theta_W}[(1 - 4\sin^2\theta_W) + \gamma_5]\right]u(p) \right\} \qquad (16.1)$$

At low energy one has $|\mathbf{q}|/M_Z \ll 1$, and the momentum-dependent terms can be neglected in the Z-propagator. Take the standard model values

$$e^2 = 4\pi\alpha$$

$$\frac{g^2}{8m_Z^2 \cos^2\theta_W} = \frac{G}{\sqrt{2}} = \frac{1.024 \times 10^{-5}}{\sqrt{2}\, m_p^2}$$

$$a = -(1 - 4\sin^2\theta_W) \qquad ; \sin^2\theta_W = 0.2315$$

$$b = -1 \qquad (16.2)$$

Then

$$S_{fi} = \frac{-(2\pi)^4 i}{\Omega^2}\delta^{(4)}(k_1 + p - k_2 - p')T_{fi}$$

$$T_{fi} = -\frac{4\pi\alpha}{q^2}\left\{ \bar{u}(k_2)\gamma_\mu u(k_1)\bar{u}(p')\gamma_\mu u(p) - \frac{Gq^2}{4\pi\alpha\sqrt{2}}\bar{u}(k_2)\gamma_\mu[a + b\gamma_5]u(k_1) \right.$$

$$\left. \times \bar{u}(p')\gamma_\mu\left[\frac{1}{2}(1 + \gamma_5) - 2\sin^2\theta_W\right]u(p) \right\} \qquad (16.3)$$

This result is easily extended to point neutrons using the Feynman rules of [Wa95] through the replacement

$$
\begin{aligned}
T_{fi} &= -\frac{4\pi\alpha}{q^2}\left\{ \bar{u}(k_2)\gamma_\mu u(k_1)\bar{u}(p')\gamma_\mu \frac{1}{2}(1+\tau_3)u(p) \right.\\
&\quad -\frac{Gq^2}{4\pi\alpha\sqrt{2}}\bar{u}(k_2)\gamma_\mu[a+b\gamma_5]u(k_1) \\
&\quad \left. \times \bar{u}(p')\gamma_\mu[(1+\gamma_5)\frac{1}{2}\tau_3 - 2\sin^2\theta_W \frac{1}{2}(1+\tau_3)]u(p) \right\}
\end{aligned}
\tag{16.4}
$$

At this juncture one can redefine things so that the result is more general than for just point nucleons

$$
\begin{aligned}
S_{fi} &= \frac{-(2\pi)^4 i}{\Omega}\delta^{(4)}(k_1+p-k_2-p')T_{fi} \\
T_{fi} &= \frac{4\pi\alpha}{q^2}\left\{ i\bar{u}(k_2)\gamma_\mu u(k_1)\langle p'|J_\mu^\gamma(0)|p\rangle \right.\\
&\quad \left. -\frac{Gq^2}{4\pi\alpha\sqrt{2}}i\bar{u}(k_2)\gamma_\mu(a+b\gamma_5)u(k_1)\langle p'|\mathscr{J}_\mu^{(0)}(0)|p\rangle \right\}
\end{aligned}
\tag{16.5}
$$

Now these are single-nucleon matrix elements of the full electromagnetic and weak neutral current densities taken between exact Heisenberg states; for point nucleons, this expression reduces to Eq. (16.4).

The dimensionless ratio $Gq^2/4\pi\alpha\sqrt{2}$ forms the small parameter in these nuclear physics parity-violation calculations.

The first term in Eq. (16.5) leads to the electron scattering cross section derived in chapter 11

$$
\begin{aligned}
d\sigma &= \frac{4\alpha^2}{q^4}\frac{d^3k_2}{2\varepsilon_2}\frac{1}{\sqrt{(k_1\cdot p)^2}}\eta_{\mu\nu}W_{\mu\nu} \\
\eta_{\mu\nu} &= -2\varepsilon_1\varepsilon_2\frac{1}{2}\sum_{s_1}\sum_{s_2}\bar{u}(k_1)\gamma_\nu u(k_2)\bar{u}(k_2)\gamma_\mu u(k_1) \\
&= k_{1\mu}k_{2\nu} + k_{1\nu}k_{2\mu} - (k_1\cdot k_2)\delta_{\mu\nu} \\
W_{\mu\nu} &= (2\pi)^3\overline{\sum_i}\sum_f \delta^{(4)}(q+p'-p)\langle p|J_\nu^\gamma(0)|p'\rangle\langle p'|J_\mu^\gamma(0)|p\rangle(\Omega E_p) \\
&= W_1^\gamma(q^2, q\cdot p)\left(\delta_{\mu\nu} - \frac{q_\mu q_\nu}{q^2}\right) \\
&\quad + W_2^\gamma(q^2, q\cdot p)\frac{1}{M_T^2}\left(p_\mu - \frac{p\cdot q}{q^2}q_\mu\right)\left(p_\nu - \frac{p\cdot q}{q^2}q_\nu\right)
\end{aligned}
\tag{16.6}
$$

It is important to note that at this point we have again generalized the target response tensor to include the possibility of inelastic processes.

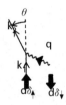

Fig. 16.2.   Cross sections for right- and left-handed electrons.

From appendix D and chapter 15 we know that the following are projections for right- and left-handed (massless) Dirac electrons

$$P_\uparrow = \frac{1}{2}(1 - \gamma_5) \qquad\qquad P_\downarrow = \frac{1}{2}(1 + \gamma_5) \qquad (16.7)$$

To calculate the cross sections for such particles (Fig. 16.2) one simply modifies $\eta_{\mu\nu}$ with the appropriate insertion of these projections and removes the average over the initial helicities[1]

$$\text{for } d\sigma_\uparrow : \quad \eta_{\mu\nu}^\uparrow = \ldots \overbrace{\left(\frac{1}{2}\right)}^{\text{omit}} \sum_{s_1}\sum_{s_2} \bar{u}(k_1) \ldots \frac{1}{2}(1 - \gamma_5)u(k_1)$$

$$\text{for } d\sigma_\downarrow : \quad \eta_{\mu\nu}^\downarrow = \ldots \sum_{s_1}\sum_{s_2} \bar{u}(k_1) \ldots \frac{1}{2}(1 + \gamma_5)u(k_1)$$

$$\text{for } d\sigma_\uparrow - d\sigma_\downarrow : \quad \eta_{\mu\nu}^{(-)} = \ldots \sum_{s_1}\sum_{s_2} \bar{u}(k_1) \ldots (-\gamma_5)u(k_1)$$

$$\text{for } d\sigma_\uparrow + d\sigma_\downarrow : \quad \eta_{\mu\nu}^{(+)} = \ldots \sum_{s_1}\sum_{s_2} \bar{u}(k_1) \ldots (1)u(k_1) \qquad (16.8)$$

Thus one now has either $(-\gamma_5)$ or (1) in the lepton trace. Since *all common factors cancel in the ratio* the asymmetry is given by

$$\mathscr{A} \equiv \frac{d\sigma_\uparrow - d\sigma_\downarrow}{d\sigma_\uparrow + d\sigma_\downarrow} = -\frac{Gq^2}{4\pi\alpha\sqrt{2}} \frac{\eta_{\mu\nu}^{(1)} W_{\mu\nu}^{(1)} + \eta_{\mu\nu}^{(2)} W_{\mu\nu}^{(2)}}{2\eta_{\mu\nu} W_{\mu\nu}} \qquad (16.9)$$

Here

$$\eta_{\mu\nu}^{(1)} = -2\varepsilon_1\varepsilon_2 \sum_{s_1}\sum_{s_2} \bar{u}(k_1)\gamma_\nu u(k_2)\bar{u}(k_2)\gamma_\mu(a + b\gamma_5)(-\gamma_5)u(k_1) \qquad (16.10)$$

$$W_{\mu\nu}^{(1)} = (2\pi)^3 \overline{\sum_i}\sum_f \delta^{(4)}(q + p' - p)\langle p|J_\nu^\gamma(0)|p'\rangle\langle p'|\mathscr{J}_\mu^{(0)}(0)|p\rangle(\Omega E_p)$$

$$\eta_{\mu\nu}^{(2)} = -2\varepsilon_1\varepsilon_2 \sum_{s_1}\sum_{s_2} \bar{u}(k_1)\gamma_\nu(a + b\gamma_5)u(k_2)\bar{u}(k_2)\gamma_\mu(-\gamma_5)u(k_1)$$

$$W_{\mu\nu}^{(2)} = (2\pi)^3 \overline{\sum_i}\sum_f \delta^{(4)}(q + p' - p)\langle p|\mathscr{J}_\nu^{(0)}(0)|p'\rangle\langle p'|J_\mu^\gamma(0)|p\rangle(\Omega E_p)$$

---

[1] Note $d\sigma^\uparrow + d\sigma^\downarrow = 2d\sigma_{\text{unpolarized}}$.

The lepton traces have been evaluated in chapter 15. The result is[2]

$$\eta_{\mu\nu}^{(1)} = \eta_{\mu\nu}^{(2)} = -2(b\eta_{\mu\nu} + a\varepsilon_{\mu\nu\rho\sigma}k_{1\rho}k_{2\sigma}) \tag{16.11}$$

Thus in the numerator of Eq. (16.9) one needs $\eta_{\mu\nu}^{(1)}(W_{\mu\nu}^{(1)} + W_{\mu\nu}^{(2)})$ and

$$
\begin{aligned}
W_{\mu\nu}^{(1)} + W_{\mu\nu}^{(2)} = (2\pi)^3 \overline{\sum_i}\sum_f \delta^{(4)}(q+p'-p) &\left[\langle p|J_\nu^\gamma(0)|p'\rangle\langle p'|\mathscr{J}_\mu^{(0)}(0)|p\rangle \right.\\
&\left. + \langle p|\mathscr{J}_\nu^{(0)}(0)|p'\rangle\langle p'|J_\mu^\gamma(0)|p\rangle \right](\Omega E_p)
\end{aligned} \tag{16.12}
$$

Now separate the weak neutral current into its Lorentz vector and axial vector parts

$$\mathscr{J}_\mu^{(0)} = J_\mu^{(0)} + J_{\mu 5}^{(0)} \qquad ; \text{V} - \text{A} \tag{16.13}$$

Since the asymmetry is already explicitly of order $Gq^2/4\pi\alpha\sqrt{2}$, one can then use the good parity of the nuclear states to write

$$W_{\mu\nu}^{(1)} + W_{\mu\nu}^{(2)} = W_{\mu\nu}^{\text{int}} + W_{\mu\nu}^{\text{V}-\text{A}} \tag{16.14}$$

Here the first term $W_{\mu\nu}^{\text{int}}$ comes from $J_\mu^{(0)}$; it has the same general structure as $W_{\mu\nu}^\gamma$ in Eq. (16.6)[3]

$$
\begin{aligned}
W_{\mu\nu}^{\text{int}} = W_1^{\text{int}}(q^2, q\cdot p)&\left(\delta_{\mu\nu} - \frac{q_\mu q_\nu}{q^2} \right) \\
&+ W_2^{\text{int}}(q^2, q\cdot p)\frac{1}{M_T^2}\left(p_\mu - \frac{p\cdot q}{q^2}q_\mu \right)\left(p_\nu - \frac{p\cdot q}{q^2}q_\nu \right)
\end{aligned} \tag{16.15}
$$

The second term in Eq. (16.14), coming from $J_{\mu 5}^{(0)}$, is a *pseudotensor*; the only pseudotensor that can be constructed from the two four-vectors (p_μ, q_μ) is[4]

$$W_{\mu\nu}^{\text{V}-\text{A}} = W_8(q^2, q\cdot p)\frac{1}{M_T^2}\varepsilon_{\mu\nu\rho\sigma}p_\rho q_\sigma \tag{16.16}$$

Now combine these expressions with Eq. (16.11). The result follows from simple algebra and kinematics of the type carried out previously. The only non-zero terms are [see Eq. (11.35)]

$$2\eta_{\mu\nu}W_{\mu\nu} = 4\varepsilon_1\varepsilon_2[W_2^\gamma\cos^2\frac{\theta}{2} + 2W_1^\gamma\sin^2\frac{\theta}{2}]$$

$$-2b\eta_{\mu\nu}W_{\mu\nu}^{\text{int}} = (-b)4\varepsilon_1\varepsilon_2[W_2^{\text{int}}\cos^2\frac{\theta}{2} + 2W_1^{\text{int}}\sin^2\frac{\theta}{2}] \tag{16.17}$$

---

[2] Note that the first term is symmetric in $\mu \leftrightarrow \nu$, while the second is antisymmetric.

[3] The proof of this result uses the fact that the current $J_\mu^{(0)}$ is conserved.

[4] Note that this expression is antisymmetric in $\mu \leftrightarrow \nu$.

and

$$(-2a\varepsilon_{\mu\nu\rho\sigma}k_{1\rho}k_{2\sigma})\left[W_8(q^2, q\cdot p)\frac{1}{M_T^2}\varepsilon_{\mu\nu\alpha\beta}p_\alpha q_\beta\right]$$

$$= -\left(\frac{4a}{M_T^2}W_8\right)(k_1\cdot p\,k_2\cdot q - k_1\cdot q\,k_2\cdot p)$$

$$= -\left(\frac{2a}{M_T^2}W_8\right)q^2\,p\cdot(k_1+k_2)$$

$$= \left(\frac{2a}{M_T}W_8\right)4\varepsilon_1\varepsilon_2\sin\frac{\theta}{2}\left(q^2\cos^2\frac{\theta}{2}+\mathbf{q}^2\sin^2\frac{\theta}{2}\right)^{1/2} \quad (16.18)$$

The ERL is assumed with $q = k_2 - k_1$, and the results are written in the laboratory frame. The last line follows from the following manipulations in that frame

$$\begin{aligned}
q^2\cos^2\frac{\theta}{2}+\mathbf{q}^2\sin^2\frac{\theta}{2} &= \mathbf{q}^2 - q_0^2\cos^2\frac{\theta}{2}\\
&= \varepsilon_2^2 + \varepsilon_1^2 - 2\varepsilon_1\varepsilon_2\cos\theta - (\varepsilon_2-\varepsilon_1)^2\cos^2\frac{\theta}{2}\\
&= (\varepsilon_1+\varepsilon_2)^2\sin^2\frac{\theta}{2} \quad (16.19)
\end{aligned}$$

The final result is

$$\left[\frac{d\sigma_\uparrow - d\sigma_\downarrow}{d\sigma_\uparrow + d\sigma_\downarrow}\right]\left[W_2^\gamma\cos^2\frac{\theta}{2}+2W_1^\gamma\sin^2\frac{\theta}{2}\right] = \frac{Gq^2}{4\pi\alpha\sqrt{2}}$$

$$\times\left\{b\left[W_2^{\text{int}}\cos^2\frac{\theta}{2}+2W_1^{\text{int}}\sin^2\frac{\theta}{2}\right]\right.$$

$$\left.-a\left(\frac{2W_8}{M_T}\right)\sin\frac{\theta}{2}\left(q^2\cos^2\frac{\theta}{2}+\mathbf{q}^2\sin^2\frac{\theta}{2}\right)^{1/2}\right\} \quad (16.20)$$

Several features of this result are of interest:

- This is the general expression for the parity-violating asymmetry in relativistic polarized electron scattering from a hadronic target arising from the interference of one-photon and one-Z exchange (Fig. 16.1).[5]

- The left hand side is the product of the asymmetry $\mathscr{A}$ [Eq. (16.9)] and the basic (e, e') response [Eqs. (16.6) and (16.17)].

---

[5] Additional contributions to the parity-violating asymmetry can arise from parity admixtures in the nuclear states coming from weak parity-violating nucleon–nucleon interactions. These contributions are generally negligible, except perhaps at very small q^2 [Se79, Dm92].

- The characteristic scale of parity violation in nuclear physics from the process $(\tilde{e}, e')$ is set by the dimensionless parameter $Gq^2/4\pi\alpha\sqrt{2}$ appearing on the right hand side.

- The parameter b characterizes the lepton axial-vector weak neutral current [Eq. (16.2)]; its coefficient here arises from the interference of the vector part of the weak neutral and electromagnetic hadronic currents [Eqs. (16.12), (16.13), and (16.15)]

$$
\begin{aligned}
W_{\mu\nu}^{\text{int}} &= (2\pi)^3 \overline{\sum_i \sum_f} \delta^{(4)}(q + p' - p) \left[\langle p|J_\nu^\gamma(0)|p'\rangle\langle p'|J_\mu^{(0)}(0)|p\rangle \right. \\
&\quad \left. + \langle p|J_\nu^{(0)}(0)|p'\rangle\langle p'|J_\mu^\gamma(0)|p\rangle \right] (\Omega E_p) \\
&= W_1^{\text{int}}(q^2, q \cdot p) \left(\delta_{\mu\nu} - \frac{q_\mu q_\nu}{q^2} \right) \\
&\quad + W_2^{\text{int}}(q^2, q \cdot p)\frac{1}{M_T^2} \left(p_\mu - \frac{p \cdot q}{q^2}q_\mu \right) \left(p_\nu - \frac{p \cdot q}{q^2}q_\nu \right) (16.21)
\end{aligned}
$$

- The parameter a characterizes the lepton vector weak neutral current [Eq. (16.2)]; its coefficient here arises from the interference of the axial vector part of the weak neutral and electromagnetic hadronic currents [Eqs. (16.12)–(16.14) and (16.16)]

$$
\begin{aligned}
W_{\mu\nu}^{\text{A}-\text{V}} &= (2\pi)^3 \overline{\sum_i \sum_f} \delta^{(4)}(q + p' - p) \left[\langle p|J_\nu^\gamma(0)|p'\rangle\langle p'|J_{\mu 5}^{(0)}(0)|p\rangle \right. \\
&\quad \left. + \langle p|J_{\nu 5}^{(0)}(0)|p'\rangle\langle p'|J_\mu^\gamma(0)|p\rangle \right] (\Omega E_p) \\
&= W_8(q^2, q \cdot p)\frac{1}{M_T^2}\varepsilon_{\mu\nu\rho\sigma}p_\rho q_\sigma
\end{aligned} \tag{16.22}
$$

- The three response functions on the right hand side of Eq. (16.20) can be separated by varying the electron scattering angle θ at fixed $(q^2, q \cdot p)$.[6]

- The parity violation arises from the interference of the transition matrix element of the electromagnetic and the weak neutral currents. If the electromagnetic matrix elements have been measured, then *parity violation in $(\tilde{e}, e)$ and $(\tilde{e}, e')$ provides a measurement of the matrix elements of the weak neutral current in nuclei at all q^2.*

We give one example [Wa84, Wa95]. Consider elastic scattering from a 0^+ target (Fig. 16.3a). Then from Lorentz covariance and current con-

---

[6] This is known as a *Rosenbluth* separation.

$$\overline{}\ 0^+ \qquad \overline{}\ 0^+.0$$
$$\text{(a)} \qquad\qquad \text{(b)}$$

Fig. 16.3. Example of parity-violating asymmetry in scattering from (a) $J^\pi = 0^+$, and (b) $(J^\pi, T) = (0^+, 0)$ target.

servation the transition matrix elements of the electromagnetic and weak neutral currents must have the form[7]

$$\langle p'|J_\mu^\gamma(0)|p\rangle = \left(\frac{M_T^2}{EE'\Omega^2}\right)^{1/2} F_0^\gamma(q^2)\frac{1}{M_T}\left(p_\mu - \frac{p\cdot q}{q^2}q_\mu\right)$$

$$\langle p'|J_\mu^{(0)}(0)|p\rangle = \left(\frac{M_T^2}{EE'\Omega^2}\right)^{1/2} F_0^{(0)}(q^2)\frac{1}{M_T}\left(p_\mu - \frac{p\cdot q}{q^2}q_\mu\right)$$

$$\langle p'|J_{\mu 5}^{(0)}(0)|p\rangle = 0 \tag{16.23}$$

The last relation follows since it is impossible to construct an axial vector from only two four-vectors (p_μ, q_μ).

Insertion of these relations in the defining equations yields

$$W_1^{\text{int}} = W^{A-V} = 0$$

$$\mathscr{A} = \frac{Gq^2}{4\pi\alpha\sqrt{2}}b\frac{2F_0^{(0)}(q^2)}{F_0^\gamma(q^2)} \tag{16.24}$$

Hence

$$\mathscr{A} = -\frac{Gq^2}{2\pi\alpha\sqrt{2}}\frac{F_0^{(0)}(q^2)}{F_0^\gamma(q^2)} \tag{16.25}$$

This expression allows one to measure the ratio of the weak neutral current and electromagnetic form factors — the latter measures the distribution of electromagnetic charge in the 0^+ target, and the former the distribution of weak neutral charge.

Now suppose that, in addition, the target has isospin $T = 0$ (Fig. 16.3b). Then only isoscalar operators can contribute to the matrix elements. In the *nuclear domain* of (u, d) quarks and antiquarks, the only isoscalar piece of the weak neutral current in the standard model arises from the electromagnetic current itself, and hence in this case (see chapter 26)

$$J_\mu^{(0)} \doteq -2\sin^2\theta_W J_\mu^\gamma \tag{16.26}$$

This implies

$$F_0^{(0)}(q^2) = -2\sin^2\theta_W F_0^\gamma(q^2) \tag{16.27}$$

---

[7] Hermiticity of the current implies that the form factors, as defined here, are real.

The ratio of form factors is then the constant $-2\sin^2\theta_W$ at all q^2 — a truly remarkable prediction![8] Insertion of this equality in the expression for the asymmetry leads to [Fe75]

$$\mathscr{A} = \frac{Gq^2}{\pi\alpha\sqrt{2}}\sin^2\theta_W \tag{16.28}$$

Several comments are of interest:

- It is important to note that this result holds *to all orders in the strong interactions* (QCD);

- This expression is linear in q^2 with a coefficient that depends only on fundamental constants;

- It can be used to measure $\sin^2\theta_W$ in the low-energy quark sector, complementing other measurements of this quantity;

- It can be used to test the remarkable prediction in Eq. (16.27) that holds in the nuclear domain.

A measurement of this parity-violating asymmetry for elastic scattering from ^{12}C at $q = 150$ MeV has been carried out in a tour de force experiment at the Bates Laboratory [So90]. Take

$$q = 150\text{ MeV} \qquad \sin^2\theta_W = 0.2315$$
$$\alpha^{-1} = 137.0 \qquad G = \frac{1.024\times10^{-5}}{m_p^2}$$
$$\mathscr{A} = 1.868\times10^{-6} \tag{16.29}$$

Then, with an electron beam polarization P_e, one has [So90, Mo90]

$$\mathscr{A}P_e = 0.691\times10^{-6} \qquad ;\text{ theory }(P_e = 0.37)$$
$$\mathscr{A}P_e = 0.60\pm0.14\pm0.02\times10^{-6} \;;\text{ experiment} \tag{16.30}$$

The first error is statistical. Note that the systematic error, the key to these experiments, has been reduced to 2×10^{-8}. This experiment provides the prototype for the next generation of electron scattering parity-violation studies.

Consider next the *extended domain* of (u,d,s,c) quarks and their antiquarks. The standard model then has an additional isoscalar term in the weak neutral current (see chapter 26)

$$\delta\mathscr{I}_\mu^{(0)} = \frac{i}{2}[\bar{c}\gamma_\mu(1+\gamma_5)c - \bar{s}\gamma_\mu(1+\gamma_5)s] \tag{16.31}$$

---

[8] This result depends on the assumption of isospin invariance that is broken to $O(\alpha)$ in nuclei.

Table 16.1. Quark sector used in discussion of parity-violating deep-inelastic electron scattering from the nucleon (see chapter 26).

	u	d	s
$Q_i^{(0)}$	$1/2 - (4/3)\sin^2\theta_W$	$-1/2 + (2/3)\sin^2\theta_W$	$-1/2 + (2/3)\sin^2\theta_W$
$Q_i^{(05)}$	$1/2$	$-1/2$	$-1/2$

This leads to an additional contribution $\delta F_0^{(0)}$ in the form factor in Eq. (16.27); the asymmetry for elastic scattering of polarized electrons on a $(0^+, 0)$ nucleus such as ^{4}He then takes the form

$$\mathscr{A} = \frac{Gq^2}{\pi\alpha\sqrt{2}}\sin^2\theta_W\left[1 - \frac{\delta F_0^{(0)}(q^2)}{2\sin^2\theta_W F_0^\gamma(q^2)}\right] \qquad (16.32)$$

The additional weak neutral current form factor comes from the vector current in Eq. (16.31), and is expected to arise predominantly from the much lighter strange quarks. Hence one has a direct measure of the *strangeness current* in nuclei. The total strangeness of this nucleus must vanish in the strong and electromagnetic sector, and hence $\delta F_0^{(0)}(0) = 0$; however, just as with electromagnetic charge in the neutron, there can be a strangeness *distribution*, which is determined in this experiment.

The *quark–parton model* predictions for parity violation in deep-inelastic scattering from the nucleon follow directly from the previous analysis. Go back to the intermediate step in Eq. (14.34) and identify in the quark response tensor

$$Q_i^2[p_\mu'p_\nu + p_\nu'p_\mu - (p\cdot p')\delta_{\mu\nu}] \rightarrow \frac{Q_i^2}{4}\text{trace}\,[\gamma_\nu(\gamma_\rho p_\rho')\gamma_\mu(\gamma_\sigma p_\sigma)] \qquad (16.33)$$

In the response tensor arising from the interference of the electromagnetic and vector weak neutral currents, one has instead

$$\frac{Q_iQ_i^{(0)}}{4}\text{trace}\,[\gamma_\nu(\gamma_\rho p_\rho')\gamma_\mu(\gamma_\sigma p_\sigma) + \gamma_\nu(\gamma_\rho p_\rho')\gamma_\mu(\gamma_\rho p_\rho)] =$$
$$2Q_iQ_i^{(0)}[p_\mu'p_\nu + p_\nu'p_\mu - (p\cdot p')\delta_{\mu\nu}] \qquad (16.34)$$

Here $Q_i^{(0)}$ is the weak neutral charge of the quarks, shown for the first few quarks in Table 16.1. The arguments proceed precisely as those following Eq. (14.34), with the result that the following combinations of response functions are predicted to satisfy Bjorken scaling

$$2W_1^{int}(v, q^2) \rightarrow H_1(x) = 2\sum_i Q_iQ_i^{(0)}f_i(x)$$
$$\left(\frac{v}{m}\right)W_2^{int}(v, q^2) \rightarrow H_2(x) = xH_1(x) = 2x\sum_i Q_iQ_i^{(0)}f_i(x) \qquad (16.35)$$

For the interference term between the axial vector and electromagnetic currents, the corresponding replacement in Eq. (16.33) is

$$\frac{Q_i Q_i^{(05)}}{4} \text{trace} \left[\gamma_\nu (\gamma_\rho p'_\rho) \gamma_\mu \gamma_5 (\gamma_\sigma p_\sigma) + \gamma_\nu \gamma_5 (\gamma_\rho p'_\rho) \gamma_\mu (\gamma_\sigma p_\sigma) \right] =$$
$$-2 Q_i Q_i^{(05)} \varepsilon_{\mu\nu\rho\sigma} p'_\rho p_\sigma \qquad (16.36)$$

Hence a repetition of the arguments following Eq. (14.34) allows one to conclude that the following combination must scale

$$-\left(\frac{\nu}{m}\right) W_8(\nu, q^2) \rightarrow H_8(x) = 2 \sum_i Q_i Q_i^{(05)} f_i(x) \qquad (16.37)$$

Here $Q_i^{(05)}$ are the axial vector couplings of the quarks, also shown for the first few quarks in Table 16.1. Note that if Q_i and $f_i(x)$ are known from DIS through the electromagnetic interaction, then the parity violation measurements allow one to determine the weak neutral current couplings of the quarks.[9]

We will return to the subject of parity violation in the discussion of applications and future directions.

---

[9] Parity violation in DIS from the nucleon is further analyzed in [Ka78].

Part 3
Quantum electrodynamics

17

Basic elements

Quantum Electrodynamics (QED) is the most accurate physical theory we have. Its development in the late 1940's by Dyson, Feynman, Schwinger, Tomonaga and others is one of man's great intellectual triumphs. The key original papers are collected in a volume edited by Schwinger [Sc58]; particularly influential are [Fe49, Fe49a, Dy49]. QED is the culmination of the development of electrodynamics, special relativity, and quantum mechanics. The understanding of QED, in terms of covariance, local gauge invariance, renormalization, and Feynman diagrams laid the basis for all modern relativistic quantum field theories of the fundamental interactions. Since electron scattering involves the electromagnetic interaction of relativistic (massless) Dirac particles, QED plays a central role in the analysis.

The content of QED can be expressed in terms of a set of *Feynman diagrams* with corresponding *Feynman rules* for the S-matrix. We will not derive these here, as that takes us too far afield; their derivation can be found in any standard text [Bj65, Fe71], or course (e.g. [Wa91]). The components of the diagrams are shown in Fig. 17.1. The rules, in the conventions used in this book, are as follows:

1. Draw all topologically distinct connected diagrams;

2. Include a factor of $(-i)(-ie) = -e$ for each order of perturbation theory. Here e is algebraic, and for an electron $e = -|e|$;

3. Include a factor of γ_μ for each vertex [Fig. 17.1(a)];

4. Include a factor of

$$\frac{-i}{(2\pi)^4} \frac{1}{i\gamma_\mu p_\mu + m_e} \tag{17.1}$$

for each fermion (i.e. electron) propagator [Fig. 17.1(b)];

129

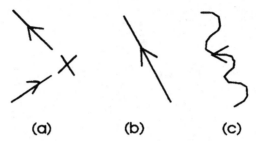

Fig. 17.1.   Basic elements of Feynman rules for the S-matrix in QED: (a) vertex; (b) fermion propagator; (c) effective photon propagator.

5. Include a factor of

$$\frac{-i}{(2\pi)^4}\frac{\delta_{\mu\nu}}{q^2} \tag{17.2}$$

for each "effective" photon propagator [Fig. 17.1(c)];[1]

6. Include a wave function for each of the external particles, e.g.

$$\frac{1}{\sqrt{\Omega}}u(p) \text{ incoming fermion}; \qquad \frac{1}{\sqrt{2\omega\Omega}}\varepsilon^{(\lambda)}_\mu \text{ incoming photon} \tag{17.3}$$

For a photon with polarization λ in the Coulomb gauge, $\varepsilon^{(\lambda)}_\mu = (\boldsymbol{\varepsilon}^{(\lambda)},0)$ and $\boldsymbol{\varepsilon}^{(\lambda)}\cdot\mathbf{k}=0$;

7. Read along fermion lines;

8. Include a factor of $(2\pi)^4\delta^{(4)}(\Delta p)$ at each vertex;

9. Integrate over all internal momenta $\int d^4q \equiv \int d^3q\,dq_0$;

10. Include a factor of (-1) for each closed fermion loop.

Here we simply treat the hadronic target as an external field, bringing an electromagnetic interaction into the electron line, which we represent by a wavy line ending in a cross. For this component:

11. Include a factor for the external field

$$\frac{a_\mu(q)}{(2\pi)^4} \tag{17.4}$$

---

[1] This result can be obtained by starting in the Coulomb gauge and then combining the terms coming from the Coulomb interaction (each interaction of order e^2) with those coming from transverse photon exchange (each of order e) in the S-matrix. Terms in q_μ or q_ν in the photon propagator do not contribute to the S-matrix because of current conservation [Bj65, Wa91].

where the external vector potential has the four-dimensional Fourier transform

$$A_\mu^{\text{ext}}(x) = \int e^{iq \cdot x} \frac{a_\mu(q)}{(2\pi)^4} d^4 q \tag{17.5}$$

We shall be content here to work to first order in the external field.[2]

Since the electron is light, it can easily radiate as it accelerates, which it does when scattering from a hadronic target. In computing the lowest order radiative corrections to the process of electron scattering, one can consistently confine the analysis to the electron line since it carries charge and runs completely through a diagram from beginning to end without termination. Thus the class of third order diagrams which are of first order in the external field, and which consist of all radiative corections of order $\alpha = e^2/4\pi$ along the electron line, provide a current conserving, gauge invariant set. Vacuum polarization in the external photon line can also be included in this set.

The Feynman diagrams giving the lowest order radiative corrections in electron scattering are then those shown in Fig. 17.2. Here a term in δm_e has been added and subtracted from the starting lagrangian (mass renormalization) so that the free lagangian represents fermions of the correct mass, and an additional interaction lagangian is then present of the form

$$\delta \mathscr{L} = \delta m_e : \bar{\psi}\psi : \tag{17.6}$$

The contribution of this mass counterterm must then also be included consistently in the Feynman rules.[3] The processes in Fig. 17.2 constitute the radiative corrections through order $\alpha = e^2/4\pi$. We will use the Feynman rules to set up each expression. The Dirac algebra is straightforward. The actual evaluation of the resulting integrals follows from the techniques of Feynman parameterization and four-dimensional momentum integration.[4] These methods are also now discussed in standard texts [Bj65], or courses

---

[2] The dominant contribution from terms of higher order in the external field consists of Coulomb interactions on the incident and outgoing electron lines. These *Coulomb corrections* imply that one should really use solutions to the Dirac equation in the Coulomb field of the target instead of plane waves for the electron. Though technically complicated, this can be done [Da51, Fe51, Ra54, Gr62, On63, Cu66, Tu68] (an updated version of the appropriate code is available from [He00]). Contributions of second order in the external field where a nuclear target is virtually excited and then de-excited, the so-called *dispersion corrections*, are much more difficult to estimate reliably [Sc55, de66, Fr72b, Do75].

[3] It is assumed that δm_e is normal ordered [Bj65, Fe71] and has a power series expansion $\delta m_{(2)} e^2 + \delta m_{(4)} e^4 + \cdots$.

[4] Or integration in $n = 4 + \epsilon$ dimensions if one uses dimensional regularization.

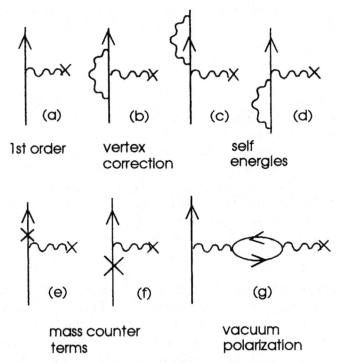

Fig. 17.2. Feynman diagrams for lowest-order radiative corrections in electron scattering.

[Wa91], and it is not the intent to reproduce the derivations. We are primarily concerned here with the results, how they fit together, how they enter into electron scattering, and their interpretation.

Let us consider each component in turn. Consider first the *electron self-energy*. The photon loop and mass counter term corrections to the S-matrix for a free electron are illustrated in Fig. 17.3. The Feynman rules give the S-matrix as[5]

$$S_{fi} = -\frac{(2\pi)^4 i}{\Omega} \delta^{(4)}(k' - k) \bar{u}(k)(\Sigma - \delta m_e)u(k) \tag{17.7}$$

Here the self-energy insertion is defined as

$$\Sigma - \delta m_e = -\frac{ie^2}{(2\pi)^4} \int \frac{d^4q}{q^2} \gamma_\mu \frac{1}{i\gamma_\lambda(k-q)_\lambda + m_e} \gamma_\mu - \delta m_e \tag{17.8}$$

From Lorentz covariance and power counting, this expression can be

---

[5] For the mass counter term one has the factors $(-1)(-i)\delta m_e$.

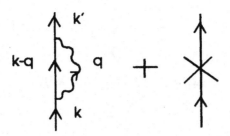

Fig. 17.3. Electron self-energy.

put in the following form

$$\Sigma - \delta m_e \ = \ A - \delta m_e$$
$$+(ik_\lambda\gamma_\lambda + m_e)B + (ik_\lambda\gamma_\lambda + m_e)\Sigma_f(ik_\sigma\gamma_\sigma + m_e) \quad (17.9)$$

Here A and B are (infinite) constants independent of k, and Σ_f is finite.

To give mathematical definition to the divergent integral in Eq. (17.8) we introduce a covariant Pauli–Villars regulator which amounts here to replacing the photon propagator by

$$\frac{1}{q^2} \longrightarrow \frac{1}{q^2} - \frac{1}{q^2 + \Lambda^2} \quad (17.10)$$

For very large Λ^2 the second term is negligible, while at fixed Λ^2 the asymptotic behavior of the photon propagator is changed to $\Lambda^2/q^2(q^2 + \Lambda^2)$, and one picks up enough convergence to make the integral finite. Explicit evaluation of the resulting integral on the mass shell, that is for $ik_\lambda\gamma_\lambda + m_e = 0$, yields the mass counter term [Bj65, Wa91]

$$A \equiv \delta m_e = \frac{3\alpha}{2\pi}m_e\left(\ln\frac{\Lambda}{m_e} + \frac{1}{4}\right) \quad (17.11)$$

Consider next *vacuum polarization*. The lowest order vacuum polarization correction to the S-matrix for a free photon as illustrated in Fig. 17.4. The analytic expression is given by

$$S_{fi} = -\frac{(2\pi)^4 i}{\Omega}\delta^{(4)}(l' - l)\frac{1}{\sqrt{4\omega\omega'}}\varepsilon_\mu^f(-\Pi_{\mu\nu})\varepsilon_\nu^i \quad (17.12)$$

The polarization part is defined by

$$\Pi_{\mu\nu} = -\frac{ie^2}{(2\pi)^4}\int d^4k \text{ trace}\left[\frac{1}{i\gamma_\lambda(k - l/2)_\lambda + m_e}\gamma_\mu\frac{1}{i\gamma_\sigma(k + l/2)_\sigma + m_e}\gamma_\nu\right]$$
$$= (l_\mu l_\nu - l^2\delta_{\mu\nu})C(l^2) \quad (17.13)$$

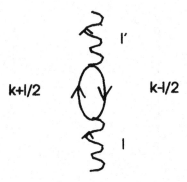

Fig. 17.4.   Lowest order vacuum polarization correction.

The second relation follows from Lorentz covariance and current conservation. One can write

$$C(l^2) = C(0) - l^2 \Pi_f(l^2) \tag{17.14}$$

In order to produce a mathematically well-defined expression, while maintaining current conservation, one can use a more general Pauli–Villars regulator on the *loop integral* in Eq. (17.13)

$$\Pi_{\mu\nu}(l, m_e^2) \rightarrow \int g(\lambda^2) d\lambda^2 [\Pi_{\mu\nu}(l, m_e^2) - \Pi_{\mu\nu}(l, m_e^2 + \lambda^2)] \tag{17.15}$$

with $g(\lambda^2)$ receiving contributions only from very large $\lambda^2 \approx \Lambda^2$ and

$$\int g(\lambda^2) d\lambda^2 = 1$$
$$\int \lambda^2 g(\lambda^2) d\lambda^2 = 0 \tag{17.16}$$

One argument in justification of this regularization procedure is that equating higher moments of λ^2 to zero, thereby obtaining additional convergence, will not change the answer. Evaluation of the integrals now results in [Bj65, Wa91]

$$C(0) = \frac{2\alpha}{3\pi} \ln \frac{\Lambda}{m_e} \tag{17.17}$$

Let us denote by e_0 the electric charge used up to this point, i.e. the "bare charge" appearing in the initial lagrangian. If one now combines the lowest order contribution with the vacuum polarization contribution for scattering of an electron in an external field [Fig. 17.2(a), (g)], the result is to change the amplitude in the limit $q^2 \rightarrow 0$ from $e_0^2/q^2 \rightarrow e^2/q^2$ where the *renormalized charge* is given by

$$e^2 = e_0^2[1 - C(0)] = e_0^2 \left(1 - \frac{2\alpha_0}{3\pi} \ln \frac{\Lambda}{m_e} \right) \tag{17.18}$$

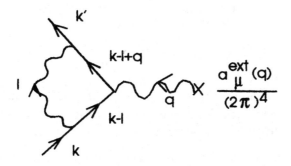

Fig. 17.5.   Vertex correction. Here $k \equiv k_1$ and $k' \equiv k_2$.

To this order, *in the radiative corrections*, one can replace $e_0^2 \to e^2$. The vacuum polarization contribution to the above process is now obtained by making the following replacement in the lowest order term

$$\frac{1}{q^2} \longrightarrow \frac{1}{q^2}[1 + q^2 \Pi_f(q^2)] \qquad (17.19)$$

Here $\Pi_f(q^2)$ is calculated with e^2.

The remaining finite momentum integrals can be evaluated to give the answer in terms of an integral over the Feynman parameter x [Bj65, Wa91]

$$l^2 \Pi_f(l^2) = \frac{2\alpha}{\pi} \int_0^1 x(1-x) \ln\left[1 + x(1-x)\frac{l^2}{m_e^2}\right] dx \qquad (17.20)$$

$$\to \frac{\alpha}{15\pi} \frac{l^2}{m_e^2} \qquad\qquad ; \; l^2 \ll m_e^2$$

$$\to \frac{\alpha}{3\pi}\left[\ln\frac{l^2}{m_e^2} - \frac{5}{3}\right] \qquad ; \; l^2 \gg m_e^2$$

Consider next the *vertex correction* in Fig. 17.5. The analytic expression for the contribution to the S-matrix is given by

$$S_{fi} = -\frac{e}{\Omega}\bar{u}(k_2)\Lambda_\mu(k_2,k_1)u(k_1)a_\mu^{\text{ext}}(q) \qquad (17.21)$$

$$\Lambda_\mu(k_2,k_1) = \frac{ie^2}{(2\pi)^4}\int \frac{d^4l}{l^2}\gamma_\nu \frac{1}{i\gamma_\lambda(k_1 - l + q)_\lambda + m_e}\gamma_\mu \frac{1}{i\gamma_\sigma(k_1 - l)_\sigma + m_e}\gamma_\nu$$

The general form of this vertex follows from Lorentz covariance and power counting as

$$\Lambda_\mu = L\gamma_\mu + \Lambda_{\mu C}(k_2,k_1) \qquad (17.22)$$

Here L is a (infinite) constant and the diagonal matrix element of the remaining convergent term, taken between Dirac spinors, vanishes. Regularization of the photon propagator as in Eq. (17.10) again eliminates the *ultraviolet divergence* at high momenta (short wavelengths), and explicit evaluation gives [Bj65, Wa91]

$$L = \frac{\alpha}{2\pi}\left[\ln\frac{\Lambda}{m_e} + \frac{9}{4} - 2\ln\frac{m_e}{\lambda}\right] \tag{17.23}$$

Here, to protect against the *infrared divergence* at low momenta (long wavelengths), the photon has been given a tiny, fictitious mass and the photon propagator has been replaced by

$$\frac{1}{q^2} \longrightarrow \frac{1}{q^2 + \lambda^2} \tag{17.24}$$

Note that no physical result can depend on the fictitious photon mass λ^2.

It is relatively easy to evaluate the matrix element of the remaining term in Eq. (17.22) between Dirac spinors $\bar{u}(k_2)\Lambda_{\mu C}(k_2, k_1)u(k_1)$ with the result [Bj65, Wa91]

$$\Lambda_{\mu C} \doteq F_E(q^2)\gamma_\mu - F_M(q^2)\frac{1}{2m_e}\sigma_{\mu\nu}q_\nu \tag{17.25}$$

$$F_M(q^2) = \frac{\alpha}{\pi}\int_0^1 dx \int_0^x dy\, \frac{m_e^2\, x(1-x)}{m_e^2\, x^2 + q^2 y(x-y)}$$

$$F_E(q^2) = -\frac{\alpha}{2\pi}\int_0^1 dx \int_0^x dy\left\{\ln\left(1 + \frac{q^2 y(x-y)}{m_e^2\, x^2 + \lambda^2(1-x)}\right)\right.$$

$$+ 2m_e^2\left(1 - x - \frac{x^2}{2}\right)$$

$$\times\left[\frac{1}{m_e^2\, x^2 + \lambda^2(1-x) + q^2 y(x-y)} - \frac{1}{m_e^2\, x^2 + \lambda^2(1-x)}\right]$$

$$\left. + q^2(1-x+y)(1-y)\left[\frac{1}{m_e^2\, x^2 + \lambda^2(1-x) + q^2 y(x-y)}\right]\right\}$$

Here $q = k_2 - k_1$, and $\doteq$ means "taken between Dirac spinors."

The limiting cases of these results are as follows

$$F_M(0) = \frac{\alpha}{2\pi}$$

$$F_E(q^2) = \frac{\alpha}{3\pi}\frac{q^2}{m_e^2}\left(\frac{3}{8} - \ln\frac{m_e}{\lambda}\right) \qquad ; q^2 \ll m_e^2$$

$$= -\frac{\alpha}{2\pi}\ln\frac{q^2}{m_e^2}\ln\frac{q^2}{\lambda^2} \qquad ; q^2 \gg m_e^2 \tag{17.26}$$

Note that the remaining finite part of the vertex $F_E(q^2)$ is infrared divergent; therefore it is *not an observable*.

By direct calculation, one can now establish to $O(\alpha)$ that

$$B = L \qquad (17.27)$$

In fact, this relation holds to all orders. It was proven by Ward who observed the general result known as *Ward's Identity* [Wa50]

$$\frac{\partial}{\partial k_\mu} \Sigma^\star(k) = i\Lambda_\mu(k, k) \qquad (17.28)$$

Here $\Sigma^\star$ is the proper self-energy and Λ_μ the proper vertex [Bj65, Fe71]. In second order, this result follows immediately from Eqs. (17.8, 17.21). Equation (17.27) can then be derived from it by taking matrix elements between Dirac spinors of identical four-momentum.[6]

When the self-energy insertion $\Sigma - \delta m_e$ is on an external line, the resulting expression obtained from Eqs. (17.8, 17.9) is ambiguous since, for example,

$$-B(i\gamma_\lambda k_\lambda + m_e) \frac{1}{i\gamma_\lambda k_\lambda + m_e} u(k) = -\frac{0}{0} B\, u(k) \qquad (17.29)$$

A proper adiabatic limiting procedure says that here the correct answer is to retain $-(B/2)u(k)$, and similarly for the other leg [Bj65, Wa91].

The use of the Fourier transform of Maxwell's equations for the external field allows one to relate that field to its source[7]

$$a_\mu^{\text{ext}}(q) = \frac{e_0}{q^2} j_\mu^{\text{ext}}(q) \qquad (17.30)$$

In *summary*, the addition of all the diagrams in Fig. 17.2 yields to $O(e_0^4)$

$$
\begin{aligned}
S_{fi} ={}& -\frac{e_0^2}{\Omega} \bar{u}(k_2) \left\{ \gamma_\mu \left[1 + L - \frac{B}{2} - \frac{B}{2} - C \right] + \gamma_\mu q^2 \Pi_f(q^2) \right. \\
& \left. + \Lambda_{\mu C}(k_2, k_1) \right\} u(k_1) \frac{1}{q^2} j_\mu^{\text{ext}}(q)
\end{aligned} \qquad (17.31)
$$

Ward's identity now leads to an exact cancellation of the term $L - B = 0$. The remaining constant C, *arising entirely from vacuum polarization*, serves to renormalize the charge according to Eq. (17.18). As above, one can then replace $\alpha_0 = \alpha + O(e_0^4)$ to this order in the radiative corrections.

---

[6] Ward's Identity follows in general by looking at all the Feynman diagrams involved, letting the external electron momentum flow along the electron line, and then differentiating with respect to this momentum.

[7] This relation explicitly exhibits the one additional power of e_0 in the process — i.e., both ends of the vacuum polarization insertion end up on a charge.

The result is

$$
\begin{aligned}
S_{fi} &= -\frac{e^2}{\Omega}\bar{u}(k_2)\left\{\gamma_\mu[1 + q^2\Pi_f(q^2)] + \Lambda_{\mu C}(k_2, k_1)\right\}u(k_1)\frac{1}{q^2}j_\mu^{\text{ext}}(q) \\
&= -\frac{e}{\Omega}\bar{u}(k_2)\left\{\gamma_\mu[1 + F_E(q^2) + q^2\Pi_f(q^2)] - F_M(q^2)\frac{1}{2m_e}\sigma_{\mu\nu}q_\nu\right\} \\
&\quad\times u(k_1)a_\mu^{\text{ext}}(q)
\end{aligned}
\tag{17.32}
$$

Several comments are relevant:

- This amplitude is to be computed with the renormalized charge;

- This result is finite as $\Lambda \to \infty$; there is no longer any ultraviolet divergence;

- The exact second-order (integral) expressions for the quantities appearing in this result are given in Eqs. (17.20, 17.25);

- The presence of form factors in this expression indicates that the electron does indeed have an internal structure; it arises from the interaction with the virtual photon field and is completely calculable within the framework of QED;[8]

- This expression is still infrared divergent in that it depends on the fictitious photon mass λ^2 — hence, as it stands, it is *unobservable*.

---

[8] There is further internal structure of the electron at much shorter distance scales arising from the weak interactions.

18

Radiative corrections

Let us now calculate the observable QED modification of the electron scattering cross section following from the analysis in the previous chapter. For this purpose, assume a static (time-independent) external field, in which case

$$a_\mu^{\text{ext}}(q) = -2\pi i \delta(W_f - W_i) \tilde{a}_\mu^{\text{ext}}(\mathbf{q}) \tag{18.1}$$

One can identify the T-matrix and cross section corresponding to Eq. (17.32) from the general relations

$$
\begin{aligned}
S_{fi} &= -2\pi i \delta(W_f - W_i) T_{fi} \\
d\sigma &= 2\pi |T_{fi}|^2 \delta(W_f - W_i) \frac{d\rho_f}{\text{Flux}}
\end{aligned} \tag{18.2}
$$

For illustration, we here confine the discussion to scattering where the target is left in its ground state. It follows that

$$\frac{d\rho_f}{\text{Flux}} = \frac{\Omega d^3 k_2}{(2\pi)^3} \frac{1}{v_1/\Omega} \tag{18.3}$$

Let the superscript denote the order in e, then to $O(e^4)$ one has for the square of the T-matrix

$$|T_{fi}|^2 = |T_{fi}^{(1)} + T_{fi}^{(3)}|^2 = |T_{fi}^{(1)}|^2 + 2\text{Re}\, T_{fi}^{(1)*} T_{fi}^{(3)} \tag{18.4}$$

If the explicit magnetic moment contribution is suppressed for the time being, then, since the QED amplitude in Eq. (17.32) contains only a real modification of the coefficient of γ_μ, one finds to this order

$$\left(\frac{d\sigma}{d\Omega}\right)_{\text{el}} \doteq \left(\frac{d\sigma}{d\Omega}\right)_0 \{1 + 2[F_E(q^2) + q^2 \Pi_f(q^2)] + \cdots\} \tag{18.5}$$

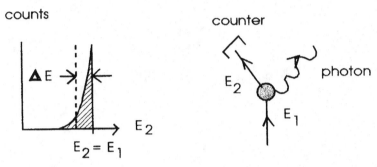

Fig. 18.1.   Observation in an electron scattering experiment.

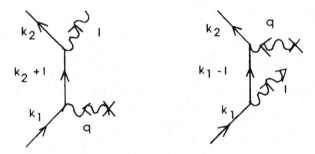

Fig. 18.2.   Bremsstrahlung in an external field. The photon polarization is ε_ν.

Here $(d\sigma/d\Omega)_0$ is the lowest order cross section, and the dots denote the magnetic moment contribution.

One now has to think carefully about what is actually *observed* in the experiment. Since an electron can always radiate a photon of arbitrarily long wavelength (or low energy) during the scattering process, what one will observe in an electron scattering experiment is illustrated in Fig. 18.1. Experimentally, all one can observe is the *sum of these elastic and inelastic electromagnetic cross sections.*

$$d\sigma = d\sigma_{\text{el}} + d\sigma_{\text{in}} \tag{18.6}$$

One is therefore required to also calculate the cross section for radiation of a photon, the bremsstrahlung cross section. The two Feynman diagrams for bremsstrahlung in the same external field are shown in Fig. 18.2. The analytic expression follows from the previous Feynman rules as

$$
\begin{aligned}
S_{fi} =& -\frac{ie^2}{\Omega}\frac{1}{\sqrt{2\omega_l\Omega}}\bar{u}(k_2)\left[\gamma_\nu\varepsilon_\nu\frac{1}{i\gamma_\lambda(k_2+l)_\lambda+m_e}\gamma_\mu\right.\\
&\left.+\gamma_\mu\frac{1}{i\gamma_\sigma(k_1-l)_\sigma+m_e}\gamma_\nu\varepsilon_\nu\right]u(k_1)a_\mu^{\text{ext}}(q)
\end{aligned}
\tag{18.7}
$$

This amplitude will give the bremsstrahlung cross section to $O(e^4)$, which is of exactly the same order as the last term in Eq. (18.4).

First, rationalize the term in brackets in Eq. (18.7) and use the Dirac equation to the left and right

$$[\cdots] = -\left[\frac{2ik_2 \cdot \varepsilon + i(\gamma_\lambda \varepsilon_\lambda)(\gamma_\sigma l_\sigma)}{2k_2 \cdot l + l^2}\gamma_\mu + \gamma_\mu \frac{2ik_1 \cdot \varepsilon - i(\gamma_\lambda l_\lambda)(\gamma_\sigma \varepsilon_\sigma)}{-2k_1 \cdot l + l^2}\right] \quad (18.8)$$

Now let the photon energy become very small

$$|\mathbf{l}| \equiv \Delta E \to 0 \quad (18.9)$$

Then

$$S_{fi} \doteq -\frac{e^2}{\Omega}\bar{u}(k_2)\gamma_\mu u(k_1)a_\mu^{\text{ext}}(q)\left[\frac{1}{\sqrt{2\omega_l\Omega}}\left(\frac{k_2 \cdot \varepsilon}{k_2 \cdot l} - \frac{k_1 \cdot \varepsilon}{k_1 \cdot l}\right)\right] \quad (18.10)$$

Note that at this stage strict current conservation (gauge invariance) is still maintained since if one replaces $\varepsilon_\mu \to l_\mu$ this amplitude vanishes.

Assume again a static external field as in Eq. (18.1) and read off the T-matrix as in Eq. (18.2). The bremsstrahlung cross section is then

$$d\sigma_{\text{in}} = 2\pi\delta(W_f - W_i)|T_{fi}|^2\frac{\Omega d^3l}{(2\pi)^3}\frac{\Omega d^3k_2}{(2\pi)^3}\frac{1}{v_1/\Omega} \quad (18.11)$$

Under the condition in Eq. (18.9), one can replace

$$\tilde{a}_\mu^{\text{ext}}(\mathbf{q}) \approx \tilde{a}_\mu^{\text{ext}}(\mathbf{k_2} - \mathbf{k_1})$$
$$E_2 + \omega_l = W_f \approx E_2 \quad (18.12)$$

Since these quantities are now the same as in elastic scattering, $d\sigma_{\text{in}}$ will again be proportional to $d\sigma_0$! It follows from the above that

$$\frac{d\sigma}{d\Omega} = \left(\frac{d\sigma}{d\Omega}\right)_{\text{el}} + \left(\frac{d\sigma}{d\Omega}\right)_{\text{in}} \quad (18.13)$$

$$= \left(\frac{d\sigma}{d\Omega}\right)_0\left\{1 + 2[F_E(q^2) + q^2\Pi_f(q^2)]\right.$$

$$\left. +\frac{\alpha}{4\pi^2}\sum_{\text{pol}}\int_0^{\Delta E}\frac{d^3l}{\omega_l}\left(\frac{k_2 \cdot \varepsilon}{k_2 \cdot l} - \frac{k_1 \cdot \varepsilon}{k_1 \cdot l}\right)^2 + \cdots\right\}$$

In this express ΔE is the resolution of the electron detector, and one must include all inelastic processes that give an electron in the detector within this resolution. A correct calculation of radiative corrections thus *depends on the geometry of the experiment*. The dots again denote the additional magnetic moment contribution.

Since a, albeit tiny, mass has been assumed for the photon, the bremsstrahlung term must be evaluated consistently for this situation. The polarization sum for a massive vector meson yields[1]

$$\sum_\sigma \varepsilon_\mu^{(\sigma)} \varepsilon_\nu^{(\sigma)} = \delta_{\mu\nu} + \frac{l_\mu l_\nu}{\lambda^2} \tag{18.14}$$

Since the bremsstrahlung amplitude satisfies strict current conservation, the terms in $l_\mu l_\nu$ in this expression give a vanishing contribution. Hence

$$\sum_{\text{pol}} \left(\frac{k_2 \cdot \varepsilon}{k_2 \cdot l} - \frac{k_1 \cdot \varepsilon}{k_1 \cdot l} \right)^2 = -\frac{m_e^2}{(k_2 \cdot l)^2} - \frac{m_e^2}{(k_1 \cdot l)^2} - \frac{2k_1 \cdot k_2}{(k_1 \cdot l)(k_2 \cdot l)} \tag{18.15}$$

Here $l_\mu = (\mathbf{l}, i\omega_l)$ where $\omega_l = \sqrt{\mathbf{l}^2 + \lambda^2}$.

One must now do the remaining $\int d^3l / \omega_l$, with the limiting results

$$\frac{\alpha}{4\pi^2} \sum_{\text{pol}} \int_0^{\Delta E} \frac{d^3 l}{\omega_l} \left(\frac{k_2 \cdot \varepsilon}{k_2 \cdot l} - \frac{k_1 \cdot \varepsilon}{k_1 \cdot l} \right)^2$$

$$= \frac{2\alpha}{3\pi} \frac{q^2}{m_e^2} \left(\ln \frac{2\Delta E}{\lambda} - \frac{5}{6} \right) \qquad k^2 \ll m_e^2 \; ; \; q^2 \ll m_e^2$$

$$= \frac{2\alpha}{\pi} \ln \frac{q^2}{m_e^2} \ln \frac{\Delta E}{\lambda} \qquad\qquad q^2 \gg m_e^2 \tag{18.16}$$

Note that this bremsstrahlung cross section is also infrared divergent so that it, by itself, is *unobservable*; however, when adding the two results in Eq. (18.13), *the infrared divergent terms in* $\ln \lambda$ *cancel identically in the observable cross section*!

A combination of the above results then yields, for the scattering of an electron in a static Coulomb field

$$\frac{d\sigma}{d\Omega} = \left(\frac{d\sigma}{d\Omega} \right)_{\text{Mott}} (1 - \delta) \tag{18.17}$$

$$\delta \approx \frac{2\alpha}{3\pi} \frac{q^2}{m_e^2} \left(\ln \frac{m_e}{\Delta E} + \frac{5}{6} - \frac{1}{5} - \frac{3}{8} + \frac{3}{8} \right) \qquad k^2 \ll m_e^2$$

$$\delta \approx \frac{2\alpha}{\pi} \ln \frac{q^2}{m_e^2} \ln \frac{E}{\Delta E} \qquad\qquad q^2 \gg m_e^2 \; ; \quad E \gg \Delta E$$

Here we have identified $(d\sigma/d\Omega)_0 = (d\sigma/d\Omega)_{\text{Mott}}$ for scattering in a static Coulomb field. The last $+3/8$ in the second line, canceling the term before it, is the hitherto suppressed magnetic moment contribution; the $-1/5$

---

[1] Use Lorentz covariance, $l \cdot e = 0$, and $l^2 = -\lambda^2$.

comes from vacuum polarization. We have also written $E_1 \equiv E$, and ΔE is the experimental resolution.[2]

These results on the radiative corrections are originally due to Schwinger [Sc49], who argued that the correct result for the infrared divergent series, to all orders in α, is really

$$1 - \delta + \frac{\delta}{2!} + \cdots = e^{-\delta} \qquad (18.18)$$

When $\Delta E \to 0$, one then has $e^{-\delta} \to 0$, and there is *no perfectly elastic scattering.*[3]

Note that while the ultraviolet divergences truly reflect a lack of knowledge of the physics at very short distances, the treatment of the infrared divergences is basically a technical problem. The emission of very long wavelength radiation (photons) is essentially governed by classical physics. This is a problem first treated in detail by Bloch and Nordsieck [Bl37]. The difficulty arises because analyzing the emission photon-by-photon (i.e. as a power series in e) is not an efficient way of attacking this problem.

---

[2] Note that $\ln q^2 \approx \ln E^2$ as $E \to \infty$.
[3] For applications of radiative corrections see [Ma69, Mo69a].

Part 4
Selected examples

19

Basic nuclear structure

The goal is to compare experimental electron scattering data with a theoretical picture of the hadronic target, and in so doing, develop an understanding of that system. We start the discussion within the traditional non-relativistic many-body description of the nucleus.

If the nucleus is modeled as a quantum mechanical system of point nucleons with intrinsic magnetic moments, then one knows how to construct the charge density, convection current density, and intrinsic magnetization density from basic quantum mechanics. In first quantization these quantities are given by

$$
\hat{\rho}_N(\mathbf{x}) = \sum_{j=1}^{A} e(j)\delta^{(3)}(\mathbf{x} - \mathbf{x}_j)
$$

$$
\hat{\mathbf{J}}_c(\mathbf{x}) = \sum_{j=1}^{A} e(j)\{\frac{\mathbf{p}(j)}{m}, \delta^{(3)}(\mathbf{x} - \mathbf{x}_j)\}_{\text{sym}}
$$

$$
\hat{\boldsymbol{\mu}}(\mathbf{x}) = \sum_{j=1}^{A} \mu(j)\frac{\boldsymbol{\sigma}(j)}{2m}\delta^{(3)}(\mathbf{x} - \mathbf{x}_j) \tag{19.1}
$$

Here $\mathbf{p} \equiv (1/i)\nabla$ and $\boldsymbol{\sigma} \equiv (\sigma_x, \sigma_y, \sigma_z)$ are the Pauli matrices. Thus for a single particle, for example

$$
\langle\hat{\rho}_N(\mathbf{x})\rangle = \int \psi^*(\mathbf{x}_p)\delta^{(3)}(\mathbf{x} - \mathbf{x}_p)\psi(\mathbf{x}_p)\,d^3x_p = |\psi(\mathbf{x})|^2 \tag{19.2}
$$

and also

$$
\langle\hat{\mathbf{J}}_c(\mathbf{x})\rangle = \int \psi^*(\mathbf{x}_p)\frac{1}{2im}[\nabla_p\delta^{(3)}(\mathbf{x} - \mathbf{x}_p) + \delta^{(3)}(\mathbf{x} - \mathbf{x}_p)\nabla_p]\psi(\mathbf{x}_p)\,d^3x_p
$$

$$
= \frac{1}{2im}\{\psi^*(\mathbf{x})\nabla\psi(\mathbf{x}) - [\nabla\psi(\mathbf{x})]^*\psi(\mathbf{x})\} \tag{19.3}
$$

147

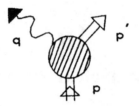

Fig. 19.1. Electromagnetic vertex for a free nucleon.

A partial integration has been used in obtaining the last equality. The charge and magnetic moment of the nucleon are given by

$$e(j) \equiv \frac{1}{2}[1 + \tau_3(j)]$$

$$\mu(j) \equiv \lambda_p \frac{1}{2}[1 + \tau_3(j)] + \lambda_n \frac{1}{2}[1 - \tau_3(j)] \tag{19.4}$$

The *anomalous* magnetic moment $\lambda'(j)$ of the nucleon is defined by

$$\lambda'(j) = \lambda'_p \frac{1}{2}[1 + \tau_3(j)] + \lambda_n \frac{1}{2}[1 - \tau_3(j)]$$

$$\mu(j) = e(j) + \lambda'(j) \tag{19.5}$$

This discussion presents a consistent non-relativistic treatment in a picture where the nucleus is made up of point nucleons with appropriate charges and intrinsic magnetic moments; however, a central goal of nuclear physics is the measurement and calculation of nuclear electromagnetic transition densities out to momentum transfers $q^2 = O(m^2)$ and well beyond. It is essential to consider corrections to the non-relativistic current operator as one moves into this regime. In order to do this, a fully relativistic treatment of the interacting many-body system is required, and the next section is devoted to this topic. For the present, we simply consider the nuclear current density arising from the full relativistic electromagnetic vertex of a free nucleon [Mc62].

The relativistic electromagnetic vertex of a free nucleon is illustrated in Fig. 19.1. The most general structure of the matrix element of the current for a free nucleon is given by [Bj65, Wa95]

$$\langle \mathbf{p}'\sigma'\rho'|J_\mu(0)|\mathbf{p}\sigma\rho\rangle = \frac{i}{\Omega}\bar{u}(\mathbf{p}',\sigma')\eta^\dagger_{\rho'}[F_1\gamma_\mu + F_2\sigma_{\mu\nu}q_\nu]\eta_\rho u(\mathbf{p},\sigma) \tag{19.6}$$

Here the spin and isospin quantum numbers have been made explicit; $\bar{u}, u$ are Dirac spinors and η_p, η_n are two-component Pauli isospinors. The four-momentum transfer is defined by $q = p - p'$, and the form factors $F_{1,2}$ are functions of q^2. The isospin structure of the form factors must be of the form

$$F_i = \frac{1}{2}(F_i^S + \tau_3 F_i^V) \qquad\qquad ; i = 1, 2 \qquad (19.7)$$

Relevant numerical values are

$$\begin{aligned}
F_1^S(0) &= F_1^V(0) = 1 \\
2m F_2^S(0) &= \lambda_p' + \lambda_n = -0.120 \\
2m F_2^V(0) &= \lambda_p' - \lambda_n = +3.706
\end{aligned} \qquad (19.8)$$

To construct the *nuclear* current density one carries out the following steps:

1. Substitute the explicit form of the Dirac spinors for a free nucleon

$$u(\mathbf{p}, \sigma) = \left(\frac{E_p + m}{2E_p}\right)^{1/2} \left[\begin{array}{c} \chi_\sigma \\[2mm] \dfrac{\boldsymbol{\sigma} \cdot \mathbf{p}}{E_p + m} \chi_\sigma \end{array}\right] \qquad (19.9)$$

Here $\chi_\uparrow, \chi_\downarrow$ are two-component Pauli spinors for spin up and down along the z-axis. Now expand the matrix element in Eq. (19.6) consistently to order $1/m^2$. The result is[1]

$$\langle \mathbf{p}'\sigma'\rho'|J_\mu(0)|\mathbf{p}\sigma\rho\rangle = \frac{1}{\Omega} \eta_{\rho'}^\dagger \chi_{\sigma'}^\dagger \mathscr{M}_\mu \chi_\sigma \eta_\rho \qquad (19.10)$$

$$\mathscr{M} = F_1 \frac{1}{2m}(\mathbf{p} + \mathbf{p}') + (F_1 + 2m F_2)\left[\frac{-i\boldsymbol{\sigma} \times \mathbf{q}}{2m}\right] + O\left(\frac{1}{m^3}\right)$$

$$\mathscr{M}_0 = F_1 - (F_1 + 4m F_2)\left[\frac{\mathbf{q}^2}{8m^2} - \frac{i\mathbf{q} \cdot (\boldsymbol{\sigma} \times \mathbf{p})}{4m^2}\right] + O\left(\frac{1}{m^3}\right)$$

Here $\mathscr{M}_\mu = (\mathscr{M}, i\mathscr{M}_0)$.

2. Take the prescription for constructing the nuclear current density operator at the origin, in second quantization, to be

$$\hat{J}_\mu(0) = \sum_{\mathbf{p}'\sigma'\rho'} \sum_{\mathbf{p}\sigma\rho} c_{\mathbf{p}'\sigma'\rho'}^\dagger \langle \mathbf{p}'\sigma'\rho'|J_\mu(0)|\mathbf{p}\sigma\rho\rangle c_{\mathbf{p}\sigma\rho} \qquad (19.11)$$

where the single-particle matrix element is that of Eq. (19.6).

3. Use the general procedure for passing from first quantization to second quantization [Fe71]. If, in first quantization the one-body nuclear density operator has the form

$$\hat{J}_\mu(\mathbf{x}) = \sum_{i=1}^A \{J_\mu^{(1)}(i)\delta^{(3)}(\mathbf{x} - \mathbf{x}_i)\} \qquad (19.12)$$

---

[1] It is assumed that both q_0 and F_2 are $O(1/m)$.

then in second quantization the operator density is

$$\hat{J}_\mu(\mathbf{x}) = \sum_{\mathbf{p}'\sigma'\rho'} \sum_{\mathbf{p}\sigma\rho} c^\dagger_{\mathbf{p}'\sigma'\rho'} \langle \mathbf{p}'\sigma'\rho' | J_\mu(\mathbf{x}) | \mathbf{p}\sigma\rho \rangle c_{\mathbf{p}\sigma\rho} \qquad (19.13)$$

with

$$\langle \mathbf{p}'\sigma'\rho' | J_\mu(\mathbf{x}) | \mathbf{p}\sigma\rho \rangle = \int d^3y \, \phi^\dagger_{\mathbf{p}'\sigma'\rho'}(\mathbf{y}) \{ J^{(1)}_\mu(\mathbf{y})\delta^{(3)}(\mathbf{x} - \mathbf{y}) \} \phi_{\mathbf{p}\sigma\rho}(\mathbf{y})$$
$$(19.14)$$

4. The discussion in chapter 9 shows that physical rates and cross sections are expressed in terms of the Fourier transform of the transition matrix element of the current

$$\int e^{-i\mathbf{q}\cdot\mathbf{x}} \langle f | \hat{J}_\mu(\mathbf{x}) | i \rangle \, d^3x \qquad (19.15)$$

Here $q = p - p'$, and in electron scattering $q = k_2 - k_1$. Define

$$\langle f | \hat{J}_\mu(\mathbf{x}) | i \rangle \equiv J_\mu(\mathbf{x})_{fi} \qquad (19.16)$$

and observe that by partial integration in Eq. (19.15), with localized densities, one can make the replacement

$$\nabla \leftrightarrow i\mathbf{q} \qquad (19.17)$$

The presence of terms in $i\mathbf{q}$ in the elementary nucleon amplitudes are then anticipated by defining

$$\begin{aligned} \mathbf{J}(\mathbf{x})_{fi} &\equiv \mathbf{J}_c(\mathbf{x})_{fi} + \nabla \times \boldsymbol{\mu}(\mathbf{x})_{fi} \\ \rho(\mathbf{x})_{fi} &\equiv \rho_N(\mathbf{x})_{fi} + \nabla \cdot \mathbf{s}(\mathbf{x})_{fi} + \nabla^2 \phi(\mathbf{x})_{fi} \end{aligned} \qquad (19.18)$$

The use of Eq. (19.13) evaluated at $\mathbf{x} = 0$ now permits the identification of the nuclear density operators in first quantization, which give rise to the required result in second quantization of Eq. (19.10). The operators take the form

$$\begin{aligned} \hat{\mathbf{J}}(\mathbf{x}) &= \hat{\mathbf{J}}_c(\mathbf{x}) + \nabla \times \hat{\boldsymbol{\mu}}(\mathbf{x}) \\ \hat{\rho}(\mathbf{x}) &= \hat{\rho}_N(\mathbf{x}) + \nabla \cdot \hat{\mathbf{s}}(\mathbf{x}) + \nabla^2 \hat{\phi}(\mathbf{x}) \end{aligned} \qquad (19.19)$$

Here the densities are defined by Eqs. (19.1), (19.4), (19.5), and

$$\hat{\phi}(\mathbf{x}) = \sum_{j=1}^A s(j) \frac{1}{8m^2} \delta^{(3)}(\mathbf{x} - \mathbf{x}_j) \qquad (19.20)$$

$$\hat{\mathbf{s}}(\mathbf{x}) = \sum_{j=1}^A s(j) \frac{1}{4m^2} \boldsymbol{\sigma}(j) \times \{ \mathbf{p}(j), \delta^{(3)}(\mathbf{x} - \mathbf{x}_j) \}_{\text{sym}}$$

where in the static limit

$$s(j) \equiv e(j) + 2\lambda'(j) \qquad (19.21)$$

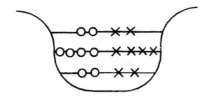

Fig. 19.2.   Basis of Hartree–Fock states.

5. It is an empirical result that in the nuclear domain[2]

$$\frac{F_1(q^2)}{F_1(0)} \approx f_{SN}(q^2) \approx \frac{F_2(q^2)}{F_2(0)}$$

$$f_{SN}(q^2) = \frac{1}{(1 + q^2/0.71 \text{ GeV}^2)^2} \qquad (19.22)$$

At finite q^2, the quantity $f_{SN}(q^2)$ enters as an overall factor in the electromagnetic vertex, and it can be included by using an overall *effective* Mott cross section $\bar{\sigma}_M$ in the (e, e') cross section

$$\bar{\sigma}_M \equiv \sigma_M |f_{SN}(q^2)|^2 \qquad (19.23)$$

The use of this effective Mott cross section represents an approximate way of taking into account in the nuclear domain the spatial extent of the *internal* charge and magnetization densities of a single constituent nucleon.

The present analysis gives the leading relativistic corrections to the nuclear current, assuming it is a one-body operator. It *neglects*, among other things: meson exchange currents, other multibody currents, relativistic terms in the *wave functions*, and off-shell corrections to the nucleon vertex in the nuclear medium. We shall return to many of these points.

Consider next the many-body matrix elements of the current [Fe71, Wa95]. Introduce a complete basis of Hartree–Fock states as illustrated in Fig. 19.2. Assume a central field and label the quantum numbers by

$$\alpha = (nl\frac{1}{2}jm_j; \frac{1}{2}m_t)$$
$$\equiv (a; m_j, m_t) \qquad (19.24)$$

---

[2] A more accurate representation of the experimental data for the proton and neutron out to very large q^2 is given by [Ba73, Wa84]

$$G_M(q^2) \equiv F_1 + 2mF_2 = f_{SN}(q^2)G_M(0)$$
$$G_E(q^2) \equiv F_1 - (q^2/2m)F_2 = f_{SN}(q^2)G_E(0)$$

although $G_E^n(q^2)$ remains to be measured well. Elastic scattering from the nucleon is discussed in more detail in chapter 20.

Then

$$-\alpha \equiv (a; -m_j, -m_t) \qquad (19.25)$$

We shall also need the phase defined by

$$S_\alpha \equiv (-1)^{j_\alpha - m_{j_\alpha}}(-1)^{1/2 - m_{t_\alpha}} \qquad (19.26)$$

Any many-body multipole operator of the above current can now be written in second quantization as [Fe71]

$$\hat{\mathcal{T}}_{JM_J;TM_T}(\kappa) = \sum_\alpha \sum_\beta c_\alpha^\dagger \langle \alpha | T_{JM_J;TM_T}(\kappa) | \beta \rangle c_\beta \qquad (19.27)$$

$c_\alpha^\dagger, c_\beta$ are creation and destruction operators for the single-particle Hartree–Fock states. The single-particle matrix elements are calculated using the wave functions of this basis, the current densities above, and the appropriate multipole projections of chapter 9.

Within the present framework, an arbitrary matrix element between exact eigenstates $|\Psi_i\rangle$ and $|\Psi_f\rangle$ of the nuclear many-body system can be written

$$\langle \Psi_f | \hat{\mathcal{T}}_{JM_J;TM_T}(\kappa) | \Psi_i \rangle = \sum_\alpha \sum_\beta \langle \alpha | T_{JM_J;TM_T}(\kappa) | \beta \rangle \, \psi_{\alpha\beta}^{fi}$$

$$\psi_{\alpha\beta}^{fi} = \langle \Psi_f | c_\alpha^\dagger c_\beta | \Psi_i \rangle \qquad (19.28)$$

The quantities $\psi_{\alpha\beta}^{fi}$ are simply *numerical coefficients*. This result has the following features:

- It assumes the current is a one-body operator — exchange currents, for example, are neglected;

- Any shell-model calculation, no matter how complicated, must give an answer of this form. The exact many-body matrix element is a sum of single-particle matrix elements with numerical coefficients;

- This is an *exact* statement within the traditional non-relativistic nuclear many-body problem.

Let us extract the angular momentum properties of the operators involved in the above. Suppress isospin for the moment; it will be restored at the end. The angular momentum operator for the system is

$$\hat{\mathbf{J}} = \sum_\alpha \sum_\beta c_\alpha^\dagger \langle \alpha | \mathbf{J} | \beta \rangle c_\beta$$

$$= \sum_{nlj} \sum_{m'} \sum_m c_{nljm'}^\dagger \langle jm' | \mathbf{J} | jm \rangle c_{nljm} \qquad (19.29)$$

Note that the single-particle matrix elements of **J** are diagonal in (nl) and independent of (nl). Now make use of the basic anti-commutation relations

$$\{c_\alpha, c_\beta^\dagger\} = \delta_{\alpha\beta}$$
$$\{c_\alpha, c_\beta\} = \{c_\alpha^\dagger, c_\beta^\dagger\} = 0 \tag{19.30}$$

It is then a matter of algebra to establish the relation[3]

$$[\hat{\mathbf{J}}, c_{nljm}^\dagger] = \sum_{m'} \langle jm'|\mathbf{J}|jm\rangle c_{nljm'}^\dagger \tag{19.31}$$

Hence $c_\alpha^\dagger$ is an irreducible tensor operator (ITO) of rank j [Ed74].

Use of the Wigner–Eckart theorem allows one to establish the following relations [Ed74]

$$\langle jm|J_{1q}|jm'\rangle = (-1)^{m'-m+1}\langle j,-m'|J_{1q}|j,-m\rangle$$
$$\sum_m \langle jm|J_{1q}|jm\rangle = 0 \tag{19.32}$$

This permits the angular momentum operator to be written in an equivalent form [recall Eq. (19.26)]

$$\hat{\mathbf{J}} = \sum_\alpha \sum_\beta (S_{-\alpha}c_{-\alpha})\langle\alpha|\mathbf{J}|\beta\rangle(S_{-\beta}c_{-\beta})^\dagger \tag{19.33}$$

$$= \sum_{nlj}\sum_{m'}\sum_m [(-1)^{j+m'}c_{nlj,-m'}]\,\langle jm'|\mathbf{J}|jm\rangle\,[(-1)^{j+m}c_{nlj,-m}^\dagger]$$

Hence one concludes the $S_{-\alpha}c_{-\alpha}$ is an ITO by the same proof as above.

Now restore isospin (treated in an exactly analogous fashion) and assume the initial and final many-body target states are eigenstates of angular momentum and isospin. Use of Eq. (19.27) and the Wigner–Eckart theorem on both the many-body and single-particle matrix elements, and a change of dummy indices, leads to[4]

$$\langle J_f M_f T_f \bar{M}_f|\hat{\mathscr{T}}_{JM_J;TM_T}|J_i M_i T_i \bar{M}_i\rangle =$$

$$(-1)^{J_f-M_f}\begin{pmatrix} J_f & J & J_i \\ -M_f & M_J & M_i \end{pmatrix} \times [J \rightleftharpoons] \times \langle J_f T_f \vdots\hat{\mathscr{T}}_{J,T}\vdots J_i T_i\rangle$$

$$= \sum_{a,b}\langle a\vdots\hat{\mathscr{T}}_{J,T}\vdots b\rangle\langle J_f M_f T_f \bar{M}_f|\left\{\sum_{m_{j_\alpha}m_{j_\beta}}\langle j_\alpha m_{j_\alpha}j_\beta m_{j_\beta}|j_\alpha j_\beta JM_J\rangle\right.$$

---

[3] See [Fe71, Wa95].

[4] The symbol $\langle\vdots O\vdots\rangle$ indicates a matrix element reduced with respect to both angular momentum and isospin.

$$\times \sum_{m_{t_\alpha} m_{t_\beta}} \langle t_\alpha m_{t_\alpha} t_\beta m_{t_\beta} | \frac{1}{2} \frac{1}{2} T M_T \rangle (-1)^{j_\beta + m_{j_\beta}} (-1)^{1/2 + m_{t_\beta}}$$

$$\times \frac{1}{\sqrt{(2J+1)(2T+1)}} c_\alpha^\dagger c_{-\beta} \Big\} |J M_i T \bar{M}_i \rangle \tag{19.34}$$

One now identifies the *tensor product* $[c_\alpha \odot S_{-\beta} c_{-\beta}]_{J M_J; T M_T}$ of two ITO [Ed74]. Use of the Wigner–Eckart on that quantity gives, upon cancellation of common factors, the following expression for the matrix element of a multipole operator

$$\langle J_f T_f \| \hat{\mathscr{T}}_{J,T}(\kappa) \| J_i T_i \rangle = \sum_{a,b} \langle a \| \mathscr{T}_{J,T}(\kappa) \| b \rangle \psi_{J,T}^{fi}(a,b) \tag{19.35}$$

$$\psi_{J,T}^{fi}(a,b) = \frac{1}{\sqrt{(2J+1)(2T+1)}} \langle J_f T_f \| \hat{\zeta}^\dagger (ab; JT) \| J_i T_i \rangle$$

$$\hat{\zeta}^\dagger (ab; J M_J, T M_T) \equiv \sum_{m_{j_\alpha} m_{j_\beta}} \langle j_\alpha m_{j_\alpha} j_\beta m_{j_\beta} | j_\alpha j_\beta J M_J \rangle$$

$$\times \sum_{m_{t_\alpha} m_{t_\beta}} \langle t_\alpha m_{t_\alpha} t_\beta m_{t_\beta} | \frac{1}{2} \frac{1}{2} T M_T \rangle c_\alpha^\dagger [S_{-\beta} c_{-\beta}]$$

This is our principal result for the many-body matrix element. It has the following features:

- It is doubly reduced with respect to angular momentum and isospin;

- It expresses the many-body matrix element as a sum of single-particle matrix elements;

- It assumes a one-body current;

- It is *exact* within the traditional non-relativistic nuclear many-body problem.

Consider the isospin dependence in more detail. The previous single-particle densities, and hence the single-particle multipole operators, all have the form

$$\mathscr{T}_{J M_J} = \frac{1}{2} \mathscr{T}_{J M_J}^{(0)} + \frac{1}{2} \tau_3 \mathscr{T}_{J M_J}^{(1)}$$

$$\equiv I_{00} \mathscr{T}_{J M_J}^{(0)} + I_{10} \mathscr{T}_{J M_J}^{(1)} \tag{19.36}$$

It follows from this definition that $\langle \frac{1}{2} \| I_T \| \frac{1}{2} \rangle = [(2T+1)/2]^{1/2}$ [Ed74]. The many-body multipole operators thus have the corresponding isospin structure

$$\hat{\mathscr{T}}_{J M_J} = \hat{\mathscr{T}}_{J M_J;00} + \hat{\mathscr{T}}_{J M_J;10} \tag{19.37}$$

In electron scattering (e, e′), as well as real photon transitions, the third component of isospin of the target cannot change, and hence $\bar{M}_f = \bar{M}_i$. Now use the Wigner–Eckart theorem on the isospin dependence of the many-particle matrix elements to obtain the doubly reduced matrix elements. The basic result in Eq. (19.35) can then be employed. The isospin dependence of the single-particle matrix elements in Eq. (19.36) *factors*, and thus the doubly reduced single-particle matrix elements are particularly simple. It follows that the many-body matrix elements that enter into the cross sections and rates must take the form

$$\langle \Psi_f || \hat{\mathscr{T}}_J(\kappa) || \Psi_i \rangle = \langle J_f; T_f \bar{M}_i || \hat{\mathscr{T}}_J(\kappa) || J_i; T_i \bar{M}_i \rangle \qquad (19.38)$$

$$= (-1)^{T_f - \bar{M}_i} \begin{pmatrix} T_f & 0 & T_i \\ -\bar{M}_i & 0 & \bar{M}_i \end{pmatrix} \sum_{a,b} \frac{1}{\sqrt{2}} \langle a || \mathscr{T}_J^{(0)} || b \rangle \psi_{J,0}^{fi}(ab)$$

$$+ (-1)^{T_f - \bar{M}_i} \begin{pmatrix} T_f & 1 & T_i \\ -\bar{M}_i & 0 & \bar{M}_i \end{pmatrix} \sum_{a,b} \sqrt{\frac{3}{2}} \langle a || \mathscr{T}_J^{(1)} || b \rangle \psi_{J,1}^{fi}(ab)$$

This is the basic multipole matrix element for the transition $|J_i; T_i \bar{M}_i\rangle \to |J_f; T_f \bar{M}_i\rangle$. It relates the many-body reduced matrix element to a sum of single-particle reduced matrix elements; the isospin dependence of the transition multipoles has now been explicitly exhibited. The many-body physics is in the numerical coefficients $\psi_{J,T}^{fi}(ab)$. Again, this is a *general result* within the current framework.

Once one has a set of coefficients $\psi_{J,T}^{fi}(ab)$ from the many-body analysis (several examples are discussed in chapter 20), the problem is reduced to computation of the single-particle reduced matrix elements of the multipole operators [Wi63]. The tables in [Do79, Do80] are a substantial aid here since all the angular momentum algebra of computing the reduced matrix element of a tensor product in a coupled basis [Ed74] has already been carried out. There are two sets of tables. The first [Do79] is in a harmonic oscillator single-particle basis where the wave functions can be written in analytic form [Fe71, Wa95]. In this case, the required radial integrals can all be done analytically in terms of hypergeometric functions [de66]. The result is of the form $\exp(-y) \times$ polynomial in y where

$$y \equiv \left(\frac{\kappa b_{\mathrm{osc}}}{2} \right)^2$$

$$\hbar \omega_{\mathrm{osc}} = \frac{\hbar^2}{m b_{\mathrm{osc}}^2} \qquad (19.39)$$

In the second set of tables [Do80], the calculation is carried out for arbitrary radial wave functions up to the point where a final radial integral must be done numerically.

If one inserts the single-particle densities from Eqs. (19.1, 19.4, 19.5, 19.20), then through $O(1/m)$ the single-particle multipole operators take the form

$$
\begin{aligned}
M_{JM_J}(\kappa) &= M_J^{M_J}(\kappa\mathbf{x}) \frac{1}{2}(1+\tau_3) \\
iT_{JM_J}^{\text{mag}}(\kappa) &= \frac{\kappa}{m}\left\{ \mathscr{M}_{JJ}^{M_J}(\kappa\mathbf{x}) \cdot \frac{1}{\kappa}\nabla \frac{1}{2}(1+\tau_3) \right. \\
&\quad \left. -\frac{1}{2}\left[\frac{1}{i\kappa}\nabla \times (\mathscr{M}_{JJ}^{M_J})\right] \cdot \boldsymbol{\sigma} \left[\frac{1}{2}(\mu_p+\mu_n) + \frac{1}{2}\tau_3(\mu_p-\mu_n)\right] \right\} \\
T_{JM_J}^{\text{el}}(\kappa) &= \frac{\kappa}{m}\left\{ \left[\frac{1}{i\kappa}\nabla \times (\mathscr{M}_{JJ}^{M_J})\right] \cdot \frac{1}{\kappa}\nabla \frac{1}{2}(1+\tau_3) \right. \\
&\quad \left. +\frac{1}{2}(\mathscr{M}_{JJ}^{M_J}) \cdot \boldsymbol{\sigma} \left[\frac{1}{2}(\mu_p+\mu_n) + \frac{1}{2}\tau_3(\mu_p-\mu_n)\right] \right\}
\end{aligned}
\tag{19.40}
$$

Here

$$
\begin{aligned}
M_J^{M_J}(\kappa\mathbf{x}) &\equiv j_J(\kappa x) Y_{JM_J}(\Omega_x) \\
\mathscr{M}_{JJ}^{M_J} &\equiv j_J(\kappa x) \mathscr{Y}_{JJ1}^{M_J}(\Omega_x)
\end{aligned}
\tag{19.41}
$$

Note the isospin dependence is now explicit and one can read off $\mathscr{T}_{JM_J}^{(0)}$ and $\mathscr{T}_{JM_J}^{(1)}$ in Eq. (19.36). Furthermore, it is no longer necessary to symmetrize the convection current since $\nabla \cdot \mathscr{M}_{JJ}^{M_J} = \nabla \cdot [\nabla \times \mathscr{M}_{JJ}^{M_J}] = 0$.

A notation which identifies the various pieces of the multipole operators in Eqs. (19.40) is introduced in [Do79]

$$
M_{JM_J}(\kappa) = \frac{1}{2}(1+\tau_3) M_J^{M_J}(\kappa\mathbf{x})
\tag{19.42}
$$

$$
iT_{JM_J}^{\text{mag}}(\kappa) \equiv \frac{\kappa}{m}\left\{ \frac{1}{2}(1+\tau_3) \Delta_J^{M_J} \right.
$$
$$
\left. + \left[\frac{1}{2}(\mu_p+\mu_n) + \frac{1}{2}\tau_3(\mu_p-\mu_n)\right] \left(-\frac{1}{2}\right) \Sigma_J'^{M_J} \right\}
$$

$$
T_{JM_J}^{\text{el}}(\kappa) \equiv \frac{\kappa}{m}\left\{ \frac{1}{2}(1+\tau_3) \Delta_J'^{M_J} \right.
$$
$$
\left. + \left[\frac{1}{2}(\mu_p+\mu_n) + \frac{1}{2}\tau_3(\mu_p-\mu_n)\right] \left(\frac{1}{2}\right) \Sigma_J^{M_J} \right\}
$$

The quantities $(\Delta, \Delta', \Sigma, \Sigma')$ follow from comparison with Eqs. (19.40). One can now directly employ the tables in [Do79, Do80].

20

Some applications

We proceed to some applications of the analysis in chapter 19. Start with closed j-shells filled up to the energy F as illustrated in Fig. 20.1, and define particle and hole operators by [Fe71]

$$
\begin{aligned}
a_\alpha^\dagger &\equiv c_\alpha^\dagger & ; \alpha > F \\
b_\alpha^\dagger &= S_{-\alpha} c_{-\alpha} & ; \alpha < F
\end{aligned}
\qquad (20.1)
$$

Recall that the phase S_α is defined by

$$
S_\alpha \equiv (-1)^{j_\alpha - m_{j_\alpha}} (-1)^{\frac{1}{2} - m_{t_\alpha}} \qquad (20.2)
$$

Equations (20.1) form a *canonical transformation* since they leave the anti-commutation relations in Eqs. (19.30) unchanged. Denote by $|0\rangle$ the non-interacting ground state illustrated in Fig. 20.1. Both particle and hole destruction operators annihilate this state

$$
a_\alpha |0\rangle = b_\alpha |0\rangle = 0 \qquad (20.3)
$$

Furthermore, by Eq. (19.31) and the discussion following Eq. (19.33), both $a^\dagger$ and $b^\dagger$ are irreducible tensor operators (ITO).

To illustrate the physics and theoretical techniques, we consider some model cases. Equations (19.35) form the starting point of this analysis.

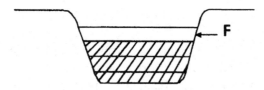

Fig. 20.1. Closed j-shells as a basis for defining particle and hole operators.

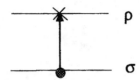

Fig. 20.2.   A particle–particle transition.

Consider first a particle–particle transition from the single-particle state σ to the state ρ as illustrated in Fig. 20.2. Take

$$
\begin{aligned}
|\Psi_i\rangle &= a_\sigma^\dagger|0\rangle \\
|\Psi_f\rangle &= a_\rho^\dagger|0\rangle
\end{aligned}
\tag{20.4}
$$

Here $\rho = (n_r l_r \frac{1}{2} j_r m_{j_\rho}; \frac{1}{2} m_{t_\rho}) \equiv (r; m_{j_\rho} m_{t_\rho})$ with a similar relation for σ. In this case one has the very simple result that the quantum numbers a must refer to the final state and b to the initial state

$$
\psi_{J,T}^{fi}(ab) = \delta_{ar}\delta_{bs} \qquad ; \text{ particle–particle}
\tag{20.5}
$$

All the angular momentum and isospin algebra has now been dealt with in arriving at the doubly reduced matrix elements. Note that here and henceforth we adopt the convention that the selection rules on J and T are implicitly contained in the reduced matrix elements.

The proof of Eq. (20.5) goes as follows. The only term in the operator $\hat{\zeta}^\dagger$ which contributes to the transition in this case is

$$
\hat{\zeta}^\dagger(ab; JM_J, TM_T) \doteq \delta_{ar}\delta_{bs}[a_\rho^\dagger \odot S_{-\sigma}a_{-\sigma}]_{JM_J TM_T}
\tag{20.6}
$$

Now take the matrix element

$$
\begin{aligned}
\langle\Psi_f|\hat{\zeta}^\dagger(rs; JM_J, TM_T)|\Psi_i\rangle &= S_\sigma\,\langle j_\rho m_{j_\rho} j_\sigma, -m_{j_\sigma}|j_\rho j_\sigma JM_J\rangle \\
&\quad \times\langle\tfrac{1}{2}m_{t_\rho}\tfrac{1}{2}, -m_{t_\sigma}|\tfrac{1}{2}\tfrac{1}{2}TM_T\rangle
\end{aligned}
\tag{20.7}
$$

This is just one form of the Wigner–Eckart theorem [Ed74]

$$
\begin{aligned}
\text{l.h.s.} &= S_\sigma\,\langle j_\rho m_{j_\rho} j_\sigma, -m_{j_\sigma}|j_\rho j_\sigma JM_J\rangle\langle\tfrac{1}{2}m_{t_\rho}\tfrac{1}{2}, -m_{t_\sigma}|\tfrac{1}{2}\tfrac{1}{2}TM_T\rangle \\
&\quad \times \frac{1}{\sqrt{(2J+1)(2T+1)}}\langle\Psi_f\,\dvdots\,\hat{\zeta}^\dagger(rs; JT)\,\dvdots\Psi_i\rangle
\end{aligned}
\tag{20.8}
$$

A comparison of these two equations gives the stated result.

Consider next the somewhat more complicated case of a particle–hole transition as illustrated in Fig. 20.3. Here a particle makes a transition out of the filled ground state, leaving a hole behind. In this case one has

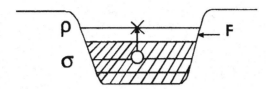

Fig. 20.3. A particle–hole transition.

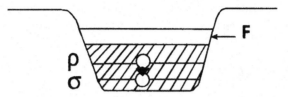

Fig. 20.4. A hole–hole transition.

$$|\Psi_i\rangle = |0\rangle$$
$$|\Psi_f\rangle = [a_\rho^\dagger \odot b_\sigma^\dagger]_{JM_J;TM_T}|0\rangle \qquad (20.9)$$

In this case, the only contributing term in the operator $\hat{\zeta}^\dagger$ is

$$\hat{\zeta}^\dagger(ab; JM_J, TM_T) \doteq \delta_{ar}\delta_{bs}\,[a_\rho^\dagger \odot b_\sigma^\dagger]_{JM_JTM_T} \qquad (20.10)$$

Now take matrix elements and use the orthonormality of the Clebsch–Gordon (C–G) coefficients [Ed74]

$$\langle \Psi_f|\hat{\zeta}^\dagger(rs; JM_J, TM_T)|\Psi_i\rangle = 1 \qquad (20.11)$$

$$= \frac{1}{\sqrt{(2J+1)(2T+1)}}\langle J, T \mathbin{\vdots\vdots} \hat{\zeta}^\dagger(rs; JT)\mathbin{\vdots\vdots} 0\rangle$$

The Wigner–Eckart theorem has again been used to obtain the second equality. Thus, in this case also, one has the simple result

$$\psi_{J,T}^{fi}(ab) = \delta_{ar}\delta_{bs} \qquad ; \text{ particle–hole} \qquad (20.12)$$

Consider the most complicated case of a hole–hole transition as illustrated in Fig. 20.4. Here the initial state has a hole in the state ρ and the final state a hole in σ. The transition is actually accomplished, of course, by a particle going in the opposite direction. In this case

$$|\Psi_i\rangle = b_\rho^\dagger|0\rangle$$
$$|\Psi_f\rangle = b_\sigma^\dagger|0\rangle \qquad (20.13)$$

The contributing term in $\hat{\zeta}^\dagger$ is

$$\hat{\zeta}^\dagger(ab, JM_J, TM_T) \doteq \delta_{ar}\delta_{bs}[S_\rho b_{-\rho} \odot b_\sigma^\dagger]_{JM_JTM_T} \qquad (20.14)$$

One has to first turn the destruction and creation operators around in computing the matrix element.[1] Use of the basic anti-commutation relations gives

$$\hat{\zeta}^\dagger(ab;JM_J,TM_T) \;\doteq\; \delta_{ar}\delta_{bs}\left\{\delta_{J0}\delta_{T0}\sqrt{2(2j_a+1)}\,\delta_{rs}\right.$$
$$\left. -(-1)^{j_a+j_b-J}(-1)^{\frac{1}{2}+\frac{1}{2}-T}[b_\sigma^\dagger \odot S_\rho b_{-\rho}]_{JM_JTM_T}\right\} \quad (20.15)$$

Here the anti-commutation relations for the b's has been employed, along with the relation for reversing the order of coupling in the C–G coefficients [Ed74]

$$\langle j_\rho m_{j_\rho} j_\sigma m_{j_\sigma}|j_\rho j_\sigma JM_J\rangle \;=\; (-1)^{j_\rho+j_\sigma-J}\langle j_\sigma m_{j_\sigma} j_\rho m_{j_\rho}|j_\sigma j_\rho JM_J\rangle$$
$$(20.16)$$

There is a similar relation for isospin. The first term in Eq. (20.15), now simply a *c*-number, follows from

$$\sum_{m_{j_\rho}}\langle j_\rho m_{j_\rho} j_\rho,-m_{j_\rho}|j_\rho j_\rho JM_J\rangle(-1)^{j_\rho-m_{j_\rho}}$$
$$=\sum_{m_{j_\rho}}\langle j_\rho m_{j_\rho} j_\rho,-m_{j_\rho}|j_\rho j_\rho JM_J\rangle\langle j_\rho m_{j_\rho} j_\rho,-m_{j_\rho}|j_\rho j_\rho 00\rangle\sqrt{2j_\rho+1}$$
$$=\sqrt{2j_\rho+1}\,\delta_{J0} \quad (20.17)$$

Again, there is a similar relation for isospin.

Now the calculation of the remaining matrix element proceeds exactly as in the particle–particle case. Note that here $S_{-\rho}=S_\rho$.

$$\psi_{J,T}^{fi}(ab) \;=\; \delta_{ar}\delta_{bs}\delta_{rs}\delta_{J0}\,\delta_{T0}\,[2(2j_a+1)(2T_i+1)(2J_i+1)]^{1/2}$$
$$-(-1)^{j_a+j_b-J}(-1)^{\frac{1}{2}+\frac{1}{2}-T}\delta_{ar}\delta_{bs} \qquad ; \text{hole–hole} \quad (20.18)$$

The reduced matrix element has been calculated in the first (scalar) term, and we now observe that there will, in fact, be an additional contribution $\psi_{0,0}^{fi}(aa)$ of this type of term for each occupied shell $a<F$ in Fig. 20.1.

Suppose one does elastic scattering from a ground state that has neither particles nor holes as in Fig. 20.1

$$|\Psi_i\rangle = |\Psi_f\rangle = |0\rangle \quad (20.19)$$

In this case, the only non-zero contribution comes from the first term in Eq. (20.18), and from Eq. (19.38) one has

$$\langle 0||\hat{M}_0(\kappa)||0\rangle = \sum_{a<F}\sqrt{2j_a+1}\,\langle a||M_0^{(0)}(\kappa)||a\rangle \quad (20.20)$$

---

[1] They must be *normal-ordered*.

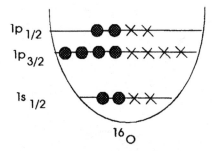

Fig. 20.5. Shell model for $^{16}_{8}$O ground state.

Now consider the case of elastic scattering from a $(J^\pi, T) = (0^+, 0)$ target. In this case, the only contributing multipole is $\hat{M}_0(\kappa)$, and one evidently measures the Fourier transform of the ground-state charge density

$$\sqrt{4\pi}\,\langle\Psi_0|\hat{M}_{00}(\kappa)|\Psi_0\rangle = \int d^3x\,\left[\frac{\sin\kappa x}{\kappa x}\right]\rho_{00}(\mathbf{x}) \qquad (20.21)$$

As an example, consider $^{16}_{8}$O modeled as indicated in Fig. 20.5. In this case, Eq. (20.20) gives

$$\langle 0||\hat{M}_0(\kappa)||0\rangle = \sqrt{2}\,\langle 1s_{1/2}||M_0^{(0)}(\kappa)||1s_{1/2}\rangle + \sqrt{4}\,\langle 1p_{3/2}||M_0^{(0)}(\kappa)||1p_{3/2}\rangle$$
$$+\sqrt{2}\,\langle 1p_{1/2}||M_0^{(0)}(\kappa)||1p_{1/2}\rangle \qquad (20.22)$$

Use of the tables in [Do79] then gives for harmonic oscillator wave functions

$$\sqrt{4\pi}\,\langle\Psi_0||\hat{M}_0(\kappa)||\Psi_0\rangle = 8\left(1 - \frac{1}{2}y\right)e^{-y} \qquad (20.23)$$

Note that at $\kappa = 0$ one must obtain Z, the total charge on the nucleus, and this result is indeed recovered. Since s and p radial wave functions contribute here, the coefficient of e^{-y} is a first-order polynomial in y. Thus there is (at most) one diffraction zero as a function of y. Figure 20.6 shows the fit to the experimental charge form factor in this model [Do75]. The data is from [Mc69]. There is one parameter in this fit, $b_{\mathrm{osc}} = 1.77\,\mathrm{fm}$. Also shown is the result obtain with Woods–Saxon radial wave functions, where a second diffraction minimum is observed [Do69]. Note that the harmonic oscillator wave functions provide an excellent description of the gross distribution of charge, and only break down when one seeks a more detailed description at higher momentum transfer.

Figure 4.2 shows the quality of elastic (e, e) data presently available. Figure 21.5 in the next section illustrates the quality of the nuclear charge distribution which can now be extracted from elastic (e, e) charge scattering. Note the small uncertainty band on the experimental result in that figure.

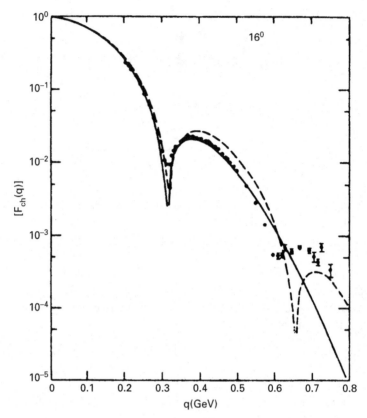

Fig. 20.6. Elastic charge form factor for $^{16}_{8}O$ showing results using harmonic oscillator (solid curve) and Woods–Saxon (dashed curve) radial wave functions [Do69, Do75, Mc69].

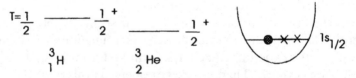

Fig. 20.7. Ground states of $^{3}_{2}$He and $^{3}_{1}$H.

Consider elastic magnetic scattering. It is proven in appendix E that only the odd magnetic multipoles contribute to elastic scattering through the transverse interaction. Consider the simplest nuclear case of elastic scattering from $^{3}_{2}$He and $^{3}_{1}$H with $(J^{\pi}, T) = (\frac{1}{2}^{+}, \frac{1}{2})$ as illustrated in Fig. 20.7. Here only $\hat{T}_{1}^{\mathrm{mag}}(\kappa)$ contributes. Although very sophisticated three-body calculations are available for this system (see later), we consider it here as a very simple illustration.

Table 20.1. Comparison of experimental and theoretical magnetic moments of the three-nucleon system calculated in the harmonic oscillator shell model.

	$\mu(^3\mathrm{H}) + \mu(^3\mathrm{He})$	$\mu(^3\mathrm{H}) - \mu(^3\mathrm{He})$
Theory	0.8795	4.7059
Experiment	0.8513	5.1064

The shell model configuration in Fig. 20.7 is a $(1s_{1/2})^{-1}$ hole in the $1s$ shell. Use of Eq. (20.18) in this case leads to

$$\psi_{JT}[(1s_{1/2})^2] = 4\delta_{J0}\delta_{T0} - (-1)^{J+T} \qquad (20.24)$$

From the general result in Eq. (19.38), and the tables in [Do79], one finds

$$\left(\frac{4\pi}{2}\right)^{1/2} \langle \tfrac{1}{2}^+ ; \tfrac{1}{2} m_t \| \hat{M}_0(\kappa) \| \tfrac{1}{2}^+ ; \tfrac{1}{2} m_t \rangle = \left(\frac{3}{2} + m_t\right) e^{-y}$$
$$= Z e^{-y}$$

$$i\left(\frac{4\pi}{2}\right)^{1/2} \langle \tfrac{1}{2}^+ ; \tfrac{1}{2} m_t \| \hat{T}_1^{\mathrm{mag}}(\kappa) \| \tfrac{1}{2}^+ ; \tfrac{1}{2} m_t \rangle$$
$$= -\sqrt{2}\left(\frac{\kappa}{2m}\right)\left[\frac{1}{2}(\mu_p + \mu_n) - m_t(\mu_p - \mu_n)\right] e^{-y}$$
$$= -\sqrt{2}\left(\frac{\kappa}{2m}\right) \mu\, e^{-y} \qquad (20.25)$$

Here Z is the charge of the target and μ its magnetic moment. The proportionality of the matrix element of $\hat{T}_1^{\mathrm{mag}}$ to the magnetic moment is a general result as shown in appendix A. At higher κ one probes the spatial distribution of the magnetic moment distribution. Table 20.1 shows the (well-known) comparison between the calculated and experimental magnetic moments for this system.

Figure 20.8 shows the low-κ elastic (e, e) data compared with this shell model calculation with an oscillator parameter chosen as $b_{\mathrm{osc}} = 1.59$ fm [Do76, Wa95]. Again, it is only the *deviations* from these simple results that show up the more sophisticated elements of nuclear structure, to which we shall return.

With higher nuclear angular momentum J_i, higher multipoles can contribute to elastic scattering. It is shown in appendix E that parity and time-reversal invariance of the strong and electromagnetic interactions imply that only the even charge multipoles and odd transverse magnetic multipoles can contribute in the elastic case. Consider a $1g_{9/2}$ proton as the single-particle shell model assignment for the ground state of $^{93}_{41}\mathrm{Nb}$ as depicted in Fig. 20.9. The *highest* magnetic multipole that can contribute in this case is $\hat{T}_9^{\mathrm{mag}}(\kappa)$. Use of the previous particle–particle results and

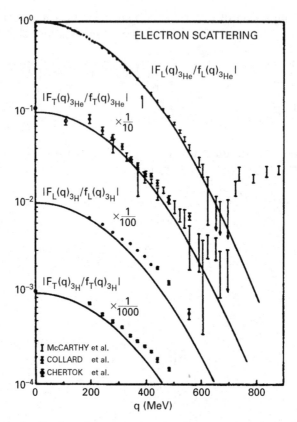

Fig. 20.8. Longitudinal and transverse elastic scattering form factors for the three-nucleon system compared with harmonic oscillator shell model results. Here $f_L = Zf_{SN}/\sqrt{4\pi}$ and $f_T = (\kappa/2m)\mu f_{SN}/\sqrt{2\pi}$. The correction factor f_{CM} has been included. [Do76, Wa84]. Data from [Co65, Ch69, Mc70].

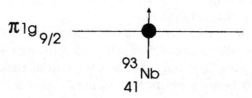

Fig. 20.9. Single-particle shell model assignment for the ground state of $^{93}_{41}$Nb.

the tables in [Do79] gives

$$\langle \frac{9}{2}^+ || \hat{T}_9^{mag}(\kappa) || \frac{9}{2}^+ \rangle = \langle \pi 1g_{9/2} || T_9(\kappa) || \pi 1g_{9/2} \rangle \qquad (20.26)$$

$$= -\frac{2^5}{3^2\sqrt{5 \cdot 11 \cdot 13 \cdot 17}} y^4 e^{-y} \left(\frac{\kappa\mu_p}{2m}\right)$$

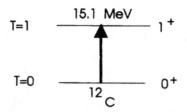

Fig. 20.10.   M1 transition in $^{12}_{6}$C as example of particle–hole transition.

Note that only the intrinsic magnetization contributes to the highest multipole in the stretched configuration $j = l + \frac{1}{2}$, and note the simplicity of this result. Note also that this multipole has the highest possible number of powers of y, which implies that it will be the dominant contribution at high momentum transfer κ. The experimental results for elastic magnetic scattering from $^{93}_{41}$Nb have already been shown in Fig. 12.4.[2] In contrast to the case with real photon interactions where the long-wavelength limit is relevant and the *lowest* multipole dominates (appendix A), here the rank of the dominant multipole increases as the momentum transfer is increased until at high κ the dominant contribution is indeed M9 — one can effectively *dial* the multipole one wants to examine.

What does one learn from this? Figure 12.5 shows the surface of half-maximum intrinsic magnetization density $\mu(x)_{\mathrm{max}}/2$ for $^{51}_{23}$V (chosen so that it would fit on a 10 fm square). Here the configuration assignment is $(1f_{7/2})_\pi$, and the nucleus is aligned so that its angular momentum points along the z-axis with $m_j = j$. The intrinsic magnetization maps the location of the valence nucleon. The nucleus is a small magnet with a current loop provided by the motion of the orbiting proton. Elastic magnetic electron scattering at all κ provides a microscope to see the spatial structure of this small current loop [Do73].

As an example of particle–hole states consider the celebrated isovector M1 transition to the $(1^+, 1)$ state at 15.11 MeV in $^{12}_{6}$C as illustrated in Fig. 20.10. Make the simplest shell model assignments

$$|\Psi_i\rangle = |0\rangle$$
$$|\Psi_f\rangle = [a^\dagger_{1p_{1/2}} \odot b^\dagger_{1p_{3/2}}]_{1^+,1} |0\rangle \qquad (20.27)$$

Then from the above discussion of particle–hole transitions and the tables

---

[2] Here $q_{\mathrm{eff}} \equiv \kappa$.

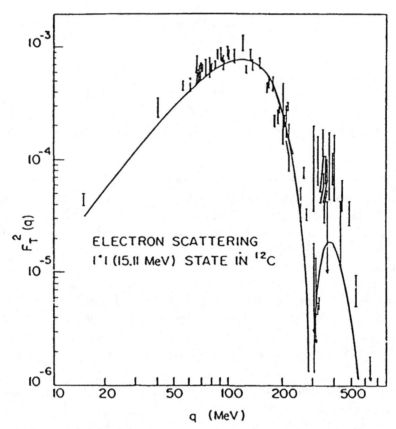

Fig. 20.11. The quantity $F_T^2(\kappa)$ for the $1^+, 1$ state in $^{12}_6\mathrm{C}(15.11\,\mathrm{MeV})$. The curve is a best fit with $b_{\mathrm{osc}} = 1.77\,\mathrm{fm}$ and $\xi = 2.25$ [Do75]. Here $q \equiv \kappa$.

in [Do79] one finds

$$i\sqrt{4\pi}\,\langle 1^+, 1\,0|| \hat{T}_1^{\mathrm{mag}}(\kappa)||0^+, 0\rangle = i\left(\frac{4\pi}{2}\right)^{1/2} \langle 1p\tfrac{3}{2}|| T_1^{\mathrm{mag}\,(1)}(\kappa)||1p\tfrac{3}{2}\rangle$$

$$= \frac{2}{3}\left(\frac{\kappa}{2m}\right)\left[1 - 2(\mu_p - \mu_n)\left(1 - \frac{1}{2}y\right)\right]e^{-y}$$

$$(20.28)$$

Now it is clear that the model of a pure particle–hole transition is an oversimplification. With configuration mixing, even within just the $1p$-shell, the amplitudes in Eq. (19.28) will be changed from the pure particle–hole value, and in general reduced from this value. A fit to the experimental data for this transition using Eq. (20.28) with an oscillator parameter $b_{\mathrm{osc}} = 1.77\,\mathrm{fm}$ and an overall reduction factor of $\xi = 2.25$ is shown in Fig. 20.11 [Do75].[3]

---

[3] In this figure $d\sigma/d\Omega \equiv 4\pi\sigma_M[(q^4/\mathbf{q}^4)F_L^2 + (q^2/2\mathbf{q}^2 + \tan^2\theta/2)F_T^2]r$.

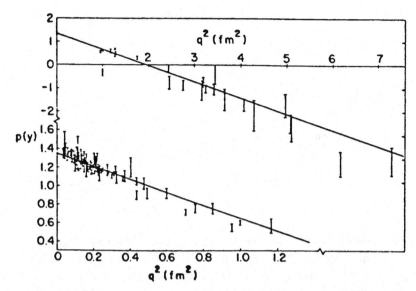

Fig. 20.12. Inelastic electron scattering to the $1^+, 1$ state in $^{12}_6C(15.11\,\text{MeV})$ plotted in terms of the polynomial $p(y)$ in Eq. (20.29). The straight line is a fit with $1p$-shell harmonic oscillator wave functions [Do79a]. Again, $q \equiv \kappa$.

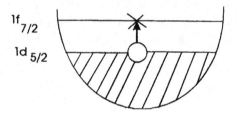

Fig. 20.13. Particle–hole configuration $(1f_{7/2})(1d_{5/2})^{-1}$.

If both the ground and excited states are configuration mixed, but still both described within the $1p$-shell oscillator model, then the polynomial in the above transition amplitude must indeed be of the form

$$p(y) = \alpha_0 - \alpha_1\, y \tag{20.29}$$

Figure 20.12 shows such a straight-line fit to the data [Do79a]. This fit determines two of the four possible one-body amplitudes $\psi^{fi}_{J,T}(ab)$ for this transition, if it is indeed described within the $1p$-shell oscillator model. Note that this straight-line plot, now on a *linear* scale, gives some credence to this description.

As an example of magnetic excitation of high-spin states, consider the isovector transition to the stretched member $(6^-, 1)$ of the particle–hole configuration in Fig. 20.13. Model the ground and excited states by

$$
\begin{aligned}
|\Psi_i\rangle &= |0\rangle \\
|\Psi_f\rangle &= [a^\dagger_{1f_{7/2}} \odot b^\dagger_{1d_{5/2}}]_{6^-,1} |0\rangle
\end{aligned} \tag{20.30}
$$

Then, as before,

$$
\begin{aligned}
i\sqrt{4\pi}\,\langle 6^-,1\,0||\hat{T}^{\mathrm{mag}}_6(\kappa)||0^+,0\rangle &= i\left(\frac{4\pi}{2}\right)^{1/2}\langle 1f\tfrac{7}{2}||T^{\mathrm{mag}\,(1)}_6(\kappa)||1d\tfrac{5}{2}\rangle \\
&= -\left(\frac{\kappa}{2m}\right)(\mu_p-\mu_n)\frac{2^3}{3\sqrt{3\cdot 11}}\,y^{5/2}\,e^{-y}
\end{aligned} \tag{20.31}
$$

Note the simplicity of this result. Again, only the magnetization contributes to this highest multipole in the stretched configuration, and this multipole gives the maximum power of y. The large isovector magnetic moment of the nucleon in Eq. (19.4) implies that isovector transitions are preferentially excited in electron scattering at large scattering angles and high momentum transfers. Figure 12.6 shows a spectrum for $^{24}_{12}\mathrm{Mg}$ taken under these conditions, and the transverse form-factor squared for the state at 15.0 MeV which dominates the spectrum; it is evidently a 6^-.[4] Such stretched particle–hole states have now been seen throughout the periodic table, including states as high as 14^- in $^{208}_{82}\mathrm{Pb}$ [Li79].

---

[4] Note that even though $^{24}_{12}\mathrm{Mg}$ is not the closed $1d_{5/2}$-shell nucleus of the shell model (that would be $^{28}_{14}\mathrm{Si}$), the $(6^-,1)$ particle–hole transition still shows up strongly in the spectrum.

21

A relativistic model of the nucleus

In current and future electron scattering studies of nuclei, one is interested in momentum transfers $q^2 \gg m^2$. For consistency, a relativistic description of the nuclear many-body system is required. One such model, *quantum hadrodynamics* (QHD) is developed in detail in [Se86, Se92, Wa95, Se97]. Relativistic mean field theory (RMFT) then includes the strong nuclear interactions in an average fashion. In its simplest version, one starts from a baryon field $\psi = \begin{pmatrix} \psi_p \\ \psi_n \end{pmatrix}$, neutral vector and scalar fields (V_μ, ϕ), and the lagrangian density

$$
\mathcal{L} = -\bar{\psi}\left[\gamma_\mu\left(\frac{\partial}{\partial x_\mu} - ig_v V_\mu\right) + (m - g_s\phi)\right]\psi - \frac{1}{2}\left[\left(\frac{\partial\phi}{\partial x_\mu}\right)^2 + m_s^2\phi^2\right]
$$
$$
- \frac{1}{4}F_{\mu\nu}F_{\mu\nu} - \frac{1}{2}m_v^2 V_\mu^2 \tag{21.1}
$$

Here

$$
F_{\mu\nu} = \frac{\partial V_\nu}{\partial x_\mu} - \frac{\partial V_\mu}{\partial x_\nu} \tag{21.2}
$$

The equations of motion for a field theory are those of continuum mechanics [Bj65a, Fe80, Wa91].

$$
\frac{\partial}{\partial x_\mu}\frac{\partial\mathcal{L}}{\partial(\partial q/\partial x_\mu)} - \frac{\partial\mathcal{L}}{\partial q} = 0 \tag{21.3}
$$

The Euler–Lagrange equations for the above fields are

$$
\left[\left(\frac{\partial}{\partial x_\mu}\right)^2 - m_s^2\right]\phi = -g_s\bar{\psi}\psi
$$

169

$$\frac{\partial}{\partial x_\nu} F_{\mu\nu} + m_v^2 V_\mu = i g_v \bar{\psi} \gamma_\mu \psi$$

$$\left[\gamma_\mu \left(\frac{\partial}{\partial x_\mu} - i g_v V_\mu \right) + (m - g_s \phi) \right] \psi = 0 \qquad (21.4)$$

The scalar field couples to the baryon scalar density, the vector field couples to the conserved baryon current, and the Dirac equation for the baryon field has the meson fields included in a minimal, linear fashion. One now imposes canonical (anti-)quantization on the dynamical variables, the fields, and this leads to a *relativistic quantum field theory* [Bj65a, Wa91].

Consider uniform nuclear matter. One can obtain an approximate solution to the field equations by replacing the meson fields by their expectation values, which are then just classical fields. By translation invariance, these quantities must be constants independent of space and time.

$$\phi \to \langle \phi \rangle \equiv \phi_0$$
$$V_\lambda \to \langle V_\lambda \rangle = i \delta_{\lambda 4} V_0 \qquad (21.5)$$

The last equality follows from the isotropy of the medium. This RMFT should become better and better as the baryon density ρ_B gets larger and the sources on the right side of the meson field equations correspondingly increase. At observed nuclear density, this is equivalent to the usual (relativistic) Hartree approximation. With this approximation, the baryon field equation is linearized.

$$\left[\gamma_\mu \frac{\partial}{\partial x_\mu} + g_v \gamma_4 V_0 + (m - g_s \phi_0) \right] \psi = 0 \qquad (21.6)$$

This linear equation can be solved exactly. Stationary state solutions to the Dirac equation lead to the eigenvalue equation

$$\varepsilon_k^{(\pm)} = g_v V_0 \pm \sqrt{\mathbf{k}^2 + m^{\star 2}}$$
$$m^\star \equiv m - g_s \phi_0 \qquad (21.7)$$

The ground state of the corresponding hamiltonian of this RMFT is obtained by filling the Dirac levels up to the Fermi momentum k_F. To be self-consistent, one determines the sources in the meson field equations by summing over the occupied levels

$$\rho_S = \langle \bar{\psi} \psi \rangle = \frac{1}{\Omega} \sum_{\mathbf{k}\lambda}^{k_F} \bar{U}(\mathbf{k}\lambda) U(\mathbf{k}\lambda)$$

$$\rho_B = \langle \psi^\dagger \psi \rangle = \frac{1}{\Omega} \sum_{\mathbf{k}\lambda}^{k_F} U^\dagger(\mathbf{k}\lambda) U(\mathbf{k}\lambda) \qquad (21.8)$$

We choose to normalize to unit probability in the box, so that $U^\dagger U = 1$. Note also the important relation, derived immediately from the Dirac equation

$$\bar{U}(\mathbf{k}\lambda)U(\mathbf{k}\lambda) = \frac{m^\star}{\sqrt{\mathbf{k}^2 + m^{\star 2}}} U^\dagger(\mathbf{k}\lambda)U(\mathbf{k}\lambda) \qquad (21.9)$$

From the field equations, for constant fields

$$\phi_0 = \frac{1}{m_s^2}\rho_S \qquad ; \quad V_0 = \frac{1}{m_v^2}\rho_B \qquad (21.10)$$

The translational invariance of nuclear matter permits ready solution of the resulting coupled, non-linear, differential equations. The quantity $m^\star$ is calculated self-consistently. Nuclear matter saturates at the right binding energy and density in RMFT if one takes

$$C_s^2 \equiv g_s^2\left(\frac{m^2}{m_s^2}\right) = 267.1 \qquad ; \quad C_v^2 \equiv g_v^2\left(\frac{m^2}{m_v^2}\right) = 195.9 \qquad (21.11)$$

The saturation and $m^\star$ curves are shown in [Wa95].

The baryon field operator in RMFT takes the following form

$$\psi(x) = \frac{1}{\sqrt{\Omega}}\sum_{\mathbf{k}\lambda}\left[A_{\mathbf{k}\lambda}U(\mathbf{k}\lambda)\exp(ik\cdot x) + B^\dagger_{\mathbf{k}\lambda}V(-\mathbf{k}\lambda)\exp(-ik\cdot x)\right]$$

$$(21.12)$$

Here $A_{\mathbf{k}\lambda}$ destroys a baryon of momentum and helicity $(\mathbf{k}, \lambda)$ in the medium and $B^\dagger_{\mathbf{k}\lambda}$ creates an antibaryon [Bj65a, Fe71, Wa91].

An *effective electromagnetic current operator* is now introduced. The internal structure of the individual nucleons, which from this hadronic point of view arises from charged meson fields, is summarized in a single-nucleon form factor and an effective Møller potential

$$\frac{1}{q^2} \rightarrow \frac{f_{SN}(q^2)}{q^2} \qquad (21.13)$$

The effective current, to be used in lowest order in the nuclear many-body problem, is then taken to be

$$J_\mu(x) = i\bar{\psi}\gamma_\mu \underline{Q}\psi + \frac{1}{2m}\frac{\partial}{\partial x_\nu}\bar{\psi}\sigma_{\mu\nu}\underline{\lambda}'\psi$$

$$\underline{Q} = \frac{1}{2}(1 + \tau_3)$$

$$\underline{\lambda}' = \lambda_p'\frac{1}{2}(1 + \tau_3) + \lambda_n\frac{1}{2}(1 - \tau_3) \qquad (21.14)$$

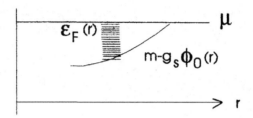

Fig. 21.1. Thomas–Fermi theory for finite systems.

If $\psi(x)$ satisfies the field Eq. (21.6), and its adjoint, then this current is both *local and conserved*

$$\frac{\partial J_\mu(x)}{\partial x_\mu} = 0 \tag{21.15}$$

Furthermore, it gives the correct result for a free nucleon.[1]

An immediate extension of RMFT to finite nuclei is through Thomas–Fermi theory [Se86]. Here one gives the meson fields a spatial variation

$$\begin{aligned}
(\nabla^2 - m_s^2)\phi_0(r) &= -\rho_S(r) \\
(\nabla^2 - m_v^2)V_0(r) &= -\rho_B(r)
\end{aligned} \tag{21.16}$$

It is then assumed that these fields vary slowly enough so that one can calculate the sources for a uniform system at the appropriate baryon density $\rho_B(r)$ parameterized by $k_F(r)$ as illustrated in Fig. 21.1.[2] The condition $k_F = 0$ determines the nuclear size and baryon number B. The electron scattering cross sections of Rosenfelder for two nuclei $^{40}_{20}$Ca(e, e$'$) and $^{208}_{82}$Pb(e, e$'$), calculated with local RMFT and then integrated over the Thomas–Fermi distributions, are shown in Figs. 21.2 and 21.3 [Ro80]. The calculations are compared with quasielastic electron scattering data from HEPL on these nuclei [Mo71]. Note the following features of

---

[1] This effective electromagnetic current assumes $F_1(q^2)/F_1(0) \approx F_2(q^2)/F_2(0) \approx f_{SN}(q^2)$. This relation breaks down at large q^2 where the data indicate that it is the Sachs form factors $G_M = F_1 + 2mF_2$ and $G_E = F_1 - q^2F_2/2m$ that scale. To incorporate this observation, make the following replacements:

$$\frac{f_{SN}(q^2)}{q^2} \to \frac{f_{SN}(q^2)}{q^2} \frac{1}{1 + q^2/4m^2}$$

$$J_\lambda \to J_\lambda - \frac{i}{4m^2} \frac{\partial}{\partial x_\nu} \frac{\partial}{\partial x_\nu} \left(\bar\psi \mu \gamma_\lambda \psi\right)$$

Here μ is the full magnetic moment [Wa95]. Applications of this improved effective current do not exist at the present time.

[2] $k_F^n(r)$ and $k_F^p(r)$ will differ if $N \neq Z$ and the Coulomb interaction is included.

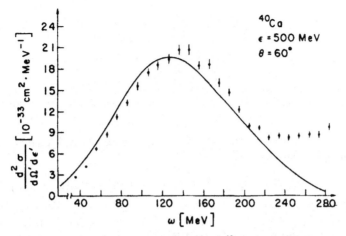

Fig. 21.2. Quasielastic electron scattering from $^{40}_{20}$Ca in RMFT compared with experimental values. The calculation assumes a local Fermi gas with the quantities $m^\star(r)$ and $\rho_B(r)$ taken from a relativistic Thomas–Fermi calculation of these quantities in QHD-I [Ro80, Wa95]. Data are from [Mo71].

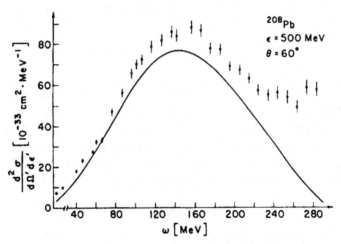

Fig. 21.3. As Fig. 21.2, but for $^{208}_{82}$Pb.

these results:

- They are calculated using the Thomas–Fermi ground state densities;

- The only parameters are those of nuclear matter and $m_s = 550\,\mathrm{MeV}$ from a fit to the mean-square charge radius of $^{40}_{20}$Ca;

- The final result involves values of $m^\star(r)$ and $\rho_B(r)$ taken over the entire nuclear density;

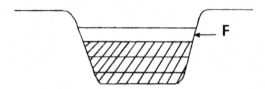

Fig. 21.4.   Occupied Dirac orbitals in relativistic Hartree theory.

- It is satisfying that the positions of the peak and values of k_F are approximately correct.[3]

A more satisfactory treatment of finite nuclei is obtained through *relativistic Hartree theory*. Here one solves the Dirac equation in the fields $[\phi_0(r), V_0(r)]$ and assumes the Dirac orbitals are occupied up to some level F in these mean fields as illustrated in Fig. 21.4.[4] The source terms in the meson field equations at a given point are then calculated self-consistently by summing over the contributions of the occupied orbitals

$$\rho_B(r) = \sum_{\alpha}^{F} \phi_{\alpha}^{\dagger}(\mathbf{r})\phi_{\alpha}(\mathbf{r})$$

$$\rho_S(r) = \sum_{\alpha}^{F} \bar{\phi}_{\alpha}(\mathbf{r})\phi_{\alpha}(\mathbf{r}) \qquad (21.17)$$

The meson field Eqs. (21.16) are then solved with these sources.

Since the relativistic Hartree calculations provide a very powerful way of dealing with the relativistic nuclear many-body problem, and since they follow from our discussion of the Dirac equation in chapter 10, we digress to develop them [Bj65, Sc68, Se86, Wa95]. Consider the hamiltonian for a Dirac particle moving in spherically symmetric vector and scalar fields.

$$h = -i\vec{\alpha} \cdot \vec{\nabla} + g_v V_0(r) + \beta[M - g_s\phi_0(r)] \qquad (21.18)$$

Define the angular momentum by

$$\vec{J} = \vec{L} + \vec{S} = -i\vec{r} \times \vec{\nabla} + \frac{1}{2}\vec{\Sigma} \qquad (21.19)$$

The wave function and spin matrix are written in two-by-two form as

$$\psi = \begin{pmatrix} \psi_A \\ \psi_B \end{pmatrix} \qquad \vec{\Sigma} = \begin{pmatrix} \vec{\sigma} & 0 \\ 0 & \vec{\sigma} \end{pmatrix} \qquad (21.20)$$

---

[3] Note there is no additional average binding energy $\bar{\varepsilon}$ required in these calculations.
[4] We here assume closed shells and spherical symmetry.

One first establishes the following commutation relations (this takes a little algebra)

$$[h, J_i] = [h, \vec{J}^2] = [h, \vec{S}^2] = 0 \qquad (21.21)$$

Note $[h, \vec{L}^2] \neq 0$. Now introduce

$$\begin{aligned} K &\equiv \beta(\vec{\Sigma} \cdot \vec{L} + 1) = \beta(\vec{\Sigma} \cdot \vec{J} - 1/2) \\ [h, K] &= 0 \end{aligned} \qquad (21.22)$$

To establish the vanishing of the second commutator again takes some algebra.

Now label the eigenvalues of K by $K\psi = -\kappa\psi$. The states can be characterized by the eigenvalues $\{j, s = 1/2, -\kappa, m\}$. One then establishes the following relations

$$\begin{aligned} K^2 &= \vec{L}^2 + \vec{\Sigma} \cdot \vec{L} + 1 = \vec{J}^2 + 1/4 \\ \kappa &= \pm(j + 1/2) \end{aligned} \qquad (21.23)$$

It follows from these relations that

$$\begin{aligned} -\kappa\psi_A &= (\vec{\sigma} \cdot \vec{L} + 1)\psi_A \\ -\kappa\psi_B &= -(\vec{\sigma} \cdot \vec{L} + 1)\psi_B \end{aligned} \qquad (21.24)$$

Hence

$$\vec{L}^2\psi_A = \left[\left(j + \frac{1}{2}\right)^2 + \kappa\right]\psi_A = l_A(l_A + 1)\psi_A$$

$$\vec{L}^2\psi_B = \left[\left(j + \frac{1}{2}\right)^2 - \kappa\right]\psi_B = l_B(l_B + 1)\psi_B \qquad (21.25)$$

Thus, although ψ is not an eigenstate of $\vec{L}^2$, the upper and lower components are separately eigenstates with eigenvalues determined from these relations. They also have fixed j and $s = 1/2$.

Now introduce spin spherical harmonics

$$\Phi_{\kappa m} = \sum_{m_l m_s} \langle l m_l \tfrac{1}{2} m_s | l \tfrac{1}{2} j m \rangle Y_{l m_l}(\theta, \phi) \chi_{m_s} \qquad (21.26)$$

Here $j = |\kappa| - 1/2$. Hence one shows that the solutions to this Dirac equation take the form

$$\psi_{n\kappa m} = \frac{1}{r}\begin{pmatrix} iG(r)_{n\kappa}\Phi_{\kappa m} \\ -F(r)_{n\kappa}\Phi_{-\kappa m} \end{pmatrix} \qquad (21.27)$$

Here $l = \kappa$ if $\kappa > 0$ and $l = -(\kappa + 1)$ if $\kappa < 0$.

Consider the relativistic Hartree equations.

$$
\begin{aligned}
(\nabla^2 - m_s^2)\phi_0 &= -g_s\rho_S(r) \\
(\nabla^2 - m_v^2)V_0 &= -g_v\rho_B(r) \\
\left(\frac{1}{i}\boldsymbol{\alpha}\cdot\nabla + g_v V_0(r) + \beta[M - g_s\phi_0(r)]\right)\psi &= i\frac{\partial\psi}{\partial t}
\end{aligned}
\tag{21.28}
$$

Label the baryon states by $\{\alpha\} = \{n\kappa t, m_\alpha\} \equiv \{a, m_\alpha\}$. Here $t = 1/2\,(-1/2)$ for protons (neutrons). Look for stationary state solutions, and insert the form of Dirac wave functions in Eq. (21.27). One needs the following identity, again established with a little work

$$
\vec{\sigma}\cdot\vec{\nabla}\,\frac{1}{r}G\Phi_{\kappa m} = -\frac{1}{r}\left(\frac{d}{dr} + \frac{\kappa}{r}\right)G\Phi_{-\kappa m}
\tag{21.29}
$$

The required relation for F is obtained with the substitution $\kappa \to -\kappa$. It then follows that the coupled radial Dirac equations reduce to

$$
\begin{aligned}
\frac{d}{dr}G_a(r) + \frac{\kappa}{r}G_a(r) - [E_a - g_v V_0(r) + M - g_s\phi_0(r)]F_a(r) &= 0 \\
\frac{d}{dr}F_a(r) - \frac{\kappa}{r}F_a(r) + [E_a - g_v V_0(r) - M + g_s\phi_0(r)]G_a(r) &= 0
\end{aligned}
\tag{21.30}
$$

The normalization condition on the Dirac wavefunctions reduces to

$$
\int_0^\infty dr(|G_a(r)|^2 + |F_a(r)|^2) = 1
\tag{21.31}
$$

Consider the relativistic Hartree Eqs. (21.28) for the meson fields. For the source terms one uses the following relation for the Dirac solutions if $\kappa = \pm\kappa'$.

$$
\sum_{m=-j}^{m=j} \Phi_{\kappa m}^\dagger \Phi_{\kappa' m} = \delta_{\kappa\kappa'}\frac{2j+1}{4\pi} \qquad ; \kappa = \pm\kappa'
\tag{21.32}
$$

Hence the meson field equations become

$$
\frac{d^2}{dr^2}\phi_0(r) + \frac{2}{r}\frac{d}{dr}\phi_0(r) - m_s^2\phi_0(r) = -g_s\sum_a^{occ}\left(\frac{2j_a+1}{4\pi r^2}\right)[|G_a(r)|^2 - |F_a(r)|^2]
$$

$$
\frac{d^2}{dr^2}V_0(r) + \frac{2}{r}\frac{d}{dr}V_0(r) - m_v^2 V_0(r) = -g_v\sum_a^{occ}\left(\frac{2j_a+1}{4\pi r^2}\right)[|G_a(r)|^2 + |F_a(r)|^2]
$$

$$
\tag{21.33}
$$

In practice, these relativistic Hartree equations are enlarged to include a condensed neutral ρ field $b_0(r)$ coupled to the third component of the

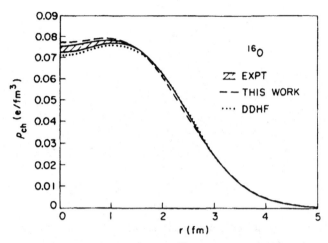

Fig. 21.5. Predicted charge density for $^{16}_{8}$O in the relativistic Hartree model [Ho81, Se86]. The solid curve and shaded area represent the fit to experimental data [Mc69]. Theoretical results are indicated by the long dashed lines [Ho81, Wa95].

isovector baryon density $g_\rho \psi^\dagger \frac{1}{2}\tau_3 \psi$ and the Coulomb field $A_0(r)$ coupled to the charge density $e_p \psi^\dagger \frac{1}{2}(1+\tau_3)\psi$. The extension of the above equations is immediate.

The resulting relativistic Hartree equations are coupled, non-linear differential equations; however, they are *local*. Fortunately, a rapidly convergent computer program has been published, available to all, that solves these equations by iteration [Ho91].

Let us return to applications of this relativistic Hartree theory. Horowitz and Serot [Ho81] make the model more realistic by adding the condensed, neutral ρ field and the Coulomb interaction. The four parameters in their theory (g_v, g_s, g_ρ, m_s) are fitted to the bulk properties of nuclear matter $(E/B, k_F, a_4)_{nm}$ and the root-mean square charge radius of $^{40}_{20}$Ca. Their results are discussed in detail in [Wa95]; the principal features are:

- One finds an excellent fit to ground-state charge densities throughout the periodic table;

- There is a strong spin–orbit interaction, and one derives the single-particle spectrum of the nuclear shell model;

- If one forms a Dirac optical potential from the relativistic Hartree densities and the empirical, Lorentz covariant N–N scattering amplitude (RIA), one obtains a quantitative description of p–A scattering, including the spin observables, up to $\approx 1\,\mathrm{GeV}$.

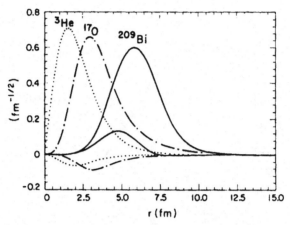

Fig. 21.6. Upper and lower component Dirac radial wave functions $G(r)$ and $F(r)$ for the three cases discussed in the text [Ki86, Wa95].

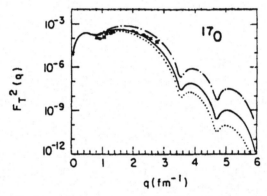

Fig. 21.7. Magnetic form factor squared for $^{17}_{8}$O. Calculated using relativistic Hartree wave functions and the effective current operator [Ki86, Wa95]. The dotted curve omits the C-M correction factor taken from the harmonic oscillator shell model, and the dashed curve omits the single-nucleon form factor. Data are from [Co65, Mc77, Ar78, Ca82].

As a first application of this relativistic Hartree model to electron scattering, consider elastic electron scattering from the extended charge distribution in oxygen $^{16}_{8}$O(e, e). The central density in $^{208}_{82}$Pb determines $(k_F)_{nm}$. The mean-square radius of $^{40}_{20}$Ca determines the one length parameter in the model. The charge distribution of $^{16}_{8}$O is then predicted [Ho81]. It is compared with the experimental determination of this charge distribution in Fig. 21.5. The agreement is all that one might hope for.

As a second application to electron scattering, consider elastic magnetic scattering from $^{17}_{8}$O(e, e). Figure 21.6 shows the Dirac radial wave functions for the valence nucleon, calculated in relativistic Hartree, for three nuclei: $^{3}_{2}$He$(\nu 1s)^{-1}_{1/2}$; $^{17}_{8}$O$(\nu 1d)_{5/2}$; $^{209}_{83}$Bi$(\pi 1h)_{11/2}$ [Ki86, Wa95]. The transverse

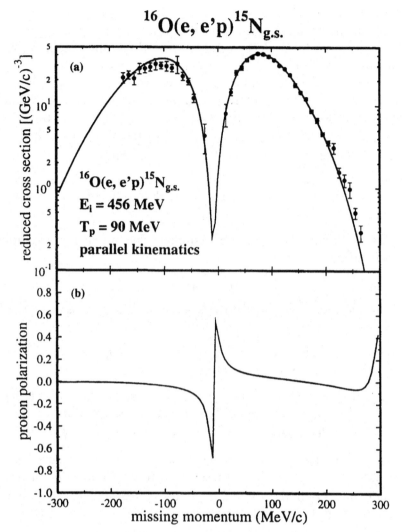

Fig. 21.8. Relativistic Hartree calculation using RIA optical potential and effective current compared with data for $^{16}_{8}\text{O}(e, e'p)^{15}_{7}\text{N}_{\text{g.s.}}$ (see text) [He95]. The author is grateful to J. I. Johansson for preparing this figure.

magnetic form factor for elastic electron scattering from $^{17}_{8}\text{O}(\nu 1d)_{5/2}$ calculated by Kim is shown in Fig. 21.7, together with the existing experimental data [Ki86]. Here we again use for the inclusive process

$$\frac{d\sigma}{d\Omega} = 4\pi\sigma_M F^2 r$$

$$F^2 \equiv \left(\frac{q_\mu^4}{\mathbf{q}^4}\right) F_L^2 + \left(\frac{q_\mu^2}{2\mathbf{q}^2} + \tan^2\frac{\theta}{2}\right) F_T^2 \tag{21.34}$$

All the parameters in the calculation are determined through relativistic Hartree and the effective current. The agreement with experiment is satisfactory (but note the scale!), and there is nothing intrinsic in the model that now limits the range of q^2 to which the calculation can apply.[5]

As a third application, consider the electron scattering coincidence reaction $^{16}_{8}O(e, e'\,p)^{15}_{7}N_{g.s.}$. Figure 21.8 shows a state-of-the-art calculation of the cross section for this (e, e' p) reaction on $^{16}_{8}O$ leading to the $(\pi 1p)^{-1}_{1/2}$ ground state of $^{15}_{7}N$. The calculation is from [He95] and the data from [La93, Le94]. The calculation has the following features:

- Relativistic Hartree wave functions [Ho81] and the Dirac RIA optical potential [Co93] are used, along with the effective electromagnetic current;[6]

- Parallel kinematics $\mathbf{q}\|\mathbf{k}$ are assumed. The missing momentum is defined by

$$\mathbf{p}_m \equiv \mathbf{q} - \mathbf{k} \qquad (21.35)$$

  The incident electron energy is 456 MeV and $|\mathbf{q}|$ is held fixed at 90 MeV;

- The ordinate is $d^5/d\varepsilon_2 d\Omega_2 d\Omega_q$ divided by $I_{inc}\,\sigma_{ep}$ where σ_{ep} is calculated under the same kinematics, but with[7]

  1. An initial proton four-momentum $p_\mu = (\mathbf{p}_m, [\mathbf{p}_m^2+(m-E_S)^2]^{1/2})$, where E_S is the separation energy of the $1p_{1/2}$ proton in $^{16}_{8}O$;
  2. A nucleon vertex $\Gamma_\mu \equiv \gamma_\mu(F_1 + 2mF_2) - (p + q)_\mu F_2$.

- The theoretical calculation has been reduced by a spectroscopic factor of S = 0.69. This represents partial occupancy of the $(1p)_{1/2}$ state in $^{16}_{8}O$ in this model. (See e.g. [Do75].) The shape of the momentum distribution is unaffected by this overall factor;

- The normal polarization of the outgoing proton is also shown. Since the most important consequence of relativistic Hartree theory and RIA is the prediction of the spin–orbit effects, it is crucially important to see how these predictions hold up for nucleons ejected from orbits in the nuclear interior;

---

[5] Except for a proper relativistic treatment of the C-M motion (see [Ki87]).

[6] To the extent that the relativistic Hartree calculations generate the real part of the empirical RIA optical potentials, this electromagnetic current is conserved. Current conservation in the presence of the imaginary part of the optical potential is a more challenging problem under active investigation.

[7] The denominator is a well-defined expression under all kinematic conditions — it yields the so-called CC1 expression of de Forest [de83].

- There is nothing intrinsic in the calculation that limits the k_μ^2 to which it can be extended, nor the $\mathbf{q}$, provided the RIA optical potential still yields a good description of the scattering.[8]

Quasielastic electron scattering for $^{40}_{20}$Ca(e, e′) is calculated in relativistic Hartree by summing over single-particle transitions, and including the random phase approximation (RPA) response, in [Ho89].

As described here, QHD is a simple model which reproduces many important aspects of nuclear structure. A much deeper basis for this relativistic quantum field theory approach to the nuclear many-body problem can be given in terms of *effective field theory*. One chooses to use hadronic degrees of freedom as generalized coordinates and writes the most general low-energy lagrangian consistent with the symmetry properties of QCD (chapter 25). This provides the density functional for the nuclear system, and minimization of that density functional leads to the relativistic Hartree equations (with additional non-linear couplings). The values of $g_s\phi_0/m$ and $g_v V_0/m$, both $\approx 1/3$, then yield expansion parameters which provide controlled approximation schemes. This approach puts the RMFT on a much firmer theoretical basis [Fu97, Se97].

---

[8] Again, except for a proper relativistic treatment of C-M motion.

22

Elastic scattering

Electron scattering studies of nuclear and nucleon structure came into their own with the beautiful experiments at the High Energy Physics Laboratory (HEPL) at Stanford University by Hofstadter and collaborators in the 1950's. These experiments measured charge distributions through elastic scattering [Ho56]. Early results throughout the periodic system from Mg to Pb were summarized in terms of a two-parameter "Fermi model" charge distribution

$$
\begin{aligned}
\rho &= \frac{\rho_0}{1 + e^{(r-R)/a}} \\
R &= r_0\, A^{1/3} \\
r_0 &\approx 1.07\,\text{fm} \\
t &\approx 2.4\,\text{fm}
\end{aligned} \tag{22.1}
$$

Here the *surface thickness* t is the 90% to 10% fall-off distance ($t \approx 2a\ln 9$). The observed distributions are illustrated in Fig. 22.1. It is difficult to overstate the impact of these experiments. One could actually *see* what the tiny nucleus at the center of the atom looks like.[1] The density of the *nuclear matter* at the center of the nucleus is approximately constant from nucleus to nucleus, as is the surface structure. As one adds nucleons, the nucleus simply grows in size. If fact, it grows exactly as a drop of water grows when more liquid is added to it. The volume of the nucleus is simply proportional to the number of nucleons

$$
V = \frac{4}{3}\pi R^3 = \frac{4}{3}\pi r_0^3\, A \tag{22.2}
$$

---

[1] Experiments on finite-size effects in the spectra of mu-mesic X-rays had previously yielded good values for the nuclear mean-square radius [Fi53].

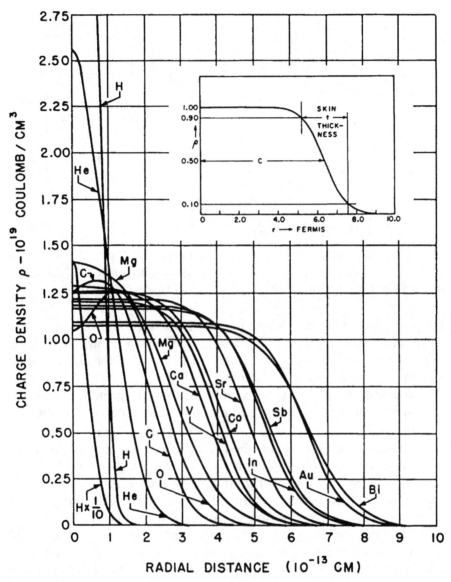

Fig. 22.1. Approximate shapes of the charge distribution of selected nuclei, including the proton and alpha particle. Note the change of scale for the proton. The insert explains the Fermi model (here $c \equiv R$) [Ho56].

Furthermore, Hofstadter and collaborators demonstrated that the charge distribution of the proton is of finite extent, with a root-mean-square radius of [Ch56, Ho56]

$$\langle r_p^2 \rangle^{1/2} \approx 0.77 \, \text{fm} \tag{22.3}$$

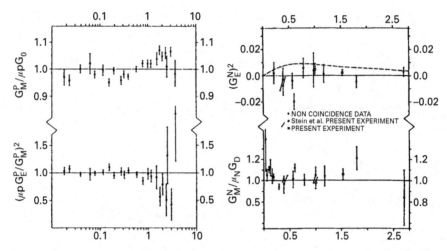

Fig. 22.2.   Compilation of the form factors of the proton. Left: $(\mu_p G_{\rm E}^p/G_{\rm M}^p)^2$ and $G_{\rm M}^p/\mu_p G_{\rm D}$ are plotted versus q^2 on a logarithmic scale. Right: $(G_{\rm E}^n)^2$ [the dashed curve is $G_{\rm E}^n(q^2) = -\tau G_{\rm M}^n(q^2)$ with $\tau \equiv q^2/4m^2$] and $G_{\rm M}^n/\mu_n G_{\rm D}$ [Ba73].

This gave direct evidence that the nucleon itself is a composite structure.[2] To extract accurate charge distributions in electron scattering from heavier nuclei, one has to deal with the Coulomb distortion of the electron wave functions, as well as other corrections [Ye65, Ra87].

Experiments on elastic magnetic scattering from the proton and neutron demonstrated the spatial extent, and measured the spatial distribution, of the magnetization in this $(J^\pi, T) = (\frac{1}{2}^+, \frac{1}{2})$ system. Free neutron targets do not exist, so one uses the next best thing, a neutron lightly bound to a proton in the deuteron ($_1^2$H). Both quasielastic scattering where the neutron is directly ejected, and elastic scattering, have been employed. The latter depends more sensitively on the ground-state deuteron wave function. Of course, the accompanying proton is not really inert, and nuclear physics comes into play in analyzing these experiments. Nevertheless, over the years reliable charge and magnetic form factors have been obtained for both the proton and neutron, and a summary of the elastic form factors for the nucleon is shown in Fig. 22.2. The *Sachs* form factors are defined in terms of the Dirac form factors according to

$$G_{\rm M}(q^2) \;\equiv\; F_1(q^2) + 2mF_2(q^2)$$

$$G_{\rm E}(q^2) \;\equiv\; F_1(q^2) - \frac{q^2}{2m}F_2(q^2) \tag{22.4}$$

The dipole form factor, to which the measured values are compared, is

---

[2] Of course, the fact that the magnetic moments of the neutron and proton already deviate so much from the Dirac values (chapter 19) strongly implies the same thing.

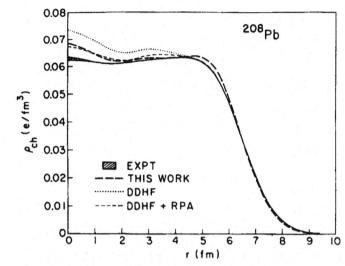

Fig. 22.3. Charge density for $^{208}_{82}$Pb. The solid curve and shaded area represent the fit to the experimental data [He69]. Relativistic mean field theory results are indicated by the long dashed lines [Ho81, Se86]. Some density-dependent Hartree–Fock calculations within the traditional picture are also shown (see [Ho81] for references).

defined by

$$G_D \equiv \frac{1}{(1 + q^2/0.71\,\mathrm{GeV}^2)^2} \tag{22.5}$$

By showing the deviations from the phenomenological dipole form, one can plot the form factors on an expanded scale. Note that at the time of publication of Fig. 22.2, the charge form factor of the neutron was not very well known at all. It is important to emphasize that even though the neutron has no net charge, it can still have a non-uniform spatial distribution of charge within it. One of the advantages of the Sachs form factors is that they have a more direct interpretation in terms of the spatial Fourier transform of the charge and total magnetic moment densities of the nucleon [Wa59].

Theoretical understanding of the charge distribution of nuclei within the traditional picture is based on self-consistent Hartree–Fock calculations of nuclear ground states [Fe71, Go79, Ne82]. The most sophisticated of these use as an interaction the local-density T-matrix calculated in the Bruekner theory of nuclear matter [Ne82]. Relativistic mean field theory (RMFT), described in the previous section, gives a very direct determination of these densities with a local interaction. In Fig. 22.3 we show the comparison of the RMFT calculation by Horowitz and Serot

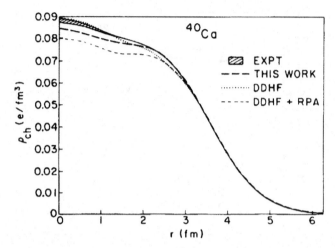

Fig. 22.4.   As Fig. 22.3, but for $^{40}_{20}$Ca [Fr79, Ho81, Se86].

of the charge density of $^{208}_{82}$Pb and the experimental determination of this quantity [Ho81, Wa95]. The error band on the experimental charge distribution arises primarily from the fact that one is always measuring only a partial Fourier transform in electron scattering, although other effects contribute [Fr73]. It is from this figure that the density of nuclear matter is determined [Ho81]. Figure 22.4 shows a similar result for $^{40}_{20}$Ca [Ho81, Wa95]. Here the half-density radius determines the scalar mass [Ho81]. All other nuclear charge densities are then predicted in RMFT (see Fig. 21.5). In summary, although in a sense it is the simplest thing one can compute [Se97], one has a good theoretical understanding of the ground-state charge densities of nuclei.

The theoretical analysis of the elastic form factors of the nucleon proceeds most directly through the *spectral representation* of these quantities [Ch58, Fe58, Dr61, Wa95]. From very general field theory principles, one establishes that the isovector and isoscalar form factors of the nucleon have the representations

$$F_i^V(q^2) = \frac{1}{\pi} \int_{(2m_\pi)^2}^{\infty} \frac{w_i^V(\sigma^2)\,d\sigma^2}{\sigma^2 + q^2} \qquad ; i = 1, 2$$

$$F_i^S(q^2) = \frac{1}{\pi} \int_{(3m_\pi)^2}^{\infty} \frac{w_i^S(\sigma^2)\,d\sigma^2}{\sigma^2 + q^2} \tag{22.6}$$

Here m_π is the pion mass. The thresholds in the representations in Eqs. (22.6) are obtained by angular momentum, isospin, and charge-conjugation considerations.

The real spectral weight functions $w_i^{S,V}(\sigma^2)$ are related to the absorptive part of the amplitude for a time-like virtual photon to go through an intermediate hadronic state and then into a nucleon–antinucleon pair. Time-like virtual photons with $q = (0, iW)$ can be created in the laboratory through the process of electron–positron annihilation in the C-M system. The process $e^+ + e^- \rightarrow pions$ can be measured experimentally for any $W \geq 2m_\pi$. The amplitude for $e^+ + e^- \rightarrow N + \bar{N}$ can be accessed experimentally only for $W \geq 2m$; for $W < 2m$, one needs analytic continuation.

The spectral representations in Eqs. (22.6) hold in the entire q^2 plane. The representation for the charge form factors probably requires one subtraction [Dr61]

$$F_1^\alpha(q^2) = 1 - \frac{q^2}{\pi} \int \frac{w_1^\alpha(\sigma^2) \, d\sigma^2}{\sigma^2(\sigma^2 + q^2)} \qquad ; \alpha = S, V \qquad (22.7)$$

One can readily establish that in elastic electron scattering from the nucleon there is always one Lorentz frame, the so-called Breit (or brickwall) frame, where the electron undergoes no energy transfer. In this frame, the four-momentum transfer takes the form $q = (\mathbf{q}, i0)$. In this case one can define the form factor as the three-dimensional Fourier transform of a charge and magnetization density according to[3]

$$F(\mathbf{q}^2) = \int d^3x \, e^{i\mathbf{q}\cdot\mathbf{x}} \rho(r)$$

$$\rho(r) = \int \frac{d^3q}{(2\pi)^3} e^{-i\mathbf{q}\cdot\mathbf{x}} F(\mathbf{q}^2) \qquad (22.8)$$

Insertion of Eqs. (22.6) in the second relation gives

$$\rho(r) = \frac{1}{4\pi^2} \int d\sigma^2 \, w(\sigma^2) \frac{e^{-\sigma r}}{r} \qquad (22.9)$$

This relation expresses the density as a linear combination of Yukawa distributions, each of mass σ. By the uncertainty principle, the mass σ of the intermediate state determines how far it extends out from the origin. The intermediate state now occurs as a virtual one in electron scattering where the momentum transfer is space-like.

Consider a simple example of these ideas. The lightest mass hadron is the pion, and it is evident from Eq. (22.9) that charged pions are responsible for the long-range part of the electromagnetic structure of the nucleon. To evaluate the two-pion contribution to the spectral weight function for $F_2(q^2)$ in Born approximation (without pion rescattering) one can simply look at the Feynman diagram for the lowest-order vertex correction illustrated in Fig. 22.5. We calculate the contribution of this diagram to S_{fi}

---

[3] The Wigner–Eckart theorem and parity invariance imply that the ground state densities must be spherically symmetric for a spin one-half system.

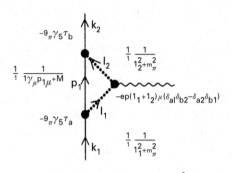

Fig. 22.5.   Two-pion contribution to S_{fi} and $F_2(q^2)$ in Born approximation.

from the following pion–nucleon and pion–photon lagrangian densities[4]

$$\mathcal{L}_{\pi N} = ig_\pi \bar{\psi}\gamma_5 \boldsymbol{\tau} \; \psi \cdot \boldsymbol{\pi}$$

$$\mathcal{L}_{\gamma\pi} = -e_p \left[\boldsymbol{\pi} \times \frac{\partial \boldsymbol{\pi}}{\partial x_\mu} \right]_3 A_\mu \tag{22.10}$$

The component contributions to the diagram are then indicated in Fig. 22.5. It is a nice exercise to show that the result from this diagram can be put into the following form

$$2mF_2(q^2) = \tau_3 \frac{g_\pi^2}{4\pi} \int_0^1 dx\, (1-x)^2 \int_0^x dy \frac{m^2}{m^2(1-x)^2 + m_\pi^2 x + q^2 y(x-y)} \tag{22.11}$$

The spectral representation and two-pion contribution to the spectral weight function follow directly. Note that this contribution is entirely isovector.

The integral in Eq. (22.11) is well-defined, and one can use it to calculate this two-pion contribution to the anomalous magnetic moment of the nucleon by simply evaluating $2mF_2(0)$. The longest range two-pion contribution to the mean-square radius of the isovector magnetic moment can be obtained through

$$\frac{F_2^V(q^2)}{F_2^V(0)} = 1 - \frac{q^2}{6} \langle r^2 \rangle_2^V + \cdots \tag{22.12}$$

The use of $g_\pi^2/4\pi = 14.4$ from pion–nucleon scattering leads to the results shown in Table 22.1. The present analysis provides a qualitative, and even semiquantitative, understanding of the anomalous magnetic moment and its mean-square radius [Ch58, Fe58].

To pursue this approach even further, it was argued *before their discovery* that vector mesons with $(J^\pi, T) = (1^-, 1)$ and $(1^-, 0)$, the ρ and ω, must be

---

[4] The absorptive part is independent of the particular form of the π–N coupling used.

Table 22.1. Two-pion contribution to the anomalous magnetic moment of the nucleon in Born approximation.

	λ'^S	λ'^V	$\langle r^2 \rangle^V_{\text{mag}}$	$(\langle r^2 \rangle^V_{\text{mag}})^{1/2}$
Theory	0	3.20	0.24 fm^2	0.49 fm
Experiment	-0.12	3.706	≈ 0.64 fm^2	≈ 0.80 fm

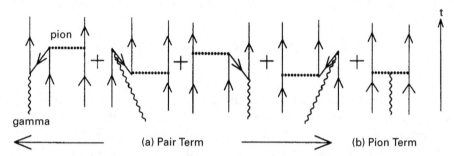

Fig. 22.6. Time-ordered Feynman diagrams retained in the one-pion exchange current calculation in [Du76].

present to make the *size* of the distributions quantitative [Na57, Fr60]. The basic idea is that a two- or three-pion resonance makes the distribution extend out further.[5]

Of course, the internal quark structure of the nucleon plays an essential role in determining the electromagnetic structure of the nucleon (chapter 24); however, it is clear from Eq. (22.9) that pions are responsible for the long-range contribution to this structure. Both elements of the internal structure clearly play a role.[6] Effective chiral lagrangians that reflect the underlying symmetry structure of QCD, and chiral perturbation theory, place the calculation of the the long-range low-q^2 pion contribution to the nucleon form factors on a firmer theoretical foundation [Be98, Ku01].

A prime example of the need for an explicit hadronic description of nuclei is provided by the additional two-body currents arising from the exchange of charged mesons between nucleons. Although many exchange current calculations exist, for concreteness we briefly describe those of Dubach, Koch, and Donnelly [Du76]. These authors keep the static limit [leading $O(1/m)$] of the time-ordered Feynman diagrams shown in Fig. 22.6. Each of these processes clearly represents an additional contribution

---

[5] In chapter 21 we were content to include the contribution of charged mesons to the internal structure of the nucleon in a phenomenological fashion, through a single-nucleon form factor $f_{\text{SN}}(q^2) = G_D(q^2)$.

[6] "Bag models" and "chiral soliton models" attempt to incorporate both elements of this internal structure [Wa95].

to the current in the traditional picture, which is now extended to

$$\hat{J}_\mu(\mathbf{x}) = \sum_{i=1}^{A} J_\mu^{(1)}(\mathbf{x}_i; \mathbf{x}) + \sum_{i<j=1}^{A} J_\mu^{(2)}(\mathbf{x}_i, \mathbf{x}_j; \mathbf{x}) \qquad (22.13)$$

The two-body current can be identified through reproduction of the S-matrix as follows.

The free Dirac propagator can be decomposed according to [Fe71, Wa95]

$$\frac{1}{i\gamma_\mu p_\mu + m} \equiv \left[\frac{1}{2E_p} \frac{\boldsymbol{\alpha} \cdot \mathbf{p} + \beta m + E_p}{E_p - p_0 - i\eta} + \frac{1}{2E_p} \frac{\boldsymbol{\alpha} \cdot \mathbf{p} + \beta m - E_p}{E_p + p_0 - i\eta} \right] \beta \qquad (22.14)$$

The first term yields the usual non-relativistic result [Fe71]; the second term gives rise to backward propagation in time. The Feynman rules from the lagrangian in Eq. (22.10) allow one to evaluate the contribution to the S-matrix from the graphs in Fig. 22.6, retaining just the second piece of the baryon propagator. An equivalent S-matrix can be constructed from the current in Eq. (22.13), and one can then identify the additional two-body current. Define

$$J_\mu(\mathbf{x}_1, \mathbf{x}_2; \mathbf{x}) = \int e^{i\mathbf{k}\cdot\mathbf{x}} J_\mu(\mathbf{x}_1, \mathbf{x}_2; \mathbf{k}) \frac{d^3k}{(2\pi)^3} \qquad (22.15)$$

Then to leading order in $1/m$, and with the neglect of k_0, the pair contribution to the pion-exchange current in Fig. 22.6(a) is given by

$$\mathbf{J}^{\text{pair}}(\mathbf{x}_1, \mathbf{x}_2; \mathbf{k}) = -e_p f_\pi^2 [\boldsymbol{\tau}^{(1)} \times \boldsymbol{\tau}^{(2)}]_3$$

$$\times \left\{ \boldsymbol{\sigma}_2 \left(\frac{\boldsymbol{\sigma}_1 \cdot \mathbf{r}}{r} \right) e^{-i\mathbf{k}\cdot\mathbf{x}_2} + \boldsymbol{\sigma}_1 \left(\frac{\boldsymbol{\sigma}_2 \cdot \mathbf{r}}{r} \right) e^{-i\mathbf{k}\cdot\mathbf{x}_1} \right\} \left(\frac{1 + x_\pi}{x_\pi^2} \right) e^{-x_\pi} \qquad (22.16)$$

Here

$$\mathbf{x}_\pi = \mu\mathbf{r} \qquad\qquad ; \mu \equiv m_\pi$$

$$\mathbf{r} = \mathbf{x}_1 - \mathbf{x}_2 \qquad\qquad ; \mathbf{R} = \frac{1}{2}(\mathbf{x}_1 + \mathbf{x}_2)$$

$$f_\pi^2 = \frac{g_\pi^2}{4\pi} \left(\frac{\mu}{2m} \right)^2 = 0.080 \qquad (22.17)$$

The pion contribution in Fig. 22.6(b) is

$$\mathbf{J}^{\text{pion}}(\mathbf{x}_1, \mathbf{x}_2; \mathbf{k}) = e_p \left(\frac{f_\pi}{\mu} \right)^2 [\boldsymbol{\tau}^{(1)} \times \boldsymbol{\tau}^{(2)}]_3 (\boldsymbol{\sigma}_1 \cdot \nabla_1)(\boldsymbol{\sigma}_2 \cdot \nabla_2)$$

$$\times \int_{-1/2}^{1/2} dv \, (-ir v \mathbf{k} + \mathbf{y}) \left(\frac{e^{-y}}{y} \right) \exp\{-i\mathbf{k} \cdot (\mathbf{R} - v\mathbf{r})\} \qquad (22.18)$$

Here

$$\mathbf{y} = \mathbf{r}\left[\mu^2 + \left(\frac{\mathbf{k}^2}{4}\right)(1 - 4v^2)\right]^{1/2} \tag{22.19}$$

There is no exchange contribution to the charge density to this order in $1/m$. These results are from [Du76], and the reader is now in a position to reproduce them.[7]

This exchange current has the following features to recommend it:

- If the current is taken to be the sum of the one-body current of chapter 19 and the above exchange current, and if a two-nucleon potential has the form $V = V^{\text{neutral}} + V^{\text{OPEP}}$ where the last term is the one-pion exchange potential [Wa95], then the current is differentially conserved [Du76]

$$\frac{\partial \hat{J}_\mu}{\partial x_\mu} = \nabla \cdot \hat{\mathbf{J}} + i[\hat{H}, \hat{\rho}] = 0 \tag{22.20}$$

- The threshold pion electroproduction part of the graphs in the above amplitude satisfies the Kroll–Ruderman (soft-pion) theorem;

- This one-pion contribution represents the longest-range part of the two-body exchange current; it is exact as $|\mathbf{x}_1 - \mathbf{x}_2| \to \infty$;

- The charge density operator is unmodified to leading $O(1/m)$; hence transition matrix elements of the charge density can be used to calibrate the nuclear structure in exchange-current calculations.

Assume that ^3_2He can be described by a $(v1s_{1/2})^{-1}$ harmonic oscillator shell model configuration as shown in Figs. 20.7 and 20.8. The magnetic moment calculated with the inclusion of the above exchange current is $\mu = -2.078$ n.m., now closer to the experimental value $\mu = -2.127$ n.m. than is the Schmidt value $\mu = -1.913$ n.m. in Table 20.1. (Here 1 n.m. $= e_p\hbar/2m_pc$). This gives one some confidence in the present exchange current calculation [Du76]. The effect on elastic magnetic electron scattering at modest momentum transfers, say $q^2 \le 6$ fm^{-2}, is shown in Fig. 22.7; the effect is not large. This illustrates the *marginal* role of exchange currents in the traditional nuclear physics domain.[8]

Figure 22.8 illustrates the state of the art with elastic magnetic scattering from ^3_2He [Ca82]. The measurements are from Saclay and Bates. The dashed line shows the result obtained from the best three-body calculation

---

[7] Use $(ab)^{-1} = \int_0^1 dz[az + b(1 - z)]^{-2}$.

[8] A relativistic QHD calculation of this exchange current, without the $1/m$ expansion, is contained in [Bl91].

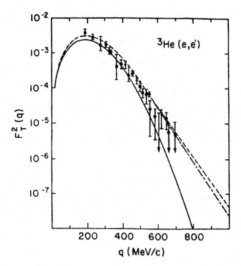

Fig. 22.7. Elastic transverse form factor for $^3_2$He(e, e′) in the harmonic oscillator model with (dashed) and without (solid) one-pion exchange currents [Du76, Wa84].

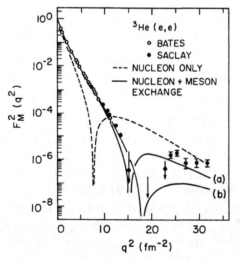

Fig. 22.8. Elastic magnetic form factor for $^3_2$He(e, e) out to high q^2 [Ca82]. Two exchange-current theories are shown: (a) from [Ha83]; (b) from [Ri80].

done in the traditional picture; the three-body wave function is obtained by solving the Faddeev equations with potentials fitted to two-body data, and the current is obtained from the properties of free nucleons. There is clear disagreement with the data as q^2 increases to $\sim 10\,\mathrm{fm}^{-2}$, not by a few percent, but by orders of magnitude. The best three-body calculation in

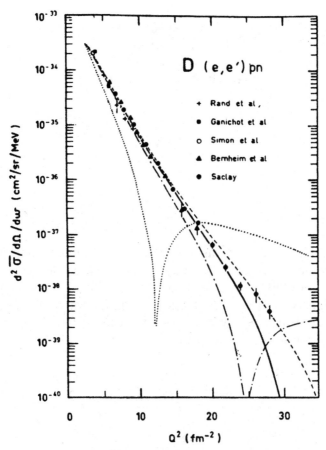

Fig. 22.9. Cross section for $^2_1$H(e, e′)pn$_{thresh}$. The dotted curve is the impulse-approximation result, the dash-dot curve includes the pion-exchange contribution, the dash curve includes also a ρ-exchange contribution, and the solid curve is the total result, which includes a Δ contribution. [Au85].

the traditional picture clearly fails at high q^2. Also shown in Fig. 22.8 are two exchange-current calculations that include the pion exchange current discussed above, as well as other hadronic contributions [Ha83, Ri80]. The difference between these two curves at high q^2 is a good measure of the present theoretical uncertainty. While the exchange current contribution is marginal at low momentum transfers, it is a *dominant* effect at large q^2. A more recent and extensive discussion of electromagnetic interactions with light nuclei is contained in [Ca91].

This pion exchange current also shows up dramatically in threshold electrodisintegration of the deuteron $^2_1$H(e, e′)pn$_{thresh}$ as shown in Fig. 22.9 [Au85].

23

Quasielastic scattering

A first description of quasielastic scattering is obtained from the electro-magnetic response of a non-interacting, non-relativistic Fermi gas. This provides a convenient, consistent picture of the dominant part of the nuclear response surface as a function of $(\mathbf{q}, \omega)$. Consider a Fermi gas of protons as illustrated in Fig. 23.1. The total charge and charge density are obtained by counting the occupied states

$$
Z = \sum_{\mathbf{k}\lambda}^{k_F} 1 \rightarrow \frac{2\Omega}{(2\pi)^3} \int_0^{k_F} d^3k = \frac{\Omega k_F^3}{3\pi^2}
$$

$$
\frac{Z}{\Omega} \equiv \rho = \frac{k_F^3}{3\pi^2} \tag{23.1}
$$

The last equality relates the proton density to the Fermi wave number. For illustration, retain just the Coulomb interaction, assuming no transverse interaction. The target response surfaces of chapter 11 then reduce to the form [compare Eq. (12.31)]

$$
W_1 = 0 \tag{23.2}
$$

$$
\frac{1}{M_T} W_2 = \frac{q_\mu^4}{\mathbf{q}^4} \overline{\sum_i} \sum_f |\langle f| \int \exp(-i\mathbf{q} \cdot \mathbf{x})\, \hat{\rho}(\mathbf{x})\, d^3x |i\rangle|^2 \delta(W_f - W_i)
$$

The notation used is indicated in Fig. 23.2.[1]

---

[1] Use of translational invariance, with eigenstates of three-momentum, allows one to do both spatial integrations which yield, in the limit of large Ω, a factor $(2\pi)^3\Omega\delta^{(3)}(\mathbf{P}_f - \mathbf{P}_i)$ — this reproduces the previous form for the response surfaces for a target initially at rest. In the present form, one can directly calculate the scattering for a stationary target.

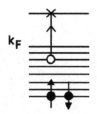

Fig. 23.1. Response of Fermi gas.

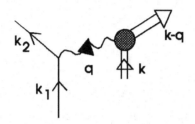

Fig. 23.2. Notation in calculation of Coulomb response of Fermi gas.

We also extract the nucleon form factor of Eq. (19.22) and write

$$\frac{1}{M_T} W_2 \equiv \frac{q_\mu^4}{\mathbf{q}^4} |f_{\mathrm{SN}}(q^2)|^2 R(\mathbf{q}^2, \omega) \tag{23.3}$$

Here $\omega \equiv \varepsilon_1 - \varepsilon_2$ is the energy loss.

The charge density operator for point protons can then be written in second quantization as

$$\hat{\rho}(\mathbf{x}) = \hat{\psi}^\dagger(\mathbf{x})\hat{\psi}(\mathbf{x}) \tag{23.4}$$

Here, the non-relativistic, two-component proton field is given by

$$\hat{\psi}(\mathbf{x}) = \frac{1}{\sqrt{\Omega}} \sum_{\mathbf{k}\lambda} a_{\mathbf{k}\lambda} \exp{(i\mathbf{k} \cdot \mathbf{x})}\, \eta_\lambda \tag{23.5}$$

Thus, upon integration

$$\int \exp{(-i\mathbf{q} \cdot \mathbf{x})}\, \hat{\rho}(\mathbf{x})\, d^3x = \sum_{\mathbf{k}\lambda} a_{\mathbf{k}-\mathbf{q},\,\lambda}^\dagger a_{\mathbf{k}\lambda} \tag{23.6}$$

Matrix elements of this expression for a Fermi gas can now be readily evaluated. One particle must be destroyed inside the Fermi sea and one created outside. Upon converting the final sum to an integral, one arrives at

$$R(\mathbf{q}^2, \omega) = \frac{2\Omega}{(2\pi)^3} \int_0^{k_F} d^3k\, \theta(|\mathbf{k} - \mathbf{q}| - k_F)\delta(\omega - \varepsilon_{\mathbf{k}-\mathbf{q}} + \varepsilon_{\mathbf{k}}) \tag{23.7}$$

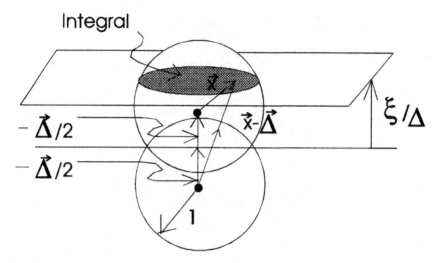

Fig. 23.3. Momentum space geometry for evaluation of response of a non-relativistic Fermi gas.

This is a general result for a Fermi gas. Now specialize to the dispersion relation for non-relativistic nucleons

$$\varepsilon_{\mathbf{k}} = \frac{\mathbf{k}^2}{2m} \tag{23.8}$$

$$R(\mathbf{q}^2, \omega) = \frac{2\Omega}{(2\pi)^3} \int_0^{k_F} d^3k \, \theta(|\mathbf{k} - \mathbf{q}| - k_F) \, \delta \left(\omega + \frac{\mathbf{k} \cdot \mathbf{q}}{m} - \frac{\mathbf{q}^2}{2m} \right)$$

Introduce dimensionless variables according to

$$\mathbf{x} \equiv \frac{\mathbf{k}}{k_F} \qquad ; \Delta \equiv \frac{\mathbf{q}}{k_F} \qquad ; \xi \equiv \frac{m\omega}{k_F^2} \tag{23.9}$$

The additional use of $\Omega = 3\pi^2 Z / k_F^3$ then leads to

$$R(\mathbf{q}^2, \omega) = \frac{3Z}{4\pi} \frac{m}{k_F^2} \int_0^1 d^3x \, \theta(|\mathbf{x} - \Delta| - 1) \, \delta \left(\xi + \Delta \cdot \mathbf{x} - \frac{\Delta^2}{2} \right) \tag{23.10}$$

This integral can be done with the aid of some simple geometric considerations [Fe71]. The situation is illustrated in Fig. 23.3. First rewrite

$$\delta \left(\xi + \Delta \cdot \mathbf{x} - \frac{\Delta^2}{2} \right) = \frac{1}{\Delta} \delta \left(\mathbf{x} \cdot \left[\frac{\Delta}{\Delta} \right] + \frac{\xi}{\Delta} - \frac{\Delta}{2} \right) \tag{23.11}$$

Energy conservation enforced by the vanishing of the argument of the

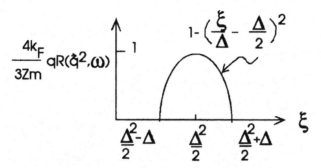

Fig. 23.4. The quantity $\zeta R(q^2, \omega)$ with $\zeta \equiv 4k_F q/3Zm$ for the case $\Delta \geq 2$.

delta function then defines the plane indicated in this figure

$$-\mathbf{x} \cdot \left[\frac{\Delta}{\Delta} \right] = \frac{\xi}{\Delta} - \frac{\Delta}{2} \tag{23.12}$$

The restrictions on the region of integration reflect the fact that the particle is initially inside the Fermi sphere (F) and must end up outside

$$
\begin{array}{lll}
|\mathbf{x} - \boldsymbol{\Delta}| & > & 1 \qquad ; \text{ outside F} \\
|\mathbf{x}| & \leq & 1 \qquad ; \text{ inside F}
\end{array} \tag{23.13}
$$

The answer for the $\Delta \times$ (integral) is the area of intersection of the plane and the restricted Fermi sphere. This is either a circle (as illustrated in the figure), or an annulus, depending on the value of ξ/Δ. Thus one can immediately write the answer in the various cases:

1) $\Delta \geq 2$ (spheres do not intersect)

$$\frac{\Delta}{2} + 1 \geq \frac{\xi}{\Delta} \geq \frac{\Delta}{2} - 1 \qquad ; \text{ plane intersects sphere}$$

$$\text{area} = \pi \left[1^2 - \left(\frac{\xi}{\Delta} - \frac{\Delta}{2} \right)^2 \right]$$

$$R(q^2, \omega) = \frac{3Z}{4\pi} \frac{m}{k_F^2} \frac{\pi}{\Delta} \left[1^2 - \left(\frac{\xi}{\Delta} - \frac{\Delta}{2} \right)^2 \right] \tag{23.14}$$

This result is sketched in Fig. 23.4, and we make two observations:

• The peak of this response occurs at $\xi = \Delta^2/2$ or

$$\omega_{\text{peak}} = \mathbf{q}^2/2m \tag{23.15}$$

This is just the free, non-relativistic kinematic relation for the energy transfer to a nucleon initially at rest, recoiling with momentum $-\mathbf{q}$. Note the position of this quasielastic peak moves with $\mathbf{q}^2$.

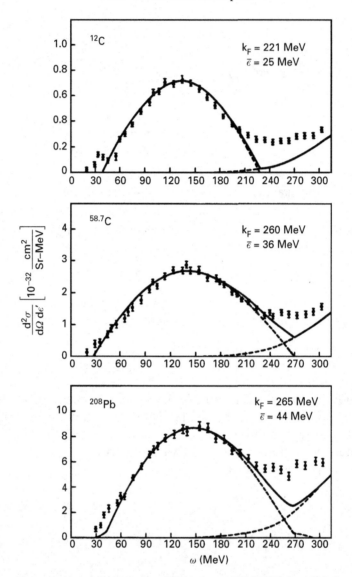

Fig. 23.5. Data on quasielastic scattering at $\varepsilon_1 = 500\,\text{MeV}, \theta = 60°$ from HEPL. Calculation includes Coulomb and transverse current interactions. $\bar{\varepsilon}$ shifts the response function by an average nuclear binding energy [Mo71, Do75].

• The width of this peak at the base is 2Δ. Thus $\delta\xi_{\text{base}} = 2\Delta$ or

$$\frac{1}{2}\delta\omega_{\text{base}} = \frac{k_F}{m}|\mathbf{q}| = v_F|\mathbf{q}| \tag{23.16}$$

There is a Doppler broadening of the quasielastic peak that increases with $|\mathbf{q}|$. This width can be used to measure the Fermi velocity as shown in Fig. 23.5.

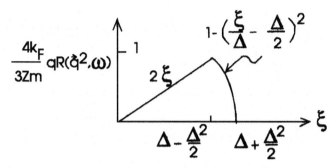

Fig. 23.6.   The quantity $\zeta R(\mathbf{q}^2, \omega)$ with $\zeta \equiv 4k_F q/3Zm$ for the case $\Delta \leq 2$.

2) $\Delta \leq 2$ (spheres intersect)

$$1 + \frac{\Delta}{2} \;\geq\; \frac{\xi}{\Delta} \geq 1 - \frac{\Delta}{2} \quad ; \text{ plane does not intersect excluded sphere}$$

$$R(\mathbf{q}^2, \omega) \;=\; \frac{3Z}{4\pi}\frac{m}{k_F^2}\frac{\pi}{\Delta}\left[1^2 - \left(\frac{\xi}{\Delta} - \frac{\Delta}{2}\right)^2\right] \tag{23.17}$$

The area and answer are the same as before.

3) $\Delta \leq 2$ (spheres intersect)

$$1 - \frac{\Delta}{2} \;\geq\; \frac{\xi}{\Delta} \geq 0 \qquad\qquad ; \text{ plane does intersect excluded sphere}$$

$$\text{area} \;=\; \pi\left\{\left[1^2 - \left(\frac{\xi}{\Delta} - \frac{\Delta}{2}\right)^2\right] - \left[1^2 - \left(\frac{\xi}{\Delta} + \frac{\Delta}{2}\right)^2\right]\right\} = 2\pi\xi$$

$$R(\mathbf{q}^2, \omega) \;=\; \frac{3Z}{4\pi}\frac{m}{k_F^2}2\pi\left(\frac{\xi}{\Delta}\right) \tag{23.18}$$

The results in the case $\Delta \leq 2$ are sketched in Fig. 23.6.

This simple model calculation provides excellent insight into quasielastic electron scattering from the nuclear many-body system.

Within the traditional framework of non-relativistic nucleons and one-body densities, it is possible to derive some *exact* results for the nuclear response functions. An integration over all energy loss in Eq. (23.2) at fixed $\mathbf{q}$ removes the energy delta function, and from Eqs. (23.3) and (23.4)[2]

$$S(\mathbf{q}^2) \equiv \int d\omega R(\mathbf{q}^2, \omega) = \sum_f |\langle \Psi_f| \int \exp(-i\mathbf{q} \cdot \mathbf{x})\,\hat{\rho}(\mathbf{x})\,d^3x|\Psi_0\rangle|^2 \tag{23.19}$$

---

[2] It is important to note that one can never fully evaluate this integral experimentally in electron scattering since there the four-momentum transfer must be space-like $q_\mu^2 = \mathbf{q}^2 - \omega^2 \geq 0$.

Here the ground state is written as $|i\rangle \equiv |\Psi_0\rangle$; it is assumed to be non-degenerate. Closure may now be used on this expression to give

$$S(\mathbf{q}^2) = \langle\Psi_0| \int\int d^3x\, d^3y \exp\{-i\mathbf{q}\cdot(\mathbf{x}-\mathbf{y})\}\, \hat{\rho}(\mathbf{y})\,\hat{\rho}(\mathbf{x})\,|\Psi_0\rangle \qquad (23.20)$$

Now make use of the canonical anti-commutation rules for the proton fields

$$\begin{aligned}
\left\{\hat{\psi}_\alpha(\mathbf{x}),\, \hat{\psi}_\beta^\dagger(\mathbf{y})\right\} &= \delta^{(3)}(\mathbf{x}-\mathbf{y})\,\delta_{\alpha\beta} \\
\{\hat{\psi}_\alpha(\mathbf{x}),\, \hat{\psi}_\beta(\mathbf{y})\} &= \left\{\hat{\psi}_\alpha^\dagger(\mathbf{x}),\, \hat{\psi}_\beta^\dagger(\mathbf{y})\right\} = 0
\end{aligned} \qquad (23.21)$$

They allow one to write

$$\begin{aligned}
\hat{\rho}(\mathbf{y})\,\hat{\rho}(\mathbf{x}) &= \hat{\psi}^\dagger(\mathbf{y})\hat{\psi}(\mathbf{y})\hat{\psi}^\dagger(\mathbf{x})\hat{\psi}(\mathbf{x}) \qquad (23.22) \\
&= \delta^{(3)}(\mathbf{x}-\mathbf{y})\hat{\psi}^\dagger(\mathbf{x})\hat{\psi}(\mathbf{x}) + \hat{\psi}^\dagger(\mathbf{y})\hat{\psi}^\dagger(\mathbf{x})\hat{\psi}(\mathbf{x})\hat{\psi}(\mathbf{y})
\end{aligned}$$

The total charge, a constant of the motion, can be identified according to

$$\int d^3x \int d^3y \exp\{-i\mathbf{q}\cdot(\mathbf{x}-\mathbf{y})\}\,\delta^{(3)}(\mathbf{x}-\mathbf{y})\,\hat{\psi}^\dagger(\mathbf{x})\,\hat{\psi}(\mathbf{x}) = \hat{Z} \qquad (23.23)$$

Hence one has the general result

$$\begin{aligned}
S(\mathbf{q}^2) &= Z \qquad (23.24) \\
&+ \int d^3x \int d^3y \exp\{-i\mathbf{q}\cdot(\mathbf{x}-\mathbf{y})\}\, \langle\Psi_0|\hat{\psi}^\dagger(\mathbf{y})\hat{\psi}^\dagger(\mathbf{x})\hat{\psi}(\mathbf{x})\hat{\psi}(\mathbf{y})|\Psi_0\rangle
\end{aligned}$$

The discussion can be focused on inelastic transitions by defining S^{in} through the restriction $\sum_{f\neq 0}$ in Eq. (23.19). This yields

$$S^{\text{in}}(\mathbf{q}^2) = S(\mathbf{q}^2) - \left|\langle\Psi_0| \int \exp(-i\mathbf{q}\cdot\mathbf{x})\,\hat{\rho}(\mathbf{x})\,d^3x\,|\Psi_0\rangle\right|^2 \qquad (23.25)$$

It follows that

$$\begin{aligned}
S^{\text{in}}(\mathbf{q}^2) &\equiv \int d\omega\, R^{\text{in}}(\mathbf{q}^2, \omega) \\
&= Z + \int\int d^3x\, d^3y \exp\{-i\mathbf{q}\cdot(\mathbf{x}-\mathbf{y})\}\, g(\mathbf{x},\mathbf{y}) \\
g(\mathbf{x},\mathbf{y}) &\equiv \langle\Psi_0|\hat{\psi}^\dagger(\mathbf{y})\hat{\psi}^\dagger(\mathbf{x})\hat{\psi}(\mathbf{x})\hat{\psi}(\mathbf{y})|\Psi_0\rangle \\
&\quad -\langle\Psi_0|\hat{\psi}^\dagger(\mathbf{x})\hat{\psi}(\mathbf{x})|\Psi_0\rangle\langle\Psi_0|\hat{\psi}^\dagger(\mathbf{y})\hat{\psi}(\mathbf{y})|\Psi_0\rangle \qquad (23.26)
\end{aligned}$$

One observes that $g(\mathbf{x},\mathbf{y})$ in the nuclear two-body charge density, and this Coulomb sum rule provides probably the only way, in principle, to un-ambiguously measure this *density–density correlation function* in nuclei. If the Fourier transform of the two-body density goes to zero as $|\mathbf{q}| \to \infty$, then

$$S^{\text{in}}(\mathbf{q}^2) \to Z \qquad ; |\mathbf{q}| \to \infty \qquad (23.27)$$

In this limit, in this traditional picture, the Coulomb sum rule simply counts the number of proton charges.

The Coulomb sum rule can be explicitly evaluated for a Fermi gas by integrating the previously derived response functions. First note that $d\xi = m \, d\omega / k_F^2$, and define $y \equiv \xi/\Delta$. Then

1) For $\Delta \geq 2$

$$S^{\text{in}}(\mathbf{q}^2) = \frac{3Z}{4} \int_{\Delta/2-1}^{\Delta/2+1} dy \left[1 - \left(y - \frac{\Delta}{2}\right)^2\right] \tag{23.28}$$

Change variables to $u \equiv y - \Delta/2$ and use $\int_{-1}^{1} du(1 - u^2) = 4/3$, thus

$$S^{\text{in}}(\mathbf{q}^2) = Z \tag{23.29}$$

2) For $\Delta \leq 2$

$$
\begin{aligned}
S^{\text{in}}(\mathbf{q}^2) &= \frac{3Z}{4} \left\{ \int_0^{1-\Delta/2} 2\Delta \, y \, dy + \int_{1-\Delta/2}^{1+\Delta/2} dy \left[1 - \left(y - \frac{\Delta}{2}\right)^2\right] \right\} \\
&= \frac{3Z}{4} \left[2\Delta \frac{1}{2} \left(1 - \frac{\Delta}{2}\right)^2 + \int_{1-\Delta}^{1} du(1 - u^2) \right] \\
&= \frac{3Z}{4} \left(\Delta - \frac{1}{12}\Delta^3\right)
\end{aligned}
\tag{23.30}
$$

Thus

$$S^{\text{in}}(\mathbf{q}^2) = Z \left(\frac{3}{4}\Delta - \frac{1}{16}\Delta^3\right) \tag{23.31}$$

In *summary*, the Coulomb sum rule defined by

$$C^{\text{in}}(q) \equiv \frac{1}{Z} S^{\text{in}}(q) \tag{23.32}$$

takes the following form for a non-relativistic Fermi gas

$$
\begin{aligned}
C^{\text{in}}(q) &= 1 && ; q \geq 2k_F \\
&= \frac{3}{2}\left(\frac{q}{2k_F}\right) - \frac{1}{2}\left(\frac{q}{2k_F}\right)^3 && ; q \leq 2k_F
\end{aligned}
\tag{23.33}
$$

This result is plotted as $C(Q)_{\text{NR}}$ in Fig. 23.7.[3]

---

[3] In the quantum field theory QHD-I, described by the lagrangian density of Eq. (21.1), the baryon field satisfies canonical anti-commutation relations; however, it contains both baryons and anti-baryons. The equations of motion imply that the local, effective current in Eq. (21.14) is conserved. A Coulomb sum rule can then be constructed in direct analogy with the derivation given in the text. The result obtained for nuclear

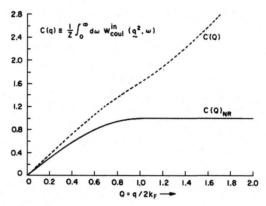

Fig. 23.7.   The Coulomb sum rule $C(Q)$ where $Q = q/2k_F$. The non-relativistic result (NR) is that of Eq. (23.33). Also shown is the RMFT result (see text).

Inspection of Eqs. (23.3) and (23.8) shows that in the non-interacting, non-relativistic Fermi gas, the Coulomb cross section can be written

$$\frac{d^2\sigma}{d\varepsilon_2 d\Omega_2} = \sigma_M \frac{q_\mu^4}{\mathbf{q}^4} |f_{SN}(q^2)|^2 R(\mathbf{q}^2, \omega) \tag{23.34}$$

$$R(\mathbf{q}^2, \omega) = \frac{3Z}{4\pi k_F^3} \int_0^{k_F} d^3k \, \theta(|\mathbf{k} - \mathbf{q}| - k_F) \delta(\omega + \frac{\mathbf{k} \cdot \mathbf{q}}{m} - \frac{\mathbf{q}^2}{2m})$$

Here we have used $\Omega = 3\pi^2 Z/k_F^3$. Now write the momentum integration region in Eq. (23.34) as $d^2k_\perp dk_\parallel$ where $\mathbf{k}_\parallel$ lies along $\mathbf{q}$, and assume that the momentum transfer $\mathbf{q}$ is large enough so that the θ function in the integrand is irrelevant. Then

$$\frac{d^2\sigma}{d\varepsilon_2 d\Omega_2} = \sigma_M \frac{q_\mu^4}{\mathbf{q}^4} |f_{SN}(q^2)|^2 \frac{m}{q} F(y) \qquad ; \mathbf{q} \to \infty$$

$$F(y) = \frac{3Z}{4\pi k_F^3} \int_0^{k_F} d^2k_\perp \, dk_\parallel \, \delta(k_\parallel - y)$$

$$y \equiv \frac{m\omega}{q} - \frac{q}{2} \tag{23.35}$$

The energy loss ω and momentum transfer $q \equiv |\mathbf{q}|$ enter the *scaling function* $F(y)$ only through the single *scaling variable* y.

---

matter, using RMFT to evaluate the two-body charge density, is shown in Fig. 23.7 [Wa83]. The Coulomb response amplitudes are more complicated in a full field theory. For example, there are other degrees of freedom that carry charge, included here in an empirical fashion in the additional anomalous magnetic moment term in the effective current [responsible for the rise in $C(\mathbf{q}^2)$]. It is also possible to produce real nucleon pairs in the time-like region [Ma83]. The nuclear Coulomb sum rule in such theories is examined in detail in [Fe94, Ko95].

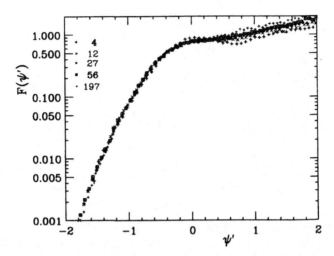

Fig. 23.8. Super-scaling analysis of Donnelly and Sick for nuclei from $A = 4-197$ [Do99]. The variable ψ', defined in that paper, is close to y/k_F.

Suppose one still has a non-interacting, non-relativistic Fermi gas but now, instead of the initial momentum distribution $(3/4\pi k_F^3)\theta(k_F - k)$, one has a more general (normalized) distribution $n(\mathbf{k}^2)$. One example would be a *thermal* Fermi distribution [Fe71]. Suppose also that the Pauli Principle is irrelevant for the final-state proton. The evident generalization of Eq. (23.35) is

$$
\begin{aligned}
F(y) &= \int d^2k_\perp \, dk_\parallel \, n(\mathbf{k}_\perp^2 + \mathbf{k}_\parallel^2)\, \delta(k_\parallel - y) \\
&= \int d^2k_\perp \, n(\mathbf{k}_\perp^2 + y^2) \\
y &\equiv \frac{m\omega}{q} - \frac{q}{2}
\end{aligned}
\tag{23.36}
$$

This result is known as *y-scaling*. It is a simple result of conservation of energy and momentum for a non-relativistic nucleon. To the extent that the nuclear density and Fermi momentum are unchanged as the size of the nucleus is increased, $F(y)$ should be a universal function independent of A. y-scaling is discussed in detail in the review article [Da90]. One of the most impressive applications of y-scaling is in the work of Donnelly and Sick shown in Fig. 23.8 [Do99], which reflects some of the extended relativistic analysis in appendix G.

The response of the relativistic Fermi gas is investigated in depth in [Mo69, Va78, Al88, Ce97]. Smith and Moniz [Sm72] have also calculated inclusive quasielastic scattering (e, e′) in a relativistic Fermi gas model of

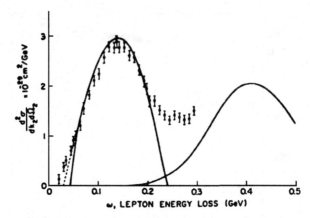

Fig. 23.9.   Relativistic calculation of quasielastic peak and $N^*(1232)$ production in electron scattering from Ni at $\varepsilon_1 = 500$ MeV, and $\theta = 60°$; here $\bar{\varepsilon} = 42$ MeV and $k_F = 271$ MeV [Sm72]. The experimental data are from Moniz *et al.* [Mo71]. The dashed line omits the Pauli principle in the final state.

the nucleus, including production of the $\Delta(1232)$ (see chapter 28). Their results for Ni are shown in Fig. 23.9.

A relativistic model which includes interactions in an average fashion is given by relativistic mean field theory (RMFT) discussed in chapter 21. Pollock has calculated the four response functions of Eq. (13.48) for coincident electron scattering $(e, e'N)$ for nuclear matter in RMFT [Po88]. He uses the current of Eq. (21.14), and his results are shown in Fig. 23.10. This is a very simple calculation, but it has the following features to recommend it:

- The RMFT provides a realistic model of nuclear matter [Wa95];

- The full nucleon vertex $\Gamma_\mu = F_1\gamma_\mu - F_2\sigma_{\mu\nu}k_\nu$ has been used; the current is conserved and gives the correct result for a free nucleon;

- The calculation is completely relativistic;

- The resulting response surfaces in Fig. 23.10 map out the complete Fermi sphere, weighted with the appropriate electromagnetic inter-action; one can examine any part of the Fermi sphere, including the deeply bound states, by looking at the appropriate region of the response surface. Correlations will modify the Fermi sphere and add a tail to the momentum distribution;

- The $(e, e' n)$ surfaces are also worth looking at [Po88].

Figures 21.2 and 21.3 show quasielastic data from HEPL on $^{40}_{20}$Ca(e, e')

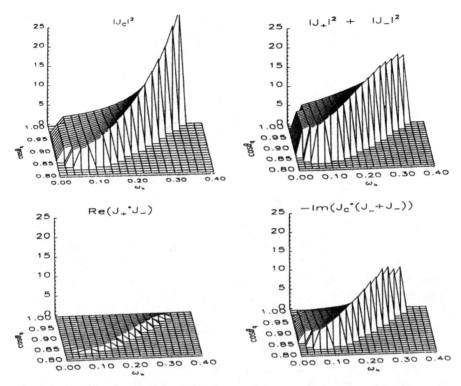

Fig. 23.10. The four proton response functions in RMFT obtained from the transition matrix elements of the current $\mathscr{J}_\mu$ and evaluated per proton are plotted as functions of energy loss and $\cos\theta_q$. Here $|\mathbf{k}| = 0.5\,\text{GeV}$ and $\phi_q = \pi/2$; the ϕ_q dependence is now in the response. Also $k_F = 0.28\,\text{GeV}$ and $m^*/m = 0.56$ (appropriate for nuclear matter). The vertical scale is $25.0\,\text{GeV}^{-1}$ for all four response functions, and ω_k is in GeV [Po88, Wa95].

and $^{208}_{82}\text{Pb}(e, e')$ compared with a calculation in RMFT [Ro80].[4] The calculation uses the relativistic densities for these nuclei, and the full, relativistic, conserved current; there are no free parameters. The position, shape, and magnitude of the peak are all well-described; it would appear that one had an understanding of nuclear quasielastic scattering. Nonetheless, the data contains both the transverse and Coulomb (longitudinal) response, and if one could isolate the Coulomb response, where the interaction is simply with the charges in the target, the understanding should be even better. Experimentalists have worked very hard to make the required Rosenbluth separation, and the result for $^{40}_{20}\text{Ca}$ is shown in Fig. 23.11. The experimental points are from Saclay; they represent the area under the Coulomb

---

[4] Quasielastic electron scattering for $^{40}_{20}\text{Ca}(e, e')$ is calculated in relativistic Hartree by summing over single-particle transitions, and including the RPA response, in [Ho89].

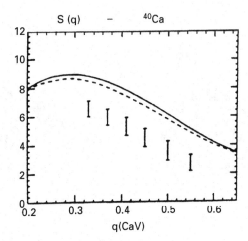

Fig. 23.11. Area under the Coulomb part of the quasielastic peak — the Coulomb sum rule — for $^{40}_{20}$Ca(e, e'). Data are from Saclay [Me84]. Theoretical curve is RMFT, with relativistic, conserved current and nucleon form factors left in [Po88, Wa95].

quasielastic peak — the Coulomb sum rule. The theoretical curve is the same RMFT calculation described above [Po88].[5] The disagreement is by almost a factor of 2 at the largest q. Several possible solutions have been proposed, including: a swelling of the nucleon in the nuclear medium [now pretty well ruled out by further (e, e') studies], modification of strong, hadronic vacuum polarization in the nuclear medium, RPA correlations, short-range correlations, and missing experimental strength.

Consider further one of these effects, the role of short-range correlations in the Coulomb sum rule. Recall from Eqs. (23.26) and (23.32) that the Coulomb sum rule, properly normalized[6] can be written [Vi77, Wa93]

$$C^{in}(q) = 1 + \tilde{\rho}^{(2)}_{pp}(q) \qquad (23.37)$$

The second term is the Fourier transform of the two-body density. Figure 23.12 [Vi77] shows the calculated quantity $|1 - C^{in}(q)|$ for infinite nuclear matter using (1) The Pauli correlations of a non-interacting Fermi gas; (2) A two-body density calculated from the Bethe–Goldstone wave function for a hard-core interaction [Fe71]; (3) A similar result with a more realistic two-body interaction. While approximately 10% correction from short-range correlations at the highest measured q above is conceivable; it is difficult to see how this could account for the factor of 2.

---

[5] The single-nucleon form factors have been left in this result.
[6] The normalization is $C^{in}(q) \to 0$ as $q \to 0$, and the single-nucleon form factor has been divided out.

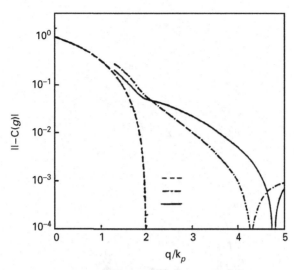

Fig. 23.12. Normalized Coulomb sum rule for nuclear matter including short-range correlations (see text) [Vi77] .

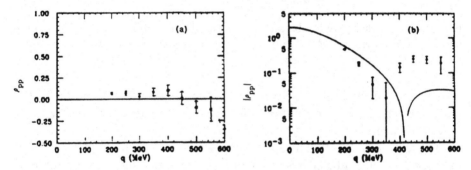

Fig. 23.13. Two-body density $\tilde{\rho}_{pp}^{(2)}(q)$ extracted from Coulomb sum rule in (a) $^3_1\text{H}(e,e')$ and (b) $^3_2\text{He}(e,e')$. The data are from Bates [Be90].

The two-body density is one of the fundamental quantities in many-particle physics. For example, the precision measurement of this quantity by inelastic neutron scattering in liquid ^4He provides the basis for much of our understanding of this quantum fluid. Despite the fact that the two-body density is the basic quantity used in the calculation of the binding energy of many-body nuclei, it had never been measured experimentally. It *has* now been measured, however, for one simple system.

A study of quasielastic scattering in both $^3_2\text{He}(e,e')$ and $^3_1\text{H}(e,e')$ has been carried out at Bates. Figure 23.13 shows the two-body density extracted from the Coulomb sum rule in these two nuclei [Be90]. This determination has a very nice self-calibration, for the two-body proton

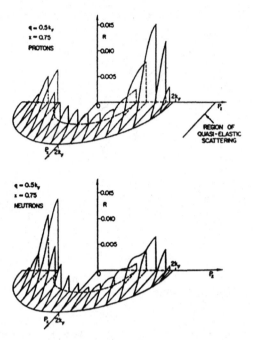

Fig. 23.14.   Nuclear Coulomb response (dimensionless units) for n.m.(e, e′ N) for hard-core Fermi gas to order $(k_F a)^2$. x is dimensionless energy transfer. Kinematics arranged so there is no quasi-elastic (e, e′ p) [de66a, de67, Wa93].

density must *vanish* in $^3_1$H, as it does. The result for $^3_2$He provides the first experimental determination of the two-body nuclear density — a significant achievement.

The resolution of the disagreement with the Coulomb sum rule in medium to heavy nuclei has been one of the most significant problems in nuclear physics [Wa95]. Jourdan has made a very important contribution here [Jo96]. By combining all the available data from the world's laboratories, he shows that it is possible to obtain a longer lever arm on the Rosenbluth plots, separating the dominant transverse scattering, which determines the slope, from the much smaller Coulomb scattering, determined from the extrapolated intercept. After an extensive analysis, he finds a ratio of experimental to theoretical value of the Coulomb sum rule of 0.97 ± 0.12 in $^{56}_{26}$Fe at $q = 570\,\text{MeV/c}$.[7]

It is interesting to investigate other effects on the electromagnetic response surfaces produced by short-range correlations. Figure 23.14 shows the Coulomb response for the reaction n.m.(e, e′ N) on a nuclear matter (n.m.) Fermi gas with hard-core interactions [Fe71]. The figure is from

---

[7] The extensive set of current experimental data on separated longitudinal and transverse quasielastic response surfaces for $^{40}_{20}$Ca is discussed in [Wi97].

Taber deForest's thesis [de66a] — done over 30 years ago in connection with the proposed program for the SCA at HEPL. The quantities $(\mathbf{q}^2, \omega)$ are arranged to be outside the allowed region for quasielastic scattering for the Fermi gas — see above and [Fe71]; thus the process can only proceed while two nucleons are in virtual collision in the nucleus.[8] The calculation of the response is *exact* to order $(k_F a)^2$ (all graphs creating 2p–2h states are retained). In Fig. 23.14 the z-axis lies along $\mathbf{q}$. While the resulting proton distribution may not look so dramatic, the backward peaking of the neutrons in the n.m.(e, e$'$ n) Coulomb response is quite spectacular.

---

[8] For quasielastic scattering, the indicated energy conservation region would have to intersect the Fermi sphere; thus these results lie on the high-energy-loss side of the quasielastic peak.

24

The quark model

The first understanding of the underlying structure and "periodic table" of the hadrons came from the quark model of Gell-Mann and Zweig [Ge64, Zw65]. We now know that quarks form the underlying fermionic degrees of freedom for QCD and a field theory of the strong interactions. Solution to the dynamics of strong-coupling QCD presents formidable problems. It is often useful to make simple dynamical models that emphasize one or another aspect of QCD and that provide physical insight and guidance for further work [Bh88, Wa95]. Models build on three features of QCD:

- Baryons have the *quantum numbers* of (qqq) systems and mesons of $(\bar{q}q)$ systems where the flavor quantum numbers of the quarks q are given in Table 24.1;

- Color and the strong color forces are *confined* to the interior of the hadrons. Quarks come in three colors (R, G, B). Lattice gauge theory calculations indicate that confinement arises from the strong nonlinear couplings of the gauge fields at large distances;

- QCD is *asymptotically free*; at short distances the renormalized coupling constant goes to zero. One can do perturbation theory at short distances.

One approach to model building is that of the M.I.T. bag which provides an extreme picture of each of the three items listed above [Ch74, Ch74a, De75, Ja76]. For baryons, three massless non-interacting quarks (correct quantum numbers), with the one-gluon-exchange interaction treated as a perturbation (asymptotic freedom), are placed inside a vacuum *bubble* of radius R (confinement). It is assumed that it takes a positive amount of internal energy density to create this bubble in the vacuum. The Dirac equation is then solved within this scalar bubble, wave functions for the

Table 24.1. Flavor quantum numbers of the lightest quarks: isospin, third component of isospin, baryon number, strangeness, charm, and electric charge, respectively.

Quark/field	T	T_3	B	S	C	$Q = T_3 + (B + S + C)/2$
u	1/2	1/2	1/3	0	0	2/3
d	1/2	−1/2	1/3	0	0	−1/3
s	0	0	1/3	−1	0	−1/3
c	0	0	1/3	0	1	2/3

nucleon are constructed, and its properties calculated. The M.I.T. bag model is discussed in detail in [Wa95].

Another approach is the non-relativistic quark model [Bh88] whose most extensive application is due to Isgur and Karl [Is77, Is80, Is85]. Here "constituent quarks" with masses of $\approx m/3$ move non-relativistically in a confining potential.[1] The confining potential is most simply taken to be that of a harmonic oscillator, which has the distinct advantage that the center-of-mass motion of the three-quark system can be treated exactly (appendix B).

Let us confine the discussion to the nuclear domain where only the lightest (u, d) quarks and their antiquarks are retained. The quark field is thus approximated by

$$\psi \doteq \begin{pmatrix} u \\ d \end{pmatrix} \qquad ; \text{ nuclear domain} \qquad (24.1)$$

To do a calculation one needs the (qqq) wave functions, including all the quantum numbers. We make an independent-quark shell model of hadrons and start with the simple case of non-relativistic quarks in a potential (where the spin and spatial wave functions decouple). In this case one can write the one-quark wave function as

$$\psi = \underbrace{\psi_{nlm_l}(\mathbf{r})}_{\text{space}} \underbrace{\chi_{m_s}}_{\text{spin}} \underbrace{\eta_{m_t}}_{\text{isospin}} \underbrace{\rho_\alpha}_{\text{color}} ; \qquad \begin{matrix} m_s = & \pm 1/2 \\ m_t = & \pm 1/2 \\ \alpha = & (R, G, B) \end{matrix} \qquad (24.2)$$

Consider the *color wave function* for the (qqq) system. The observed hadrons are color singlets. Hence the color wave function in this case is just the completely antisymmetric combination (a Slater determinant with respect to color)

$$\Psi_{\text{color}}(1, 2, 3) = \frac{1}{\sqrt{6}} \begin{vmatrix} \rho_R(1) & \rho_G(1) & \rho_B(1) \\ \rho_R(2) & \rho_G(2) & \rho_B(2) \\ \rho_R(3) & \rho_G(3) & \rho_B(3) \end{vmatrix} \qquad ; \text{ antisymmetric} \quad (24.3)$$

---

[1] The masses of these constituent quarks are presumably generated by spontaneously broken chiral symmetry in QCD.

If G_α^{color} with $\alpha = 1, \ldots, 8$ are the generators of the color transformation among the quarks, then all of the generators annihilate this wave function[2]

$$G_\alpha^{\text{color}} \Psi_{\text{color}} = 0 \qquad\qquad ; \alpha = 1, \ldots, 8 \qquad (24.4)$$

Since the total wave function must be antisymmetric in the interchange of any two fermions, the remaining space-spin-isospin wave function must be *symmetric*.

For the ground state in this shell model, the spatial wave functions $\psi_{n00}(\mathbf{r})$ will all be the same, all $1s$, and hence the spatial part of the wave function is totally symmetric

$$\Psi_{\text{space}}(1, 2, 3) = \psi_{1s}(\mathbf{r}_1)\psi_{1s}(\mathbf{r}_2)\psi_{1s}(\mathbf{r}_3) \qquad ; \text{symmetric} \qquad (24.5)$$

The spin-isospin wave function must thus be *totally symmetric*. Start with isospin. One is faced with the problem of coupling three angular momenta; however, the procedure follows immediately from the discussion of 6-j symbols in quantum mechanics [Ed74]. An eigenstate of total angular momentum can be formed as follows

$$|(j_1 j_2)j_{12}j_3 jm\rangle = \sum_{m_1 m_2 m_3 m_{12}} \langle j_1 m_1 j_2 m_2 | j_1 j_2 j_{12} m_{12}\rangle \qquad (24.6)$$
$$\times \langle j_{12}m_{12}j_3 m_3 | j_{12}j_3 jm\rangle |j_1 m_1\rangle |j_2 m_2\rangle |j_3 m_3\rangle$$

These states form a complete orthonormal basis for given (j_1, j_2, j_3). The states formed by coupling in the other order $|j_1(j_2 j_3)j_{23}jm\rangle$ are linear combinations of these with 6-j symbols as coefficients.

For *isospin* in the nuclear domain all the $t_i = 1/2$, thus there are a total of $2 \times 2 \times 2 = 8$ basis states. Consider first the states with total $T = 3/2$. Here the only possible intermediate value is $t_{12} = 1$. The state with $T_3 = 3/2$ is readily constructed from the above as $\alpha(1)\alpha(2)\alpha(3)$. Now apply the total lowering operator $T_- = t(1)_- + t(2)_- + t(3)_-$ and use $t_-\alpha = \beta$, $t_-\beta = 0$. The set of states with $T = 3/2$ follows immediately

$$\Phi\left[\left(\frac{1}{2}\frac{1}{2}\right)1\frac{1}{2}\frac{3}{2}\frac{3}{2}\right] = \alpha(1)\alpha(2)\alpha(3)$$

$$\Phi\left[\left(\frac{1}{2}\frac{1}{2}\right)1\frac{1}{2}\frac{3}{2}\frac{1}{2}\right] = \frac{1}{\sqrt{3}}[\beta(1)\alpha(2)\alpha(3) + \alpha(1)\beta(2)\alpha(3) + \alpha(1)\alpha(2)\beta(3)]$$

$$\Phi\left[\left(\frac{1}{2}\frac{1}{2}\right)1\frac{1}{2}\frac{3}{2} - \frac{1}{2}\right] = \frac{1}{\sqrt{3}}[\beta(1)\beta(2)\alpha(3) + \beta(1)\alpha(2)\beta(3) + \alpha(1)\beta(2)\beta(3)]$$

$$\Phi\left[\left(\frac{1}{2}\frac{1}{2}\right)1\frac{1}{2}\frac{3}{2} - \frac{3}{2}\right] = \beta(1)\beta(2)\beta(3) \qquad ; \text{4 symmetric states} \qquad (24.7)$$

There are four symmetric states with $T = 3/2$.

---

[2] Just as the fully occupied Slater determinant of spins has $S = 0$, or of j-shells has $J = 0$.

Consider next the states with total $T = 1/2$. Here there are two possible intermediate values in the above, $t_{12} = 0, 1$. For the first of these values one finds

$$\Phi^\rho \left[\left(\frac{1}{2} \frac{1}{2} \right) 0 \frac{1}{2} \frac{1}{2} \frac{1}{2} \right] = \frac{1}{\sqrt{2}} \left[\alpha(1)\beta(2) - \alpha(2)\beta(1) \right] \alpha(3) \qquad (24.8)$$

$$\Phi^\rho \left[\left(\frac{1}{2} \frac{1}{2} \right) 0 \frac{1}{2} \frac{1}{2} - \frac{1}{2} \right] = \frac{1}{\sqrt{2}} \left[\alpha(1)\beta(2) - \alpha(2)\beta(1) \right] \beta(3) \quad ; \text{2 states}$$

These two states have *mixed symmetry*; they are antisymmetric in the interchange of particles $(1 \leftrightarrow 2)$.

The second value $t_{12} = 1$ yields

$$\Phi^\lambda \left[\left(\frac{1}{2} \frac{1}{2} \right) 1 \frac{1}{2} \frac{1}{2} \frac{1}{2} \right] = \frac{1}{\sqrt{6}} [2\alpha(1)\alpha(2)\beta(3) - \alpha(1)\beta(2)\alpha(3) - \beta(1)\alpha(2)\alpha(3)]$$

$$\Phi^\lambda \left[\left(\frac{1}{2} \frac{1}{2} \right) 1 \frac{1}{2} \frac{1}{2} - \frac{1}{2} \right] = \qquad\qquad (24.9)$$

$$-\frac{1}{\sqrt{6}} [2\beta(1)\beta(2)\alpha(3) - \beta(1)\alpha(2)\beta(3) - \alpha(1)\beta(2)\beta(3)] \qquad ; \text{2 states}$$

These two states also have mixed symmetry; they are symmetric in the interchange of particles $(1 \leftrightarrow 2)$.

Now look at the *spin* wave functions. The analysis is exactly the same! We have a set of spin states Ξ identical to those above.

For the overall spin-isospin wave function, we must take a product of these wave functions and make the result totally symmetric. Recall first from quantum mechanics how one makes a wave function totally antisymmetric. Introduce the antisymmetrizing operator

$$\mathscr{A} = N \sum_{(P)} (-1)^P P \qquad\qquad (24.10)$$

Here the sum goes over all permutations, produced by the operator P, of a complete set of coordinates for each particle. The signature of the permutation is $(-1)^P$, and $N = 1/\sqrt{N_P}$ where N_P is the total number of permutations.

Similarly, to make a wave function totally symmetric introduce the (unnormalized) *symmetrizing operator*

$$\mathscr{S} = N \sum_{(P)} P \qquad\qquad (24.11)$$

Note that if a wave function is antisymmetric under the interchange of any two particles, the application of $\mathscr{S}$ will give zero. This result is established as follows. Use

$$P_{12}\mathscr{S} = \mathscr{S}P_{12} \qquad\qquad (24.12)$$

Table 24.2. Totally symmetric spin-isospin states for three non-relativistic quarks.

T	S	Number of states
3/2	3/2	16
1/2	1/2	4
		20

This follows since as P goes over all permutations, so does $P_{12}P$ or PP_{12}

$$\sum_{(P)} P_{12}P = \sum_{(P)} P = \sum_{(P)} PP_{12} \tag{24.13}$$

It follows that

$$P_{12}\mathscr{S}\psi = \mathscr{S}\psi = \mathscr{S}P_{12}\psi = -\mathscr{S}\psi = 0 \tag{24.14}$$

This is the stated result.

Note further that if the operator $\mathscr{S}$ is applied to the product of the totally symmetric 3/2 state and either of the 1/2 states with mixed symmetry, the result will vanish. The proof is as follows. Since $\mathscr{S}\Phi_{3/2} = \Phi_{3/2}\mathscr{S}$, one just needs to show that

$$\mathscr{S}[A\Phi^\rho + B\Phi^\lambda] = 0 \tag{24.15}$$

The first term gives zero since Φ^ρ is antisymmetric in the interchange of the first pair of particles. The second vanishes because of the nature of the sums in Eqs. (24.9) and the fact that $\mathscr{S}$ produces an identical result when applied to each term in the sum

$$\mathscr{S}(\alpha\alpha\beta) = \mathscr{S}(\alpha\beta\alpha) = \mathscr{S}(\beta\alpha\alpha) \tag{24.16}$$

It is a consequence of these two observations that *the only non-zero totally symmetric wave function will be obtained by combining the spin and isospin wave functions of the same symmetry*. Thus one must combine the two totally symmetric spin and isospin states and the other two pairs of states with the same mixed symmetry; in the latter case there is only one totally symmetric linear combination (this is proven in appendix J of [Wa95]). This leads to the set of totally symmetric spin-isospin states shown in Table 24.2 and given by

$$\Phi_{\frac{3}{2}m_t}\Xi_{\frac{3}{2}m_s}$$

$$\frac{1}{\sqrt{2}}\left(\Phi^\lambda_{\frac{1}{2}m_t}\Xi^\lambda_{\frac{1}{2}m_s} + \Phi^\rho_{\frac{1}{2}m_t}\Xi^\rho_{\frac{1}{2}m_s}\right) \tag{24.17}$$

These are all the baryons one can make in this model. Since all these states are degenerate in the model as presently formulated, one has a *supermultiplet* of baryons. The present calculation predicts the spins and isospins of the members of this supermultiplet.[3]

These arguments can be extended to the situation in the M.I.T. bag model where, in contrast to massive, non-relativistic constituents, one has massless relativistic quarks. The problem is more complicated since the space–spin parts of the wave functions are now coupled; however, if the quarks occupy a common lowest positive energy $\psi_{1s_{1/2}m_j}(\mathbf{r})$ ground state, the problem is greatly simplified. Make the following replacement in the space–spin wave functions discussed above

$$\psi_{1s}(\mathbf{r})\chi_{m_s} \rightarrow \psi_{1s_{1/2}m_j}(\mathbf{r}) \tag{24.18}$$

Instead of the spin $\mathbf{S}$, now talk about the total angular momentum $\mathbf{J}$; the angular momentum and symmetry arguments are then *exactly the same as before*.

Let us investigate some consequences of the quark model. Consider the nucleon (N) ground-state expectation value of the following operator

$$O = \sum_{i=1}^{3} O_i(\mathbf{r}_i, \sigma_i)I_i(\tau_i) \tag{24.19}$$

Assume that the isospin factor is diagonal $I_i = (1, \tau_3)_i$. Since the wave function is totally symmetric, it follows that one need evaluate the matrix element only for the third particle.[4]

$$\langle \Psi_N | \sum_{i=1}^{3} O_i I_i | \Psi_N \rangle = 3\langle \Psi_N | O_3 I_3 | \Psi_N \rangle \tag{24.20}$$

Substitution of Eq. (24.17) then yields[5] for the state of total $m_j = 1/2$

$$3\langle \Psi_N | O_3 I_3 | \Psi_N \rangle = \frac{3}{2}\langle \Phi^\rho | I_3 | \Phi^\rho \rangle \langle \tfrac{1}{2}(3) | O_3 | \tfrac{1}{2}(3) \rangle$$
$$+ \frac{3}{2}\langle \Phi^\lambda | I_3 | \Phi^\lambda \rangle \frac{1}{6} \left\{ 4\langle -\tfrac{1}{2}(3) | O_3 | -\tfrac{1}{2}(3) \rangle + 2\langle \tfrac{1}{2}(3) | O_3 | \tfrac{1}{2}(3) \rangle \right\} \tag{24.21}$$

---

[3] Define $\zeta_i \equiv \chi_{m_s}\eta_{m_t}$ with $(m_s, m_t) = (\pm 1/2, \pm 1/2)$. Then in a non-relativistic quark model with spin-independent interactions one has an internal global $SU(4)$ (flavor) symmetry — this is just Wigner's supermultiplet theory [Wi37]. Here the baryons belong to the totally symmetric irreducible representation one gets from $4 \otimes 4 \otimes 4$; this is the [20] dimensional representation with spin-isospin content worked out in the text and shown in Table 24.2.

[4] Assume the operators form the identity with respect to color; the color wave function then goes right through the matrix element, and it is normalized.

[5] Use $\langle \Phi^\rho | I_3 | \Phi^\lambda \rangle = 0$ if I_3 is diagonal; this follows immediately from the form of Eqs. (24.8) and the orthogonality of the mixed-symmetry wave functions.

Here the remaining labels on the single-particle matrix elements of O_3 are $|m_j, (\text{particle number})\rangle$. The result is

$$\langle\Psi^N_{m_t\frac{1}{2}}|\sum_{i=1}^{3}O_iI_i|\Psi^N_{m_t\frac{1}{2}}\rangle = \langle\frac{1}{2}|O|\frac{1}{2}\rangle\left[\frac{3}{2}\langle\Phi^\rho|I_3|\Phi^\rho\rangle + \frac{1}{2}\langle\Phi^\lambda|I_3|\Phi^\lambda\rangle\right]$$
$$+\langle-\frac{1}{2}|O|-\frac{1}{2}\rangle\left[\langle\Phi^\lambda|I_3|\Phi^\lambda\rangle\right] \qquad (24.22)$$

This result is for total $m_j = 1/2$; the remaining isospin operator I_3 acts only on the third particle. For an *isoscalar* operator with $I_3 = 1$ this expression reduces to

$$\langle\Psi^N_{m_t\frac{1}{2}}|\sum_{i=1}^{3}O_i|\Psi^N_{m_t\frac{1}{2}}\rangle = 2\langle\frac{1}{2}|O|\frac{1}{2}\rangle + \langle-\frac{1}{2}|O|-\frac{1}{2}\rangle \qquad (24.23)$$

This is now just a sum of single-particle matrix elements. For an *isovector* operator with $I_3 = \tau_3$, the required isospin matrix elements for the proton with $m_t = 1/2$ follow from Eqs. (24.8) and (24.9)

$$\langle\Phi^\rho|\tau_3(3)|\Phi^\rho\rangle = 1 \qquad (24.24)$$
$$\langle\Phi^\lambda|\tau_3(3)|\Phi^\lambda\rangle = \frac{1}{6}(-4+1+1) = -\frac{1}{3} \qquad ; \text{ proton } m_t = \frac{1}{2}$$

For a neutron with $m_t = -1/2$, these isovector matrix elements simply change sign. It follows that

$$\langle\Psi^N_{\frac{1}{2}\frac{1}{2}}|\sum_{i=1}^{3}O_i\tau_3(i)|\Psi^N_{\frac{1}{2}\frac{1}{2}}\rangle = \frac{4}{3}\langle\frac{1}{2}|O|\frac{1}{2}\rangle - \frac{1}{3}\langle-\frac{1}{2}|O|-\frac{1}{2}\rangle$$
$$\langle\Psi^N_{-\frac{1}{2}\frac{1}{2}}|\sum_{i=1}^{3}O_i\tau_3(i)|\Psi^N_{-\frac{1}{2}\frac{1}{2}}\rangle = -\frac{4}{3}\langle\frac{1}{2}|O|\frac{1}{2}\rangle + \frac{1}{3}\langle-\frac{1}{2}|O|-\frac{1}{2}\rangle \quad (24.25)$$

The notation here is $\Psi^N_{m_t,m_j}$.

In the nuclear domain with only (u, d) quarks the electric charge is given by

$$e_i = \left[\frac{1}{6} + \frac{1}{2}\tau_3(i)\right]e_p \qquad (24.26)$$

Hence the expectation value of an operator proportional to the charge in the composite three-quark proton and neutron ground state is given by

$$\langle p| \sum_{i=1}^{3} O_i e_i |p\rangle = e_p \left[\frac{1}{6}(2O_{1/2} + O_{-1/2}) + \frac{1}{2}(\frac{4}{3}O_{1/2} - \frac{1}{3}O_{-1/2}) \right]$$

$$= e_p \langle \frac{1}{2}|O|\frac{1}{2}\rangle$$

$$\langle n| \sum_{i=1}^{3} O_i e_i |n\rangle = e_p \left[\frac{1}{6}(2O_{1/2} + O_{-1/2}) + \frac{1}{2}(-\frac{4}{3}O_{1/2} + \frac{1}{3}O_{-1/2}) \right]$$

$$= -\frac{e_p}{3}\langle \frac{1}{2}|O|\frac{1}{2}\rangle + \frac{e_p}{3}\langle -\frac{1}{2}|O| -\frac{1}{2}\rangle \qquad (24.27)$$

Let us apply this result to compute the magnetic moment of the ground state of the nucleon in the non-relativistic quark model using for the expectation value of the single quark matrix element the Dirac magnetic moment of a point quark of mass m_q

$$\langle \frac{1}{2}|O|\frac{1}{2}\rangle = \frac{1}{2m_q} \qquad (24.28)$$

Since the magnetic moment is a vector operator, its expectation value in the state $m_j = -1/2$ must simply change sign $\langle -\frac{1}{2}|O| -\frac{1}{2}\rangle = -1/2m_q$. This yields

$$\mu_p = \frac{e_p}{2m_q} \qquad \mu_n = -\frac{2\mu_p}{3} \qquad (24.29)$$

The experimental results are

$$\mu_p = +2.79 \text{ n.m.} \qquad \mu_n = -1.91 \text{ n.m.} \qquad (24.30)$$

The calculated ratio is quite impressive, and the absolute value can be fitted in the first relation with a constituent quark mass of $m_q = m/2.79$, which is certainly in the right ballpark.

Suppose that instead of just the static magnetic moment, one wanted the matrix element of the transverse magnetic dipole operator at all momentum transfer in the constituent quark model, how would the calculation change? From Eq. (9.16) one has

$$\hat{T}_{1M}^{\text{mag}}(\kappa) = \int d^3x \left\{ j_1(\kappa x)\mathscr{Y}_{111}^{M} \cdot \hat{\mathbf{J}}_c(\mathbf{x}) + [\nabla \times j_1(\kappa x)\mathscr{Y}_{111}^{M}] \cdot \hat{\boldsymbol{\mu}}(\mathbf{x}) \right\} \qquad (24.31)$$

There is no convection current in a $1s$ state, so the first term does not contribute. For the second term use the general relation [Ed74]

$$\nabla \times [j_J(\kappa x)\mathscr{Y}_{JJ1}^M] = -i\kappa \left[j_{J+1}(\kappa x)\left(\frac{J}{2J+1}\right)^{1/2} \mathscr{Y}_{J,J+1,1}^M \right.$$
$$\left. - j_{J-1}(\kappa x)\left(\frac{J+1}{2J+1}\right)^{1/2} \mathscr{Y}_{J,J-1,1}^M \right] \quad (24.32)$$

Since there is no orbital angular momentum in the initial and final states, the first term does not contribute; retention of just the second leads to

$$\hat{T}_{1M}^{\text{mag}}(\kappa) \doteq i\kappa \left(\frac{2}{3}\right)^{1/2} \int d^3x \, j_0(\kappa x)\mathscr{Y}_{101}^M \cdot \hat{\boldsymbol{\mu}}(\mathbf{x})$$
$$= i\kappa \left(\frac{1}{6\pi}\right)^{1/2} \int d^3x \, j_0(\kappa x)\hat{\boldsymbol{\mu}}(\mathbf{x})_{1M} \quad (24.33)$$

The spatial distribution of the magnetization is that of a 1s harmonic oscillator wave function, and from the discussion is chapter 20 we know that

$$\langle 1s|j_0(\kappa x)|1s\rangle = e^{-y} \qquad ; \ y = \left(\frac{\kappa \, b_{\text{osc}}}{2}\right)^2 \quad (24.34)$$

Hence, for the nucleon

$$e_p\langle N\frac{1}{2}|\hat{T}_{10}^{\text{mag}}(\kappa x)|N\frac{1}{2}\rangle = i\left(\frac{1}{6\pi}\right)^{1/2} \kappa \, \mu_N \, e^{-y} \quad (24.35)$$

The C-M motion for particles in a harmonic oscillator is now treated as in appendix B.

Consider the transition magnetic dipole moment between the ground state (N) and the excited state (Δ) formed from the product of the totally symmetric isospin state and totally symmetric space–spin state. Since only different m_j states are involved in the latter, we are in a position to calculate this matrix element. The wave functions are given by

$$\Psi_{\frac{1}{2}\frac{1}{2}}^N = \frac{1}{\sqrt{2}} \left[\Phi_{\frac{1}{2}\frac{1}{2}}^\lambda \Xi_{\frac{1}{2}\frac{1}{2}}^\lambda + \Phi_{\frac{1}{2}\frac{1}{2}}^\rho \Xi_{\frac{1}{2}\frac{1}{2}}^\rho \right] \quad (24.36)$$
$$\Psi_{\frac{1}{2}\frac{1}{2}}^\Delta = \Phi_{\frac{3}{2}\frac{1}{2}} \Xi_{\frac{3}{2}\frac{1}{2}}$$

The subscripts on the left are (m_t, m_j) and those of the right (Tm_t, Jm_j); in detail, these wave functions are

$$\Phi_{\frac{3}{2}\frac{1}{2}} = \quad (24.37)$$
$$\frac{1}{\sqrt{3}} \left[\phi_{-\frac{1}{2}}(1)\phi_{\frac{1}{2}}(2)\phi_{\frac{1}{2}}(3) + \phi_{\frac{1}{2}}(1)\phi_{-\frac{1}{2}}(2)\phi_{\frac{1}{2}}(3) + \phi_{\frac{1}{2}}(1)\phi_{\frac{1}{2}}(2)\phi_{-\frac{1}{2}}(3) \right]$$

A similar expression holds for $\Xi_{\frac{3}{2}\frac{1}{2}}$. The transition magnetic dipole moment is now given by

$$\mu^* = \langle \Psi^\Delta_{\frac{1}{2}\frac{1}{2}} | \sum_{i=1}^{3} \mu(i) \frac{1}{2} \tau_3(i) e_p | \Psi^N_{\frac{1}{2}\frac{1}{2}} \rangle = \frac{3}{2} e_p \langle \Psi^\Delta_{\frac{1}{2}\frac{1}{2}} | \mu(3) \tau_3(3) | \Psi^N_{\frac{1}{2}\frac{1}{2}} \rangle \quad (24.38)$$

Here it has been observed that only the isovector part of the magnetic dipole operator can contribute to the transition and the total symmetry of the states has been used. It now follows from Eq. (24.37) and the previous results that

$$\langle \Phi_{\frac{3}{2}\frac{1}{2}} | \tau_3(3) | \Phi^\rho_{\frac{1}{2}\frac{1}{2}} \rangle = 0 \quad (24.39)$$

$$\langle \Phi_{\frac{3}{2}\frac{1}{2}} | \tau_3(3) | \Phi^\lambda_{\frac{1}{2}\frac{1}{2}} \rangle = \frac{1}{\sqrt{18}} \left[2\langle -\frac{1}{2} | \tau_3 | -\frac{1}{2} \rangle - 2\langle \frac{1}{2} | \tau_3 | \frac{1}{2} \rangle \right] = -\frac{4}{\sqrt{18}}$$

$$\langle \Xi_{\frac{3}{2}\frac{1}{2}} | \mu(3) | \Xi^\lambda_{\frac{1}{2}\frac{1}{2}} \rangle = \frac{1}{\sqrt{18}} \left[2\langle -\frac{1}{2} | \mu | -\frac{1}{2} \rangle - 2\langle \frac{1}{2} | \mu | \frac{1}{2} \rangle \right] = -\frac{4}{\sqrt{18}} \langle \frac{1}{2} | \mu | \frac{1}{2} \rangle$$

Use of Eqs. (24.27) allows the final result for μ^* to be expressed in terms of the ground-state magnetic moment of the proton

$$\mu^* = \frac{3}{2} \frac{1}{\sqrt{2}} \frac{16}{18} \mu_p = \frac{4}{3\sqrt{2}} \mu_p \quad (24.40)$$

This is the matrix element for $(m_j, m_t) = (\frac{1}{2}\frac{1}{2}) \rightarrow (\frac{1}{2}\frac{1}{2})$; other components follow from the Wigner–Eckart theorem. This result agrees to about 30% with experimental observations of the transition magnetic dipole matrix element obtained from electroproduction of the first nucleon resonance [Ka83].

Since only the spin is flipped in the constituent quark model, and the radial $1s$ wave functions are unchanged in the $N \rightarrow \Delta$ transition, one can simply read off from Eq. (24.35) that the transition matrix element of the transverse magnetic dipole operator is given by

$$e_p \langle \Delta^+ \frac{1}{2} | \hat{T}^{\text{mag}}_{10}(\kappa x) | p\frac{1}{2} \rangle = i \left(\frac{1}{6\pi} \right)^{1/2} \kappa \left(\frac{4}{3\sqrt{2}} \mu_p \right) e^{-y}$$

$$= i \frac{2}{3\sqrt{3\pi}} \kappa \mu_p e^{-y} \quad (24.41)$$

Particularly simple is then the *ratio* of the transition to the static matrix elements of the transverse magnetic dipole operator

$$\frac{\langle \Delta^+ \frac{1}{2} | \hat{T}^{\text{mag}}_{10}(\kappa x) | p\frac{1}{2} \rangle}{\langle p\frac{1}{2} | \hat{T}^{\text{mag}}_{10}(\kappa x) | p\frac{1}{2} \rangle} = \frac{2\sqrt{2}}{3} \quad (24.42)$$

Note that this ratio is a numerical constant independent of κ in the constituent quark model. This result is also independent of the detailed form of the single-quark wave function since the form factor cancels in the ratio.[6]

The Coulomb monopole moment for the proton simply reflects the $1s$ radial wave function of each quark, and, as in chapter 20, the elastic scattering form factor for the proton is given by

$$\langle p\tfrac{1}{2}|M_{00}(\kappa x)|p\tfrac{1}{2}\rangle = \frac{1}{\sqrt{4\pi}}\,e^{-y} \qquad (24.43)$$

The transition magnetic dipole form factor is thus proportional to the elastic form factor of the proton in this model.

$$\frac{e_p\langle\Delta^+\tfrac{1}{2}|\hat{T}_{10}^{\mathrm{mag}}(\kappa x)|p\tfrac{1}{2}\rangle}{\langle p\tfrac{1}{2}|M_{00}(\kappa x)|p\tfrac{1}{2}\rangle} = i\frac{4}{3\sqrt{3}}\kappa\mu_p \qquad (24.44)$$

This result is again independent of the form of the single-quark radial wave functions since the form factor cancels in this ratio. Since there is no orbital angular momentum in either the ground or excited state, the transition matrix elements of the Coulomb and transverse electric quadrupole operators vanish here.

To the extent that the cross section is dominated by the transverse interaction and $q_\mu^2 \approx \mathbf{q}^2 \equiv \kappa^2$, the constancy of the ratio in Eq. (24.42) is indeed manifest by the experimental data shown in Fig. 12.9. Of course, the experimental elastic form factor itself falls off as a dipole [Eq. (22.5)] and not the gaussian of the simple-harmonic oscillator model, and it is certainly inconsistent to use a non-relativistic model for $\kappa \geq m_q$.

The $N \to \Delta$ transition is particularly simple in the constituent quark model. Higher excitations of the nucleon can be constructed by promoting one of the quarks to a higher oscillator state and then constructing totally symmetric space-spin-isospin wave functions for the nucleon. Similarly, the hyperfine splitting coming from (asymptotically-free) one-gluon exchange can be readily included in the model. We refer the reader to the literature for these developments [Is77, Bh88].

---

[6] The present treatment of the C-M motion, however, only holds in the simple harmonic oscillator model (appendix B).

25

Quantum chromodynamics

The primary evidence that hadrons are composed of a simpler substructure of *quarks* is the following:

- If one assumes the baryons are composed of quark triplets (qqq) and the mesons are quark–antiquark pairs ($q\bar{q}$) then, with appropriate quantum numbers for the quarks (flavors), one can describe and predict the observed supermultiplets of hadrons;

- The assumption of interaction with point-like quarks provides a marvelously simple and accurate description of electroweak currents;

- Dynamic evidence for a point-like quark–parton substructure of hadrons is obtained from deep-inelastic electron scattering (e, e') and neutrino reactions (v_l, l^-).

Quarks come in many *flavors*; the quark field can be written as

$$\psi = \begin{pmatrix} u \\ d \\ s \\ c \\ \vdots \end{pmatrix} \tag{25.1}$$

One assigns quarks an additional intrinsic degree of freedom called *color*, which takes three values $i = R, G, B$. The quark field then becomes (we focus here on the four lightest quarks)

$$\psi = \begin{pmatrix} u_R & u_G & u_B \\ d_R & d_G & d_B \\ s_R & s_G & s_B \\ c_R & c_G & c_B \end{pmatrix} = (\psi_R, \psi_G, \psi_B) \equiv \psi_i \qquad ; i = R, G, B \tag{25.2}$$

221

It is convenient to construct a column vector from the color fields

$$\underline{\psi} \equiv \begin{pmatrix} \psi_R \\ \psi_G \\ \psi_B \end{pmatrix} \tag{25.3}$$

Matrices in this color space will be here denoted with a bar under a symbol. This is a very compact notation

- Each ψ_i has many flavors;

- Each flavor is a four-component Dirac field.

Quantum chromodynamics (QCD) is a theory of the strong interactions binding quarks into the observed hadrons. It is a Yang–Mills non-abelian gauge theory [Ya54]. It is built on the underlying color symmetry and invariance under local $SU(3)_C$.

The lagrangian density[1] for the free quark fields can be written compactly as

$$\mathscr{L} = -\underline{\bar{\psi}} \left(\gamma_\mu \frac{\partial}{\partial x_\mu} + \underline{M} \right) \underline{\psi} \tag{25.4}$$

Here the mass term is the unit matrix with respect to color. It may be *anything* with respect to flavor, for example,

$$\underline{M} = \begin{pmatrix} \underline{m} & & \\ & \underline{m} & \\ & & \underline{m} \end{pmatrix} \qquad \underline{m} = \begin{pmatrix} m_u & & & \\ & m_d & & \\ & & m_s & \\ & & & m_c \end{pmatrix} \tag{25.5}$$

The lagrangian in Eq. (25.4) has a *global* invariance with respect to unitary transformations mixing the three internal color variables [$SU(3)_C$]. We denote the generators of this transformation by $\hat{G}^a$ with $a = 1,\ldots,8$ and the eight parameters characterizing a three-by-three unitary, unimodular matrix by θ^a with $a = 1,\ldots,8$. There are eight three-by-three, traceless, hermitian, Gell-Mann matrices $\underline{\lambda}_a$ — the analogs of the Pauli matrices. These matrices satisfy the Lie algebra of $SU(3)$, the same algebra as satisfied by the generators

$$[\frac{1}{2}\underline{\lambda}^a, \frac{1}{2}\underline{\lambda}^b] = if^{abc}\frac{1}{2}\underline{\lambda}^c \tag{25.6}$$

Here the f^{abc} are the structure constants of the group; they are antisymmetric in the indices (abc). The matrices $(\lambda^a)_{ij}$ for $a = 1,\ldots,8$ are given in

---

[1] See [Fe80] for a background discussion of continuum mechanics and lagrangian densities, and [Bj65a, Fe71] for an introduction to quantum field theory.

order by

$$
\begin{pmatrix} & 1 & \\ 1 & & \\ & & \end{pmatrix}
\begin{pmatrix} & -i & \\ i & & \\ & & \end{pmatrix}
\begin{pmatrix} 1 & & \\ & -1 & \\ & & \end{pmatrix}
\begin{pmatrix} & & 1 \\ & & \\ 1 & & \end{pmatrix}
\begin{pmatrix} & & -i \\ & & \\ i & & \end{pmatrix}
$$

$$
\begin{pmatrix} & & \\ & & 1 \\ & 1 & \end{pmatrix}
\begin{pmatrix} & & \\ & & -i \\ & i & \end{pmatrix}
\begin{pmatrix} 1/\sqrt{3} & & \\ & 1/\sqrt{3} & \\ & & -2/\sqrt{3} \end{pmatrix} \qquad (25.7)
$$

The operator producing the finite color transformation is then given by

$$
\hat{R} \;=\; e^{i\theta^a \hat{G}^a} \tag{25.8}
$$

It has the following effect on the quark field

$$
\hat{R}\underline{\psi}\hat{R}^{-1} \;=\; \underline{U}(\theta)\underline{\psi} \;=\; \left[e^{-\frac{i}{2}\underline{\lambda}^a \theta^a} \right] \underline{\psi} \tag{25.9}
$$

Latin indices will now run from $1,\ldots,8$, and repeated Latin indices are summed. The transformation in Eq. (25.9) with constant, finite θ^a leaves the lagrangian in Eq. (25.4) unchanged. Here $\underline{U}(\theta)$ is a unitary, unimodular three-by-three matrix, and the quark field in Eq. (25.3) forms a basis for the fundamental representation of $SU(3)$. The symmetry is with respect to color.

One can now make this global color invariance a *local* invariance where the transformation $\theta^a(x)$ can vary from point to point in space-time by using the theory developed by Yang and Mills [Ya54, Ab73]:

1. Introduce massless vector meson fields, one for each generator

$$
A^a_\mu(x) \qquad\qquad ; \; a = 1,\ldots,8 \tag{25.10}
$$

These vector mesons are known as *gluons*;

2. Define the covariant derivative by

$$
\frac{D}{Dx_\mu}\underline{\psi} = \left[\frac{\partial}{\partial x_\mu} - \frac{i}{2}g\underline{\lambda}^a A^a_\mu(x) \right] \underline{\psi} \tag{25.11}
$$

3. Define the field tensor for the vector meson fields as

$$
\mathscr{F}^a_{\mu\nu} = \frac{\partial A^a_\nu}{\partial x_\mu} - \frac{\partial A^a_\mu}{\partial x_\nu} + gf^{abc}A^b_\mu A^c_\nu \tag{25.12}
$$

Here f^{abc} are the structure constants of $SU(3)$;

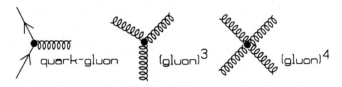

Fig. 25.1.   Processes described by the interaction terms in the QCD lagrangian.

4. Under infinitesimal local gauge transformations $\theta^a \to 0$ the vector meson fields and the field tensor transform according to

$$
\delta A_\mu^a = -\frac{1}{g}\frac{\partial \theta^a}{\partial x_\mu} + f^{abc}\theta^b A_\mu^c
$$

$$
\delta \mathscr{F}_{\mu\nu}^a = f^{abc}\theta^b \mathscr{F}_{\mu\nu}^c \qquad ; \theta^a \to 0 \qquad (25.13)
$$

5. A combination of these results leads to the *lagrangian* of QCD

$$
\mathscr{L}_{\text{QCD}} = -\underline{\bar{\psi}}\left\{\gamma_\mu\left[\frac{\partial}{\partial x_\mu} - \frac{i}{2}g\underline{\lambda}^a A_\mu^a(x)\right] + \underline{M}\right\}\underline{\psi} - \frac{1}{4}\mathscr{F}_{\mu\nu}^a \mathscr{F}_{\mu\nu}^a \qquad (25.14)
$$

The lagrangian in Eq. (25.14) can be written out explicitly in powers of the coupling constant g

$$
\mathscr{L}_{\text{QCD}} = \mathscr{L}_0 + \mathscr{L}_1 + \mathscr{L}_2 \qquad (25.15)
$$

$$
\mathscr{L}_0 = -\underline{\bar{\psi}}\left(\gamma_\mu\frac{\partial}{\partial x_\mu} + \underline{M}\right)\underline{\psi} - \frac{1}{4}F_{\mu\nu}^a F_{\mu\nu}^a
$$

$$
\mathscr{L}_1 = \frac{i}{2}g\underline{\bar{\psi}}\gamma_\mu\underline{\lambda}^a\underline{\psi}A_\mu^a(x) - \frac{g}{2}f^{abc}F_{\mu\nu}^a A_\mu^b A_\nu^c
$$

$$
\mathscr{L}_2 = -\frac{g^2}{4}f^{abc}f^{ade}A_\mu^b A_\nu^c A_\mu^d A_\nu^e
$$

Here

$$
F_{\mu\nu}^a \equiv \frac{\partial A_\nu^a}{\partial x_\mu} - \frac{\partial A_\mu^a}{\partial x_\nu} \qquad (25.16)
$$

The various processes described by the interaction terms in this lagrangian are illustrated in Fig. 25.1.

To obtain further insight into these results, it is useful to write the Yukawa interaction between the quarks and gluons in more detail. Recall, for example, the structure of the first two $\underline{\lambda}^a$ matrices

$$
\underline{\lambda}^1 = \begin{pmatrix} & 1 & \\ 1 & & \\ & & \end{pmatrix} \qquad\qquad \underline{\lambda}^2 = \begin{pmatrix} & -i & \\ i & & \\ & & \end{pmatrix} \qquad (25.17)
$$

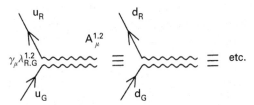

Fig. 25.2. Individual processes described by the quark–gluon Yukawa coupling in QCD.

These matrices connect the (R, G) quarks, and with explicit identification of the flavor components of the color fields, it is evident that this interaction contains the individual processes illustrated in Fig. 25.2. The quarks interact here by changing their color, which in turn is carried off by the gluons; the flavor of the quarks is unchanged and all flavors of quarks have an identical color coupling. If the gluons are represented with double lines connected to the incoming and outgoing quark lines respectively, and a color assigned to each line as indicated in this figure, then color can be viewed as running continuously through a Feynman diagram built from these components.

The Euler–Lagrange equations in continuum mechanics follow from Hamilton's principle [Fe71]

$$\delta \int \mathscr{L}\left(q, \frac{\partial q}{\partial x_\mu}\right) d^4x = 0 \tag{25.18}$$

The Euler–Lagrange equations following from the QCD lagrangian are readily derived as

$$\left\{\gamma_\mu\left[\frac{\partial}{\partial x_\mu} - \frac{i}{2}g\underline{\lambda}^a A_\mu^a(x)\right] + \underline{M}\right\}\underline{\psi} = 0$$

$$\underline{\bar\psi}\left\{\gamma_\mu\left[\frac{\overleftarrow{\partial}}{\partial x_\mu} + \frac{i}{2}g\underline{\lambda}^a A_\mu^a(x)\right] - \underline{M}\right\} = 0$$

$$\frac{\partial \mathscr{F}_{\mu\nu}^a}{\partial x_\nu} = \frac{i}{2}g\underline{\bar\psi}\gamma_\mu\underline{\lambda}^a\underline{\psi} + gf^{abc}\mathscr{F}_{\mu\nu}^b A_\nu^c \tag{25.19}$$

It follows from these equations of motion that currents built out of quark fields and a unit matrix with respect to color are *conserved*.

$$\frac{\partial}{\partial x_\mu}\left(\frac{i}{3}\underline{\bar\psi}\gamma_\mu\underline{\psi}\right) = 0 \qquad\qquad \text{; baryon current}$$

$$\frac{\partial}{\partial x_\mu}\left(i\underline{\bar\psi}\gamma_\mu\underline{\Sigma}\underline{\psi}\right) = 0 \qquad\qquad \text{; flavor current} \tag{25.20}$$

In the second line, $\underline{\Sigma}$ is a unit matrix with respect to color satisfying $[\underline{\Sigma}, \lambda^a] = 0$; the flavor submatrices are *arbitrary* as long as they commute with the mass matrix

$$\underline{\Sigma} = \begin{pmatrix} \underline{\sigma} & & \\ & \underline{\sigma} & \\ & & \underline{\sigma} \end{pmatrix} \qquad ; [\underline{\sigma}, \underline{m}] = 0 \qquad (25.21)$$

The conserved electromagnetic current for the (u, d, s, c) quarks, with charges $(2/3, -1/3, -1/3, 2/3)$ respectively, is given by the point Dirac value

$$J_\mu^\gamma = i\bar{\psi}\gamma_\mu \underline{Q}\,\psi \qquad (25.22)$$

$$\underline{Q} = \begin{pmatrix} \underline{q} & & \\ & \underline{q} & \\ & & \underline{q} \end{pmatrix} \qquad ; \underline{q} = \begin{pmatrix} 2/3 & & & \\ & -1/3 & & \\ & & -1/3 & \\ & & & 2/3 \end{pmatrix}$$

The gluons are absolutely neutral to the electromagnetic interaction.

It follows from the four-divergence of the third of Eqs. (25.19) and the antisymmetry of $\mathscr{F}_{\mu\nu}^a = -\mathscr{F}_{\nu\mu}^a$ that the color current, the source of the color field, is also conserved.

$$\frac{\partial}{\partial x_\mu} \left(\frac{i}{2} g\bar{\psi}\gamma_\mu \lambda^a \psi + gf^{abc}\mathscr{F}_{\mu\nu}^b A_\nu^c \right) = 0 \qquad (25.23)$$

The theory of QCD can again be characterized by a set of *Feynman rules*. Here we give the Feynman rules for the Green's functions, which characterize the quantum field theory [Fe71]. The quark Green's function in the vacuum sector is defined by

$$iG_{\alpha\beta}(\mathbf{x}_1 t_1, \mathbf{x}_2 t_2) \equiv \langle 0|P[\hat{\psi}_\alpha(\mathbf{x}_1 t_1), \hat{\bar{\psi}}_\beta(\mathbf{x}_2 t_2)]|0\rangle$$

$$\equiv \int \frac{d^4k}{(2\pi)^4} e^{ik\cdot(x_1-x_2)} iG_{\alpha\beta}(k) \qquad (25.24)$$

The Feynman rules for $iG(k)$ are derived in [Qu83, Ch84, Ai89, Wa91]; they are as follows:[2]

1. Draw all topologically distinct, connected diagrams;

2. Include the following factors for the quark, gluon, and *ghost* lines, respectively (Fig. 25.3):[3]

---

[2] See Ref. [Ch84] for a much more extensive discussion, including Feynman rules with other choices of gauge.

[3] All quark indices are now explicit: $i, j = R, G, B$ for color; $l, m = u, d, s, c, \cdots$ for flavor.

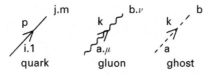

Fig. 25.3. Propagators in QCD.

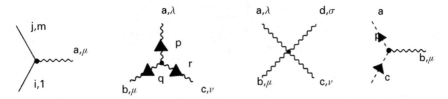

Fig. 25.4. Vertices in QCD.

$$\frac{1}{i}\frac{1}{i\gamma_\mu p_\mu}\delta_{ij}\delta_{lm} \qquad ; \text{quark (massless)}$$

$$\frac{1}{i}\delta^{ab}\frac{1}{k^2}\left(\delta_{\mu\nu} - \frac{k_\mu k_\nu}{k^2}\right) \qquad ; \text{gluon (Landau gauge)}$$

$$\frac{1}{i}\delta^{ab}\frac{1}{k^2} \qquad ; \text{ghost} \qquad (25.25)$$

The ghost is an internal element, coupled to gluons, that is required to generate the correct S-matrix in a non-abelian gauge theory;

3. Include the following factors for the vertices indicated in Fig. 25.4:

$$-g\frac{1}{2}\lambda_{ji}^a\delta_{lm}\gamma_\mu \qquad\qquad ; (\text{quark})^2-\text{gluon}$$

$$gf^{abc}[(q-r)_\lambda\delta_{\mu\nu} + (p-q)_\nu\delta_{\lambda\mu} + (r-p)_\mu\delta_{\lambda\nu}] \qquad ; (\text{gluon})^3$$

$$-ig^2[f^{abe}f^{cde}(\delta_{\lambda\nu}\delta_{\sigma\mu} - \delta_{\lambda\sigma}\delta_{\mu\nu}) + f^{ace}f^{bde}(\delta_{\lambda\mu}\delta_{\sigma\nu} - \delta_{\lambda\sigma}\delta_{\mu\nu})$$
$$+ f^{ade}f^{cbe}(\delta_{\lambda\nu}\delta_{\sigma\mu} - \delta_{\sigma\nu}\delta_{\lambda\mu})] \qquad\qquad ; (\text{gluon})^4$$

$$-gf^{abc}p_\mu \qquad\qquad ; (\text{ghost})^2-\text{gluon} \qquad (25.26)$$

4. Take the Dirac matrix product along fermion lines;

5. Conserve four-momentum at each vertex;

6. Include a factor $\int d^4q/(2\pi)^4$ for each independent internal line;

7. Include a factor of $(-1)^{F+G}$ where F is the number of closed fermion loops and G is the number of closed ghost loops;

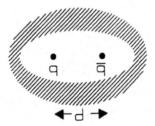

Fig. 25.5.   Confinement in QCD. Lattice gauge theory calculations indicate that the separation energy grows linearly with d.

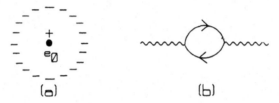

Fig. 25.6.   (a) Shielding of point charge by (b) vacuum polarization in QED.

QCD has two absolutely remarkable properties, confinement and asymptotic freedom.

Colored quarks and gluons, the basic underlying degrees of freedom in the strong interactions, are evidently never observed as free asymptotic scattering states in the laboratory; you cannot hold an isolated quark or gluon in your hand. Quarks and gluons are confined to the interior of hadrons. There are strong indications from lattice gauge theory calculations [Wi74], that *confinement* is indeed a dynamic property of QCD arising from the strong, nonlinear gluon couplings in the lagrangian. One can show in these calculations, for example, that the energy of a static $(q\bar{q})$ pair grows linearly with the distance d separating the pair (see Fig. 25.5). What actually happens as the $(q\bar{q})$ pair is separated is that another $(q\bar{q})$ pair is formed, completely shielding the individual color charges of the first pair, and producing two mesons from one.

The second remarkable property is *asymptotic freedom*. Recall from QED that vacuum polarization shields a point electric charge e_0 as indicated in Fig. 25.6(a). The renormalized charge e_2^2 changes with the distance scale, or momentum transfer λ^2, at which one measures the interior charge. The mathematical statement of this fact is the renormalization group equation of Gell-Mann and Low [Ge54]

$$\frac{de_2^2}{d\ln(\lambda^2/M^2)} = \psi(e_2^2) \qquad (25.27)$$

Fig. 25.7.   Anti-shielding of color charge in QCD by strong vacuum polarization.

The lowest order modification of the charge in QED arises from the vacuum polarization graph indicated in Fig. 25.6 (b). The renormalization group equations can be used to sum the leading logarithmic corrections to the renormalized charge to all orders. The result is that the renormalized charge measured at large $\lambda^2 \gg M^2$ is related to the usual value of the total charge e_1^2 by

$$e_2^2 \approx \frac{e_1^2}{1 - (e_1^2/12\pi^2)\ln(\lambda^2/M^2)} \tag{25.28}$$

The first term in the expansion of the denominator arises from the graph in Fig. 25.6 (b). The renormalized electric charge in QED is evidently *shielded* by vacuum polarization; the measured charge *increases* as one goes to shorter and shorter distances, or higher and higher λ^2.

Similar, although somewhat more complicated, arguments can be made in QCD. An isolated color charge g_0 is modified by strong vacuum polarization and surrounded with a corresponding cloud of color charge as indicated schematically in Fig. 25.7. In this case, the renormalization group equations lead to a sum of the leading ln corrections for $\lambda^2 \gg \lambda_1^2$ of the form [Gr73a, Gr73b, Po73, Po74]

$$g_2^2 \approx \frac{g_1^2}{1 + (g_1^2/16\pi^2)(33/3 - 2N_f/3)\ln(\lambda^2/\lambda_1^2)} \tag{25.29}$$

Here N_f is the number of quark flavors.[4] An expansion of the denominator again gives the result obtained by combining the lowest-order perturbation theory corrections to the quark and gluon propagators and quark vertex. The plus sign in the denominator in this expression is crucial. One now draws the conclusion that there is *anti-shielding*; the charge *decreases* at shorter distances, or with larger λ^2.[5] The implications are enormous, for

---

[4] With $N_f = 1$, no gluon contribution of 33/3, and the observation $\mathrm{tr}(\frac{1}{2}\lambda^a \frac{1}{2}\lambda^b) = \frac{N_f}{2}\delta^{ab}$, one recovers the result in Eq. (25.28). It is the gluon contribution that changes the sign in the denominator.

[5] The vacuum in QCD thus acts like a *paramagnetic* medium, where a moment surrounds itself with like moments, rather than the *dielectric* medium of QED where a charge surrounds itself with opposite charges.

one now concludes that it is consistent to do *perturbation theory* at very short distances, or high momentum transfer. The renormalization group equations then provide a tool for summing the leading ln's of perturbation theory. The powerful result of asymptotic freedom in QCD is due to Gross and Wilczek [Gr73a, Gr73b] and Politzer [Po73, Po74]; see also Fritzsch and Gell-Mann [Fr72a, Fr73a]. References [Ma78, Re81, Wi82, Do93] contain good background material on QCD.

While a multitude of QCD-inspired models exist [Wa95], the most ambitious attempt to solve the QCD field equations relies on lattice gauge theory where the theory is put on a finite space-time lattice [Wi74].[6] Low-energy applications can be found in terms of *effective field theory*, where hadronic degrees of freedom are the generalized coordinates of choice, and an effective lagrangian constructed which reflects the symmetry properties of QCD [Do93, Se97]. While we will not give an extensive discussion of effective field theory here, it is possible to capture the spirit of these efforts.

Consider the nuclear domain of massless (u, d) quarks. The kinetic energy term as rewritten in Eq. (26.2), and hence $\mathscr{L}_{QCD}$, is invariant under the chiral $SU(2)_L \otimes SU(2)_R$ transformation $\psi_L \to \underline{L}\psi_L$, $\psi_R \to \underline{R}\psi_R$ where $\underline{L}$ and $\underline{R}$ are global $SU(2)$ matrices. Consider the pion sector of the hadronic theory and represent the pion field $\boldsymbol{\pi}$ through the $SU(2)$ matrix

$$\underline{U} = \exp\left(i\boldsymbol{\tau} \cdot \boldsymbol{\pi}/f_\pi\right) \tag{25.30}$$

Here f_π reflects the mass scale (say m_p) at which this chiral symmetry, as manifest in nature, is spontaneously broken. The leading term in an effective lagrangian can be constructed as follows

$$\mathscr{L}_2 + \mathscr{L}_{csb} = -\frac{f_\pi^2}{4}\mathrm{tr}\left(\frac{\partial \underline{U}^\dagger}{\partial x_\lambda}\right)\left(\frac{\partial \underline{U}}{\partial x_\lambda}\right) + \frac{f_\pi^2 m_\pi^2}{4}\mathrm{tr}\left(\underline{U} + \underline{U}^\dagger - 2\right) \tag{25.31}$$

If $m_\pi = 0$, this lagrangian is invariant under the chiral transformation $\underline{U} \to \underline{L}\underline{U}\underline{R}^\dagger$ (the pion mass term reflects chiral symmetry-breaking at the lagrangian level through u, d quark masses). This effective lagrangian should be applicable in the low energy domain where $q/f_\pi \ll 1$. Higher order terms in the effective lagrangian can now be similarly constructed in terms of $\underline{U}, \partial \underline{U}/\partial x_\lambda$. An expansion of the exponential then leads to

$$\mathscr{L}_2 + \mathscr{L}_{csb} = -\frac{1}{2}\left(\frac{\partial \boldsymbol{\pi}}{\partial x_\lambda}\right)^2 - \frac{m_\pi^2}{2}\boldsymbol{\pi}^2$$

$$-\frac{1}{6f_\pi^2}\left[\left(\boldsymbol{\pi} \cdot \frac{\partial \boldsymbol{\pi}}{\partial x_\lambda}\right)^2 - \boldsymbol{\pi}^2\left(\frac{\partial \boldsymbol{\pi}}{\partial x_\lambda}\right)^2\right] + \frac{m_\pi^2}{24f_\pi^2}\boldsymbol{\pi}^4 + \cdots \tag{25.32}$$

and π–π scattering to $O(1/f_\pi^2)$ can now be calculated from this result.

---

[6] An extensive introduction to lattice gauge theory can be found in [Wa95].

26

The standard model

The development of a unified theory of the electroweak interactions surely must be regarded as one of the great intellectual achievements of our era. We assume the reader is familiar with the basic phenomenology of the weak interactions [Cu83, Wa95] and turn to the so-called *standard model* of the electroweak interactions [We67, We72, Sa64, Gl70]. The discussion follows [Wa95]

Most leptons (l, v_l) are light, or massless, and they can be created and destroyed in weak interactions. This indicates that they *must* be described with relativistic quantum fields. In the interaction representation, fermion fields take the following form [Bj65, Fe71]

$$\psi(x) = \frac{1}{\sqrt{\Omega}} \sum_{\mathbf{k}\lambda} \left[a_{\mathbf{k}\lambda} u(\mathbf{k}\lambda) e^{ik \cdot x} + b_{\mathbf{k}\lambda}^\dagger v(-\mathbf{k}\lambda) e^{-ik \cdot x} \right] \qquad (26.1)$$

In this expression a destroys a lepton, $b^\dagger$ creates an antilepton, and λ denotes the helicity with respect to the accompanying momentum variable.[1]

A spinor field can always be decomposed as follows

$$\psi = \frac{1}{2}(1 + \gamma_5)\psi + \frac{1}{2}(1 - \gamma_5)\psi \equiv \psi_L + \psi_R$$

$$\bar{\psi}\gamma_\mu \frac{\partial}{\partial x_\mu}\psi = \bar{\psi}_L\gamma_\mu\frac{\partial}{\partial x_\mu}\psi_L + \bar{\psi}_R\gamma_\mu\frac{\partial}{\partial x_\mu}\psi_R$$

$$\bar{\psi}\psi = \bar{\psi}_L\psi_R + \bar{\psi}_R\psi_L \qquad (26.2)$$

The lepton fields for the electron and electron neutrino[2] will be com-

---

[1] Hole theory implies that $v(-\mathbf{k}\lambda)$ is a negative-energy wave function with helicity λ with respect to $-\mathbf{k}$.

[2] And similarly for the other leptons.

231

bined in the following fashion:

$$\psi_l = \begin{pmatrix} \psi_{v_e} \\ \psi_e \end{pmatrix} \equiv \begin{pmatrix} v \\ e \end{pmatrix} \tag{26.3}$$

The fields (L, R) are defined by

$$L \equiv \begin{pmatrix} v_L \\ e_L \end{pmatrix} = \frac{1}{2}(1 + \gamma_5)\psi_l$$

$$R \equiv e_R = \frac{1}{2}(1 - \gamma_5)\psi_e \tag{26.4}$$

The kinetic energy of the leptons is then given by[3]

$$\mathscr{L}^0_{\text{lepton}} = -\left[\bar{\psi}_e \gamma_\mu \frac{\partial}{\partial x_\mu} \psi_e + \bar{v}_L \gamma_\mu \frac{\partial}{\partial x_\mu} v_L \right]$$

$$= -\left[\bar{L} \gamma_\mu \frac{\partial}{\partial x_\mu} L + \bar{R} \gamma_\mu \frac{\partial}{\partial x_\mu} R \right] \tag{26.5}$$

This lagrangian is invariant under a global $SU(2)_W$ symmetry — a weak (left-handed) isospin — which treats the field L as a weak isodoublet and R as a weak isosinglet. The generators for this $SU(2)_W$ symmetry can be immediately written in terms of the above fields as

$$\hat{T}^i_W = \int L^\dagger(\mathbf{x}) \frac{1}{2} \tau_i L(\mathbf{x}) d^3x$$

$$= \int \psi_l^\dagger(\mathbf{x}) \frac{1}{2} \tau_i \frac{1}{2}(1 + \gamma_5) \psi_l(\mathbf{x}) d^3x \tag{26.6}$$

It follows immediately from the canonical (anti)commutation relations that these generators satisfy an $SU(2)$ algebra

$$[\hat{T}^i_W, \hat{T}^j_W] = i\varepsilon_{ijk} \hat{T}^k_W \tag{26.7}$$

The finite symmetry transformations are given by

$$\exp\{i\boldsymbol{\theta} \cdot \hat{\mathbf{T}}_W\} L \exp\{-i\boldsymbol{\theta} \cdot \hat{\mathbf{T}}_W\} = [e^{-\frac{i}{2}\boldsymbol{\theta} \cdot \boldsymbol{\tau}}] L \quad ; \text{doublet}$$

$$\exp\{i\boldsymbol{\theta} \cdot \hat{\mathbf{T}}_W\} R \exp\{-i\boldsymbol{\theta} \cdot \hat{\mathbf{T}}_W\} = [1] R \quad ; \text{singlet} \tag{26.8}$$

These equations follow from the projection properties of $(1 \pm \gamma_5)/2$.

The mass term for the electron has the following form

$$-m_e \bar{\psi}_e \psi_e = -m_e[\bar{e}_L e_R + \bar{e}_R e_L] \tag{26.9}$$

---

[3] There is only one neutrino field in the standard model $v_L \equiv \frac{1}{2}(1 + \gamma_5)\psi_v$; it describes left-handed neutrinos and right-handed antineutrinos. This is put in by hand, as is the fact that this neutrino is massless $m_v = 0$.

This expression is *not* invariant under $SU(2)_W$. Hence if one wants to build on this symmetry, it is necessary to start with *massless fermions*.

The corresponding lagrangian for point Dirac nucleon fields is relatively simple and illustrates the general structure of the theory [We72]; we present this first. Matrix elements for physical nucleons then follow from general symmetry considerations. The somewhat more complex formulation in terms of quarks is then given [Gl70].

Proton and neutron fields are included in a manner analogous to the above

$$N_L = \begin{pmatrix} p_L \\ n_L \end{pmatrix} = \frac{1}{2}(1 + \gamma_5)\psi_N \quad ; \text{doublet} \tag{26.10}$$

$$p_R, n_R \hspace{5cm} ; \text{singlets}$$

$$\mathscr{L}^0_{\text{nucleon}} = -\left[\bar{N}_L \gamma_\mu \frac{\partial}{\partial x_\mu} N_L + \bar{p}_R \gamma_\mu \frac{\partial}{\partial x_\mu} p_R + \bar{n}_R \gamma_\mu \frac{\partial}{\partial x_\mu} n_R \right]$$

This lagrangian is now also invariant under $SU(2)_W$; again this is true only if one starts with massless fermions.

The standard model introduces an additional global $U(1)_W$ symmetry *weak hypercharge* defined so that the fields transform according to

$$\exp\{i\alpha\hat{Y}_W\} \phi \exp\{-i\alpha\hat{Y}_W\} = e^{-i\alpha Y_W}\phi \tag{26.11}$$

Now assign quantum numbers to the fields (and corresponding particles) so that the lagrangian is invariant and the *electric charge* is still given by the Gell-Mann–Nishijima relation[4]

$$Q = (T_3 + \frac{1}{2}Y)_W \tag{26.12}$$

Conservation of electric charge will be imposed as an exact symmetry of the theory. Assignments of the weak quantum numbers for the fields introduced so far are shown in Table 26.1.

As with QCD in chapter 25, this is now made into a Yang–Mills local gauge theory based on the symmetry group $SU(2)_W \otimes U(1)_W$ [Ya54, Ab73]. The only slight new complexity is that now one has the direct product of two symmetry groups with commuting generators; however, an examination of the basic concept shows that this is an inessential complication. The steps of the Yang–Mills construction are as follows:

1. Add *gauge bosons*, one for each of the generators $(\hat{T}^i_W, \hat{Y}_W)$

$$A^i_\mu(x) \hspace{3cm} ; i = 1, 2, 3$$

$$B_\mu(x) \tag{26.13}$$

---

[4] This follows for the fermions by constructing the appropriate operators in second quantization [Wa95].

Table 26.1. Weak symmetry quantum numbers in the standard model.

Particle/field	T_W	T_{3W}	Y_W	Q
$(v_e)_L$	1/2	1/2	−1	0
e_L	1/2	−1/2	−1	−1
e_R	0	0	−2	−1
p_L	1/2	1/2	1	1
n_L	1/2	−1/2	1	0
p_R	0	0	2	1
n_R	0	0	0	0
ϕ^+	1/2	1/2	1	1
ϕ^0	1/2	−1/2	1	0

2. Use the *covariant derivative* in the lagrangian

$$\frac{\partial}{\partial x_\mu} \to \frac{D}{Dx_\mu} \equiv \left(\frac{\partial}{\partial x_\mu} - \frac{i}{2}g' Y_W B_\mu - \frac{i}{2}g\tau \cdot \mathbf{A}_\mu \right) \quad ; \text{on doublets}$$

$$\equiv \left(\frac{\partial}{\partial x_\mu} - \frac{i}{2}g' Y_W B_\mu \right) \quad ; \text{on singlets} \qquad (26.14)$$

3. Include a kinetic energy term for the gauge bosons

$$\mathcal{L}_{\text{gauge}} = -\frac{1}{4}\left(\frac{\partial B_\nu}{\partial x_\mu} - \frac{\partial B_\mu}{\partial x_\nu} \right)^2 - \frac{1}{4}\left(\frac{\partial \mathbf{A}_\nu}{\partial x_\mu} - \frac{\partial \mathbf{A}_\mu}{\partial x_\nu} + g\mathbf{A}_\mu \times \mathbf{A}_\nu \right)^2 \quad (26.15)$$

Mass terms of the form $m_B^2 B_\mu B_\mu$ or $m_A^2 \mathbf{A}_\mu \cdot \mathbf{A}_\mu$ break the local gauge invariance; hence the gauge bosons must be *massless*.

The Yang–Mills lagrangian thus takes the form

$$\mathcal{L}_{\text{lepton}} = -\left[\bar{L}\gamma_\mu \left(\frac{\partial}{\partial x_\mu} - \frac{i}{2}(-1)g' B_\mu - \frac{i}{2}g\tau \cdot \mathbf{A}_\mu \right) L \right.$$

$$\left. + \bar{R}\gamma_\mu \left(\frac{\partial}{\partial x_\mu} - \frac{i}{2}(-2)g' B_\mu \right) R \right]$$

$$\mathcal{L}_{\text{nucleon}} = -\left[\bar{N}_L\gamma_\mu \left(\frac{\partial}{\partial x_\mu} - \frac{i}{2}(1)g' B_\mu - \frac{i}{2}g\tau \cdot \mathbf{A}_\mu \right) N_L \right.$$

$$\left. + \bar{p}_R\gamma_\mu \left(\frac{\partial}{\partial x_\mu} - \frac{i}{2}(2)g' B_\mu \right) p_R + \bar{n}_R\gamma_\mu \left(\frac{\partial}{\partial x_\mu} - \frac{i}{2}(0)g' B_\mu \right) n_R \right]$$

$$\mathcal{L}_{\text{gauge}} = -\frac{1}{4}B_{\mu\nu}B_{\mu\nu} - \frac{1}{4}\mathcal{F}^i_{\mu\nu}\mathcal{F}^i_{\mu\nu} \qquad (26.16)$$

The masses for the gauge bosons are now generated by *spontaneous symmetry breaking*. One proceeds to:

1. Introduce a weak isodoublet of *complex scalar mesons*

$$\underline{\phi} \equiv \begin{pmatrix} \phi^+ \\ \phi^0 \end{pmatrix} \tag{26.17}$$

2. Assign weak quantum numbers as indicated in Table 26.1;

3. Use the covariant derivative of Eq. (26.14);

4. Add a term to the lagrangian for this scalar field that is invariant under local $SU(2)_W \otimes U(1)_W$

$$\mathscr{L}_{\text{scalar}} = -\left(\frac{D\underline{\phi}}{Dx_\mu}\right)^\star \left(\frac{D\underline{\phi}}{Dx_\mu}\right) - V(\underline{\phi}^\dagger \underline{\phi}) \tag{26.18}$$

Thus[5]

$$\left(\frac{D\underline{\phi}}{Dx_\mu}\right)^\star \left(\frac{D\underline{\phi}}{Dx_\mu}\right) = \tag{26.19}$$

$$\underline{\phi}^\dagger \left(\frac{\overleftarrow{\partial}}{\partial x_\mu} + \frac{i}{2}g'B_\mu + \frac{i}{2}g\boldsymbol{\tau}\cdot\mathbf{A}_\mu\right)\left(\frac{\partial}{\partial x_\mu} - \frac{i}{2}g'B_\mu - \frac{i}{2}g\boldsymbol{\tau}\cdot\mathbf{A}_\mu\right)\underline{\phi}$$

5. Assume the most general form of the scalar self-interaction potential V for a renormalizable theory

$$V = \mu^2 \underline{\phi}^\dagger \underline{\phi} + \lambda(\underline{\phi}^\dagger \underline{\phi})^2 \tag{26.20}$$

Assume that $\mu^2 < 0$ and $\lambda > 0$ so that the potential V has the shape shown in Fig. 26.1. The minimum of the potential no longer occurs at the origin with $\underline{\phi} = 0$, but now at a finite value of $\underline{\phi}$. Hence the scalar field acquires a *vacuum expectation value*. Only the neutral component of the field can be allowed to develop a vacuum expectation value in order to preserve electric charge conservation. Furthermore, the (constant) phase of the field can always be redefined so that this vacuum expectation value is real. Thus we write

$$\langle \phi^0 \rangle = \langle \phi^{0*} \rangle \equiv \frac{v}{\sqrt{2}} \tag{26.21}$$

---

[5] The metric is not complex conjugated in $v_\mu^\star \equiv (\mathbf{v}^\dagger, +iv_0^\dagger)$.

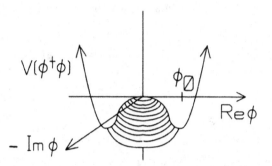

Fig. 26.1.   Form of the scalar self-interaction potential to generate mass for the gauge bosons by spontaneous symmetry breaking. The illustration is for a single, neutral, complex ϕ.

At the minimum of the vacuum expectation value of the potential one finds

$$v^2 \;=\; -\frac{\mu^2}{\lambda} \tag{26.22}$$

Without loss of generality, one can now *parameterize* the complex scalar field ϕ in terms of four real parameters $\{\xi(x), \eta(x)\}$ describing the fluctuations around the vacuum expectation value in the following fashion [Ab73]:

$$\underline{\phi} \;\equiv\; \exp\left\{\frac{-i}{2v}\xi \cdot \tau\right\} \begin{pmatrix} 0 \\ \frac{1}{\sqrt{2}}(v + \eta) \end{pmatrix} \tag{26.23}$$

The theory has been constructed to be locally gauge invariant. Make use of this fact to simplify matters. Make a gauge transformation to eliminate the first factor in this equation. Define

$$\underline{\phi}' \equiv \exp\left\{\frac{+i}{2v}\xi \cdot \tau\right\}\underline{\phi} = \underline{U}(\xi)\underline{\phi} = \frac{1}{\sqrt{2}}\begin{pmatrix} 0 \\ v + \eta \end{pmatrix} \tag{26.24}$$

Written in terms of the new field ϕ', the three scalar field variables $\{\xi(x)\}$ now no longer appear in the lagrangian; and, as we proceed to demonstrate, the free lagrangian has instead a simple interpretation in terms of massive vector and scalar particles. The lagrangian in this form is said to be written in the *unitary gauge* where the particle content of the theory is manifest. The procedure for generating the mass of the gauge bosons in this fashion is known as the *Higgs mechanism* [Cu83, Ab73].

Substitution of the expression in Eq. (26.24) in the scalar lagrangian in

Eqs. (26.18)–(26.20) leads to

$$\mathscr{L}_{\text{scalar}} = -V\left[\frac{1}{2}(v+\eta)^2\right] - \frac{1}{2}\chi_\downarrow^\dagger\left[\frac{\partial\eta}{\partial x_\mu} + \frac{ig'}{2}(v+\eta)B_\mu + \frac{ig}{2}(v+\eta)\boldsymbol{\tau}\cdot\mathbf{A}_\mu\right]$$
$$\times \left[\frac{\partial\eta}{\partial x_\mu} - \frac{ig'}{2}(v+\eta)B_\mu - \frac{ig}{2}(v+\eta)\boldsymbol{\tau}\cdot\mathbf{A}_\mu\right]\chi_\downarrow \tag{26.25}$$

Here $\chi_\downarrow \equiv \begin{pmatrix} 0 \\ 1 \end{pmatrix}$. An evaluation of the potential term, utilizing the minimization condition in Eq. (26.22), gives

$$V\left[\frac{1}{2}(v+\eta)^2\right] = \frac{\mu^2}{2}(v+\eta)^2 + \frac{\lambda}{4}(v+\eta)^4 \tag{26.26}$$
$$= v^2\left(\frac{\mu^2}{4}\right) + \eta^2(-\mu^2) + \eta^3(\lambda v) + \eta^4\left(\frac{\lambda}{4}\right)$$

Note that there is no term linear in η when one expands about the true minimum in V. The coefficient of the term linear in $\partial\eta/\partial x_\mu$ similarly vanishes in Eq. (26.25).

The remaining boson interactions in $\mathscr{L}_{\text{scalar}}$ are proportional to

$$\chi_\downarrow^\dagger(g'B_\mu + g\boldsymbol{\tau}\cdot\mathbf{A}_\mu)(g'B_\mu + g\boldsymbol{\tau}\cdot\mathbf{A}_\mu)\chi_\downarrow$$
$$= \chi_\downarrow^\dagger(g'^2B_\mu^2 + g^2\mathbf{A}_\mu^2 + 2gg'B_\mu\boldsymbol{\tau}\cdot\mathbf{A}_\mu)\chi_\downarrow$$
$$= (g'^2B_\mu^2 + g^2\mathbf{A}_\mu^2 - 2gg'B_\mu A_\mu^{(3)}) \tag{26.27}$$

Hence the scalar lagrangian in the unitary gauge is given by

$$\mathscr{L}_{\text{scalar}} = -\frac{1}{2}\left[\left(\frac{\partial\eta}{\partial x_\mu}\right)^2 + (-2\mu^2)\eta^2\right] - \frac{\lambda}{4}(4v\eta^3 + \eta^4) - \frac{1}{4}\mu^2v^2$$
$$-\frac{1}{8}(v+\eta)^2(g'^2B_\mu^2 + g^2\mathbf{A}_\mu^2 - 2gg'B_\mu A_\mu^{(3)}) \tag{26.28}$$

The term in v^2 in the second line now provides the sought-after mass for the gauge bosons. The coefficient of this term is a quadratic form in the gauge fields, which can be put on principal axes with the introduction of the following linear combinations of fields:

$$W_\mu^{(+)} \equiv W_\mu^\star \equiv \frac{1}{\sqrt{2}}(A_\mu^{(1)} + iA_\mu^{(2)})$$
$$W_\mu^{(-)} \equiv W_\mu \equiv \frac{1}{\sqrt{2}}(A_\mu^{(1)} - iA_\mu^{(2)})$$

$$Z_\mu \equiv \frac{-gA_\mu^{(3)} + g'B_\mu}{(g^2 + g'^2)^{1/2}}$$

$$A_\mu \equiv \frac{g'A_\mu^{(3)} + gB_\mu}{(g^2 + g'^2)^{1/2}} \tag{26.29}$$

The fields $(W_\mu^\star, W_\mu)$ will create particles (W_μ^+, W_μ^-), respectively, the third field describes a neutral Z_μ^0 vector boson, and the fourth is the photon field. The relation between $(B_\mu, A_\mu^{(3)})$ and (Z_μ, A_μ) is an *orthogonal transformation*. Note in particular that the weak angle is defined by

$$\sin\theta_W \equiv \frac{g'}{(g^2 + g'^2)^{1/2}} \tag{26.30}$$

The scalar lagrangian can thus finally be written in the unitary gauge as

$$\begin{aligned}
\mathscr{L}_{\text{scalar}} = & -\frac{1}{2}\left[\left(\frac{\partial\eta}{\partial x_\mu}\right)^2 + (-2\mu^2)\eta^2\right] - \frac{\lambda}{4}(4v\eta^3 + \eta^4) - \frac{1}{4}\mu^2 v^2 \\
& -\frac{1}{4}v^2(g^2 + g'^2)\frac{1}{2}Z_\mu^2 - \frac{1}{4}v^2 g^2 W_\mu W_\mu^\star \\
& -\frac{1}{8}\eta(2v + \eta)[(g^2 + g'^2)Z_\mu^2 + 2g^2 W_\mu W_\mu^\star]
\end{aligned} \tag{26.31}$$

The first term in the first line is the lagrangian for a free, neutral scalar field of mass $-2\mu^2$ — the *Higgs field*; this is the only remaining physical degree of freedom from the complex doublet of scalar fields introduced previously, in this unitary gauge. The second term describes cubic and quartic self-couplings of the Higgs field; the third term in the first line is simply an additive constant. The terms in the second line proportional to the constant v^2 represent the quadratic mass terms for the gauge bosons. Note, in particular, that no mass term has been generated for the photon field, which thus remains massless, as it must. Finally, the terms in the last line proportional to $(2v\eta + \eta^2)$ represent cubic and quartic couplings of the Higgs to the massive gauge bosons.

Since the transformation in Eqs. (26.29) is orthogonal, the quadratic part of the kinetic energy of the gauge bosons remains on principal axes and Eq. (26.15) can be rewritten as

$$\begin{aligned}
\mathscr{L}_{\text{gauge}} = & -\frac{1}{2}W_{\mu\nu}^\star W_{\mu\nu} - \frac{1}{4}Z_{\mu\nu}Z_{\mu\nu} - \frac{1}{4}F_{\mu\nu}F_{\mu\nu} \\
& -\frac{g}{2}\mathbf{F}_{\mu\nu} \cdot (\mathbf{A}_\mu \times \mathbf{A}_\nu) - \frac{g^2}{4}(\mathbf{A}_\mu \times \mathbf{A}_\nu)^2
\end{aligned} \tag{26.32}$$

Here the field tensors are defined by the linear Maxwell form $V_{\mu\nu} \equiv \partial V_\nu/\partial x_\mu - \partial V_\mu/\partial x_\nu$ and the original gauge field $\mathbf{A}_\mu$ in the nonlinear terms must still be expressed in terms of the physical fields defined through Eqs. (26.29). The second line in the above result represents cubic and quartic couplings of the physical gauge fields.

The *particle content* of the theory is now made manifest in this unitary gauge, since the free lagrangian has the required quadratic form in the kinetic energy and masses. In addition to the original (still massless!) fermions, the theory evidently now contains

1. A massive neutral weak vector meson Z^0_μ with mass given by

$$M_Z^2 = \frac{v^2(g^2 + g'^2)}{4} \tag{26.33}$$

2. Massive charged weak vector mesons $W_\mu^{(\pm)}$ with masses

$$M_W^2 = \frac{v^2 g^2}{4} = M_Z^2 \cos^2 \theta_W \tag{26.34}$$

3. A massless photon

$$M_\gamma^2 = 0 \tag{26.35}$$

The lagrangian retains the exact local $U(1)$ gauge invariance generated by the electric charge $\hat{Q}$, corresponding to QED.

The total lagrangian for the standard model as presented so far is the sum of the individual contributions discussed above

$$\mathscr{L} = \mathscr{L}_{\text{lepton}} + \mathscr{L}_{\text{nucleon}} + \mathscr{L}_{\text{gauge}} + \mathscr{L}_{\text{scalar}} \tag{26.36}$$

Note that this lagrangian now contains all the electroweak interactions [Cu83, Wa95]. The coupling of the leptons to the gauge bosons follows immediately from Eqs. (26.16) and (26.29)[6]

$$\begin{aligned}
\mathscr{L}_{\text{lepton}}^{(\pm)} &= \frac{g}{2\sqrt{2}}[j_\mu^{(+)}W_\mu + j_\mu^{(-)}W_\mu^\star] \\
\mathscr{L}_{\text{lepton}}^{(0)} &= -\frac{g}{2\cos\theta_W}j_\mu^{(0)}Z_\mu \\
\mathscr{L}_{\text{lepton}}^{\gamma} &= e_p j_\mu^\gamma A_\mu
\end{aligned} \tag{26.37}$$

Here the electric charge e_p is defined by

$$e_p \equiv \frac{gg'}{(g^2 + g'^2)^{1/2}} \tag{26.38}$$

---

[6] The details of this algebra are provided in [Wa95].

The lepton currents are given by the following expressions

$$
\begin{aligned}
j_\mu^{(\pm)} &= i\bar\psi_l\gamma_\mu(1+\gamma_5)\tau_\pm\psi_l \\
j_\mu^\gamma &= i\bar\psi_l\gamma_\mu\left[-\frac{1}{2}(1-\tau_3)\right]\psi_l \\
j_\mu^{(0)} &= i\bar\psi_l\gamma_\mu(1+\gamma_5)\frac{1}{2}\tau_3\psi_l - 2\sin^2\theta_W j_\mu^\gamma
\end{aligned}
\tag{26.39}
$$

The interaction of the point nucleons with the gauge fields takes exactly the same form as in Eqs. (26.37), with hadronic currents given by

$$
\begin{aligned}
\mathscr{J}_\mu^{(\pm)} &= i\bar\psi\gamma_\mu(1+\gamma_5)\tau_\pm\psi \\
J_\mu^\gamma &= i\bar\psi\gamma_\mu\left[\frac{1}{2}(1+\tau_3)\right]\psi \\
\mathscr{J}_\mu^{(0)} &= i\bar\psi\gamma_\mu(1+\gamma_5)\frac{1}{2}\tau_3\psi - 2\sin^2\theta_W J_\mu^\gamma
\end{aligned}
\tag{26.40}
$$

The lepton and nucleon doublets appearing in these currents are defined by

$$
\psi_l = \begin{pmatrix} \psi_{v_e} \\ \psi_e \end{pmatrix} \qquad \psi = \begin{pmatrix} \psi_p \\ \psi_n \end{pmatrix}
\tag{26.41}
$$

An analysis of the S-matrix for single, heavy weak boson exchange shows how interactions with the gauge bosons of the form in Eqs. (26.37) lead to an effective current–current lagrangian in the low-energy, nuclear domain where $\mathbf{q}^2 \ll M_W^2, M_Z^2$. In particular, comparison with that analysis immediately establishes the following relationships between the gauge couplings and masses of the standard model and the traditional weak Fermi coupling constant [Wa95]

$$
\frac{G}{\sqrt{2}} = \frac{g^2}{8M_W^2} = \frac{g^2}{8M_Z^2\cos^2\theta_W}
\tag{26.42}
$$

It is also evident that the total weak currents here receive additive contributions from the leptons and hadrons

$$
\begin{aligned}
\mathscr{J}_\lambda^{(\pm)} &= \mathscr{J}_\lambda^{(\pm)}(\text{hadrons}) + j_\lambda^{(\pm)}(\text{leptons}) \\
\mathscr{J}_\lambda^{(0)} &= \mathscr{J}_\lambda^{(0)}(\text{hadrons}) + j_\lambda^{(0)}(\text{leptons})
\end{aligned}
\tag{26.43}
$$

The semileptonic parts of this effective low-energy lagrangian form the basis of most of the nuclear applications. Formulation in terms of quarks, discussed below, simply changes the underlying structure of $\mathscr{J}_\lambda(\text{hadrons})$.

The corresponding effective four-fermion lagrangians are [Wa95]

$$\mathcal{L}_{\text{eff}}^{(\pm)} = \frac{iG}{\sqrt{2}} \left\{ [\bar{\psi}_e \gamma_\lambda (1 + \gamma_5)\psi_{v_e} + (e \leftrightarrow \mu)] \mathcal{J}_\lambda^{(+)}(\text{hadrons}) \right. \tag{26.44}$$

$$+ [\bar{\psi}_{v_e} \gamma_\lambda (1 + \gamma_5)\psi_e + (e \leftrightarrow \mu)] \mathcal{J}_\lambda^{(-)}(\text{hadrons}) \Big\}$$

$$\mathcal{L}_{\text{eff}}^{(v)} = \frac{iG}{\sqrt{2}} [\bar{\psi}_{v_e} \gamma_\lambda (1 + \gamma_5)\psi_{v_e} + (e \leftrightarrow \mu)] \mathcal{J}_\lambda^{(0)}(\text{hadrons})$$

$$\mathcal{L}_{\text{eff}}^{(l)} = -\frac{iG}{\sqrt{2}} \left[\bar{\psi}_e \gamma_\lambda (1 + \gamma_5)\psi_e - 4\sin^2\theta_W \bar{\psi}_e \gamma_\lambda \psi_e + (e \leftrightarrow \mu) \right] \mathcal{J}_\lambda^{(0)}$$

The theory as formulated assumes massless fermions. The fermion mass is now *put in by hand*. One adds Yukawa couplings of the fermions to the previously introduced complex scalar field that preserve the local $SU(2)_W \otimes U(1)_W$ local gauge symmetry. One such coupling is introduced for each fermion field. The fermions then acquire mass when the scalar field develops its vacuum expectation value. As a consequence of this procedure, each fermion also has a prescribed Yukawa coupling to the *fluctuation* of the scalar field about its vacuum expectation value — the real scalar Higgs. We illustrate the procedure in the case of leptons.[7]

Start with the following lagrangian with Yukawa couplings of the fermions to the complex scalar field and invariant under local $SU(2)_W \otimes U(1)_W$

$$\mathcal{L}_{\text{int}} = -G_e \bar{R}(\phi^\dagger \underline{L}) + \text{h.c.} \tag{26.45}$$

Each term is a weak isoscalar, and each term is neutral in weak hypercharge (Table 26.1). Now with the previously discussed spontaneous symmetry breaking, and in the unitary gauge ϕ is given by Eq. (26.24). Substitution into Eq. (26.45) and the use of Eq. (26.9) then gives

$$\mathcal{L}_{\text{int}} = -G_e \bar{e}_R \left\{ [0, \frac{1}{\sqrt{2}}(v + \eta)] \begin{pmatrix} v_L \\ e_L \end{pmatrix} \right\} + \text{h.c.}$$

$$= -\frac{v}{\sqrt{2}} G_e \bar{e}e - \frac{\eta}{\sqrt{2}} G_e \bar{e}e \tag{26.46}$$

The first term is the sought-after fermion mass (there is one adjustable coupling constant for each fermion mass in the theory). The second term is the remaining Yukawa interaction with the real scalar Higgs particle, with a prescribed coupling determined by the mass of the fermion.

The deeper formulation of the electroweak theory is in terms of quarks. At first glance, one might expect that the first quark weak isodoublet would just be that constructed from (u, d) quarks. The actual quark weak

---

[7] For point nucleons see [Wa95].

isospin doublets that couple in the electroweak interaction have a more complicated form [Ca63, Gl70]. They are

$$q_L = \begin{pmatrix} u_L \\ d_L \cos\theta_C + s_L \sin\theta_C \end{pmatrix} \equiv \begin{pmatrix} u_L \\ d_{cL} \end{pmatrix} \qquad (26.47)$$

$$Q_L = \begin{pmatrix} c_L \\ -d_L \sin\theta_C + s_L \cos\theta_C \end{pmatrix} \equiv \begin{pmatrix} c_L \\ D_{cL} \end{pmatrix} \quad ; \text{ weak doublets}$$

The fact that it is a rotated combination of fields in the charge-changing current, which includes a small strangeness-changing component, was first noted by Cabibbo [Ca63]. The discovery that one requires a second doublet with an additional c quark and the orthogonal rotated combination is due to Glashow, Iliopolous, and Maiani (GIM) [Gl70] who *predicted* the existence of the c quark on the basis of the arguments given below.[8]

As before, the right-handed quark fields form weak isosinglets

$$u_R, d_R, s_R, c_R \qquad\qquad ; \text{ weak singlets} \qquad (26.48)$$

The quarks are assigned the weak quantum numbers in Table 26.2. The assignments are again made so that the electric charge operator is

$$\hat{Q} = (\hat{T}_3 + \frac{1}{2}\hat{Y})_W \qquad (26.49)$$

Because one has two orthogonal linear combinations, the following (GIM) identity holds

$$\begin{aligned} \bar{d}_c d_c + \bar{D}_c D_c &= (\bar{d}\cos\theta_C + \bar{s}\sin\theta_C)(d\cos\theta_C + s\sin\theta_C) \\ &\quad + (-\bar{d}\sin\theta_C + \bar{s}\cos\theta_C)(-d\sin\theta_C + s\cos\theta_C) \\ &= \bar{d}d + \bar{s}s \end{aligned} \qquad (26.50)$$

No off-diagonal, strangeness-changing terms appear in this expression; as a consequence, the neutral currents generated in the standard model have no lowest-order strangeness-changing components — an empirical observation that was the primary motivation for the introduction of the c quark in [Gl70].

The GIM identity can be used to rewrite the non-interacting quark kinetic energy as

$$\mathcal{L}^0_{\text{quark}} = -\left[\bar{q}_L \gamma_\mu \frac{\partial}{\partial x_\mu} q_L + \bar{Q}_L \gamma_\mu \frac{\partial}{\partial x_\mu} Q_L \right. \qquad (26.51)$$

$$\left. + \bar{u}_R \gamma_\mu \frac{\partial}{\partial x_\mu} u_R + \bar{d}_R \gamma_\mu \frac{\partial}{\partial x_\mu} d_R + \bar{s}_R \gamma_\mu \frac{\partial}{\partial x_\mu} s_R + \bar{c}_R \gamma_\mu \frac{\partial}{\partial x_\mu} c_R \right]$$

---

[8] The extension to include still another (heavy) quark family is discussed in [Wa95].

Table 26.2. Weak isospin and weak hypercharge assignments for the quarks.

Field /particle	q_L	Q_L	u_R	d_R	s_R	c_R
T_W	1/2	1/2	0	0	0	0
Y_W	1/3	1/3	4/3	$-2/3$	$-2/3$	4/3

The covariant derivatives acting on the quark fields are as before (see Table 26.2)

$$(\frac{\partial}{\partial x_\mu} - \frac{i}{2}g' Y_W B_\mu - \frac{i}{2}g\tau \cdot \mathbf{A}_\mu) \qquad ; \text{on isodoublets}$$

$$(\frac{\partial}{\partial x_\mu} - \frac{i}{2}g' Y_W B_\mu) \qquad ; \text{on isosinglets} \quad (26.52)$$

The gauge boson and Higgs sectors of the theory are exactly the same as discussed above. The electroweak currents representing the interaction with the physical gauge bosons can also be identified exactly as before [Wa95]. The charge-changing weak current is given by

$$\begin{aligned} \mathscr{J}_\mu^{(\pm)} &= i\bar{q}\gamma_\mu(1+\gamma_5)\tau_\pm q + i\bar{Q}\gamma_\mu(1+\gamma_5)\tau_\pm Q \\ \mathscr{J}_\mu^{(+)} &= i\bar{u}\gamma_\mu(1+\gamma_5)(d\cos\theta_C + s\sin\theta_C) \\ &\quad + i\bar{c}\gamma_\mu(1+\gamma_5)(-d\sin\theta_C + s\cos\theta_C) \end{aligned} \quad (26.53)$$

Note it is the Cabibbo-rotated combination that enters into these charge-changing currents. The electromagnetic current of QED is just the point Dirac current multiplied by the correct charge

$$J_\mu^\gamma = i\left[\frac{2}{3}(\bar{u}\gamma_\mu u + \bar{c}\gamma_\mu c) - \frac{1}{3}(\bar{d}\gamma_\mu d + \bar{s}\gamma_\mu s)\right] \quad (26.54)$$

The weak neutral current is

$$\begin{aligned} \mathscr{J}_\mu^{(0)} &= i\bar{q}\gamma_\mu(1+\gamma_5)\frac{1}{2}\tau_3 q + i\bar{Q}\gamma_\mu(1+\gamma_5)\frac{1}{2}\tau_3 Q - 2\sin^2\theta_W J_\mu^\gamma \\ \mathscr{J}_\mu^{(0)} &= \frac{i}{2}[\bar{u}\gamma_\mu(1+\gamma_5)u + \bar{c}\gamma_\mu(1+\gamma_5)c - \bar{d}\gamma_\mu(1+\gamma_5)d - \bar{s}\gamma_\mu(1+\gamma_5)s] \\ &\quad - 2\sin^2\theta_W J_\mu^\gamma \end{aligned} \quad (26.55)$$

The second equality follows with the aid of the GIM identity. Terms of the form $(\bar{s}d)$ or $(\bar{d}s)$ have been eliminated; hence there are no strangeness-changing weak neutral currents in this quark-based standard model, as advertised.

The quarks can be given mass in the same fashion as were the leptons above, although the argument is somewhat more complicated in the case of quarks [Ab73, Cu83, Wa95].

How does the standard model of electroweak interactions get combined with QCD, the theory of the *strong* forces binding quarks into hadrons? Consider for simplicity the nuclear domain of (u, d) quarks. Quarks now carry an additional color index that takes three values (R, G, B), and the quark field gets extended to

$$\psi = \begin{pmatrix} u \\ d \end{pmatrix} \rightarrow \begin{pmatrix} u_R & u_G & u_B \\ d_R & d_G & d_B \end{pmatrix} \equiv (\psi_R, \psi_G, \psi_B) \qquad (26.56)$$

These get combined into a three-component (actually multicomponent) field $\underline{\psi}$

$$\underline{\psi} \equiv \begin{pmatrix} \psi_R \\ \psi_G \\ \psi_B \end{pmatrix} \qquad (26.57)$$

Let $\underline{O}$ be a matrix that is the *identity* with respect to color, but an *arbitrary* matrix O with respect to flavor

$$\underline{O} \equiv \begin{pmatrix} O & & \\ & O & \\ & & O \end{pmatrix} \qquad (26.58)$$

Then under the extension of the quark fields to include color, all *electroweak currents* are defined to be correspondingly extended to

$$\begin{aligned} \bar{\psi}\gamma_\mu O\psi & \rightarrow \bar{\psi}_R\gamma_\mu O\psi_R + \bar{\psi}_G\gamma_\mu O\psi_G + \bar{\psi}_B\gamma_\mu O\psi_B \\ & \equiv \underline{\bar{\psi}}\gamma_\mu \underline{O}\underline{\psi} \end{aligned} \qquad (26.59)$$

Such currents are evidently invariant under strong $SU(3)_C$.

The full lagrangian of the strong and electroweak interactions thus takes the form (see [Do93] for an extended discussion)

$$\mathscr{L} = \mathscr{L}^0 + \mathscr{L}^{\text{int}}_{\text{QCD}} + \mathscr{L}^{\text{int}}_{\text{EW}} \qquad (26.60)$$

This lagrangian is locally gauge invariant under the full symmetry group

$$SU(3)_C \bigotimes SU(2)_W \bigotimes U(1)_W \qquad (26.61)$$

This full theory is renormalizable. It has the following characteristic properties:

- The electroweak interactions are colorblind — they are the same, independent of the color of the quarks;

- The gluons are absolutely *neutral* to the electroweak interactions — the electroweak interactions couple to the quarks.

Let us examine the implications of this development for nuclear physics. To summarize the weak and electromagnetic quark currents in the standard model, we have

$$\begin{aligned}
\mathscr{J}_\mu^{(+)} &= i\bar{u}\gamma_\mu(1+\gamma_5)[d\cos\theta_C + s\sin\theta_C] \\
&\quad + i\bar{c}\gamma_\mu(1+\gamma_5)[-d\sin\theta_C + s\cos\theta_C] \\
\mathscr{J}_\mu^{(0)} &= \frac{i}{2}[\bar{u}\gamma_\mu(1+\gamma_5)u + \bar{c}\gamma_\mu(1+\gamma_5)c \\
&\quad - \bar{d}\gamma_\mu(1+\gamma_5)d - \bar{s}\gamma_\mu(1+\gamma_5)s] - 2\sin^2\theta_W J_\mu^\gamma \\
J_\mu^\gamma &= i\left[\frac{2}{3}(\bar{u}\gamma_\mu u + \bar{c}\gamma_\mu c) - \frac{1}{3}(\bar{d}\gamma_\mu d + \bar{s}\gamma_\mu s)\right]
\end{aligned} \qquad (26.62)$$

Each current is actually a sum over three colors $\sum_{\text{colors}}(\cdots)$ leading to an operator which is an $SU(3)_C$ - singlet as discussed above.

To a good approximation, the hadrons that make up the nucleus are composed of (u,d) quarks. As a starting point for nuclear physics, consider that subspace of the full Hilbert space consisting of any number of (u,d) quarks and their antiquarks $(\bar{u},\bar{d})$. We refer to this as the *nuclear domain*. The quark field in this sector takes the form

$$\psi \doteq \begin{pmatrix} u \\ d \end{pmatrix} \qquad ; \text{ nuclear domain} \qquad (26.63)$$

Assume that the (u,d) quarks have the *same mass* in the lagrangian; they are in fact both nearly massless. In this case, the lagrangian of the strong interactions, with the full complexity of QCD, has an *exact symmetry* — the $SU(2)$ of strong isospin. This is the familiar isotopic spin symmetry of nuclear physics. It is important to note that one still has the full complexity of strong-coupling QCD with colored quarks and gluons in this truncated flavor sector of the nuclear domain; nevertheless, one can draw conclusions that are exact to all orders in the strong interactions using this strong isospin symmetry.

The quark field ψ in Eq. (26.63) forms an isodoublet under this strong isospin. The quark currents in Eqs. (26.62) can then be written in terms of this isospinor in the nuclear domain as follows

$$\begin{aligned}
J_\mu^\gamma &= i\bar{\psi}\gamma_\mu\left(\frac{1}{6} + \frac{1}{2}\tau_3\right)\psi \\
\mathscr{J}_\mu^{(\pm)} &= i\bar{\psi}\gamma_\mu(1+\gamma_5)\tau_\pm\psi \\
\mathscr{J}_\mu^{(0)} &= i\bar{\psi}\gamma_\mu(1+\gamma_5)\frac{1}{2}\tau_3\psi - 2\sin^2\theta_W J_\mu^\gamma
\end{aligned} \qquad (26.64)$$

The properties of these currents under general symmetry properties of the theory now follow by inspection [Wa95]

$$
\begin{aligned}
\mathscr{I}_\mu &= J_\mu + J_{\mu 5} & &; V - A \\
\mathscr{I}_\mu^{(\pm)} &= \mathscr{I}_\mu^{V_1} \pm i\mathscr{I}_\mu^{V_2} & &; \text{isovector} \\
J_\mu^\gamma &= J_\mu^S + J_\mu^{V_3} & &; \text{EM current} \\
J_\mu^{(\pm)} &= J_\mu^{V_1} \pm iJ_\mu^{V_2} & &; \text{CVC} \\
\mathscr{I}_\mu^{(0)} &= \mathscr{I}_\mu^{V_3} - 2\sin^2\theta_W J_\mu^\gamma & &; \text{standard model} \quad (26.65)
\end{aligned}
$$

Here the Cabibbo angle has been absorbed into the definition of the hadronic weak charge-changing Fermi coupling constant

$$
G^{(\pm)} \equiv G\cos\theta_C \qquad\qquad \cos\theta_C = 0.974 \qquad (26.66)
$$

Note that the numerical value of $\cos\theta_C$ is, in fact, very close to 1 [Cu83].

The first of Eqs. (26.65) indicates that the weak current is the sum of a Lorentz vector and axial vector, the second that the charge-changing weak current is an isovector, and the third that the electromagnetic current is the sum of an isoscalar and third component of an isovector. The fourth equation is the statement of CVC. The conserved vector current (CVC) relation states that the Lorentz vector part of the weak charge-changing current is simply obtained from the other spherical isospin components of the same isovector operator that appears in the electromagnetic current. As a consequence, one can relate matrix elements of the Lorentz vector part of the charge-changing weak currents to those of the isovector part of the electromagnetic current by use of the Wigner–Eckart theorem applied to isospin. The resulting relations are then independent of the details of hadronic structure; they depend only on the existence of the isospin symmetry of the strong interactions. CVC is a powerful, deep, and far-reaching result, for it established the first direct relation between the electromagnetic and weak interactions which *a priori have nothing to do with each other!* All known applications of CVC are consistent with experiment. The last of Eqs. (26.65) exhibits the structure of the weak neutral current of the standard model in the nuclear domain.

If the discussion is extended to that sector of the full theory with no *net* strangeness or charm, and the electroweak interactions are treated in lowest order, then the first four of Eqs. (26.65) still hold; however, the weak neutral current is modified by the addition of an isoscalar contribution

$$
\delta\mathscr{I}_\mu^{(0)} = \frac{i}{2}[\bar{c}\gamma_\mu(1+\gamma_5)c - \bar{s}\gamma_\mu(1+\gamma_5)s] \qquad (26.67)
$$

In this sector of the theory, (s, c) quarks and their antiparticles $(\bar{s}, \bar{c})$ enter through loop processes.

27

Parity violation

We are now in a position to understand the standard model foundation of the analysis carried out in chapter 16 of parity violation in the process $A(\vec{e}, e')$, where A includes the nucleon. The Feynman rules for the diagrams shown in Fig. 16.1 follow immediately from the lepton currents in Eqs. (26.39) and quark currents in Eqs. (26.54) and (26.55). The result is the S-matrix in Eq. (16.1). The analysis in chapter 16 then leads to the general expression for the parity-violating asymmetry given in Eq. (16.20), where the response functions are defined in terms of matrix elements of the current by Eqs. (16.21, 16.22). One application has already been presented in chapter 16. Here we briefly discuss two others.

The measurement of parity violation in the scattering of longitudinally polarized electrons in deep-inelastic electron scattering from deuterium at SLAC is a classic experiment which played a pivotal role in the establishment of the weak neutral current structure of the standard model [Pr78, Pr79].

The analysis of parity violation in inclusive DIS in the quark–parton model was given in the end of chapter 16. The response functions $W_{1,2}^{\text{int}}$ and W_8^{int} are given in terms of the quark charges and momentum distributions by Eqs. (16.35) and (16.37). The electromagnetic and weak neutral charges of the quarks follow from the discussion in chapter 26; they have already been presented in Table 16.1.

Here we carry out a very simplified calculation of $^2\text{H}(\vec{e}, e')$ in the deep inelastic region [Wa95]. Assume forward angles with $\theta_e \to 0$ as in the SLAC experiment. Assume also that $\sin^2 \theta_W \approx 1/4$. It follows from Eqs. (16.2) and (16.20) that the asymmetry is then given by

$$\mathscr{A} = -\frac{G q^2}{4\pi\alpha\sqrt{2}} \left[\frac{\nu W_2(\nu, q^2)^{\text{int}}}{\nu W_2(\nu, q^2)^\gamma} \right] \tag{27.1}$$

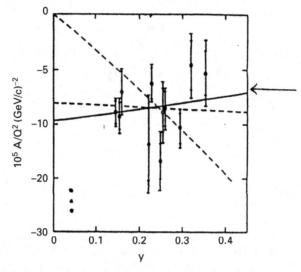

Fig. 27.1. SLAC results for parity-violating asymmetry for scattering of longitudinally polarized electrons from deuterium at forward angles in DIS region [Pr78, Pr79]. Here $y \equiv (E_0 - E')/E_0$. The result in Eq. (27.4) is also shown.

For a nucleon, assume the nuclear domain with just three valence quarks, and assume that the quark distribution functions $f_i(q^2)$ are, in fact, identical for these valence quarks. If this is the case, the required ratio of structure functions reduces simply to a ratio of charges

$$\frac{vW_2(v, q^2)^{\text{int}}}{vW_2(v, q^2)^{\gamma}} = \frac{\sum_i 2Q_i^{\gamma} Q_i^{(0)}}{\sum_i (Q_i^{\gamma})^2} \qquad (27.2)$$

The target in the initial SLAC experiment was a deuteron, which consists of a very loosely bound neutron and proton. In this case, the cross sections are just an incoherent sum of the cross sections from the nucleon constituents. An incoherent sum of the corresponding structure functions yields

$$\mathscr{A}_{^2\mathrm{H}} = -\frac{Gq^2}{4\pi\alpha\sqrt{2}} 2 \left\{ \frac{[\sum_i Q_i^{\gamma} Q_i^{(0)}]_p + [\sum_i Q_i^{\gamma} Q_i^{(0)}]_n}{[\sum_i (Q_i^{\gamma})^2]_p + [\sum_i (Q_i^{\gamma})^2]_n} \right\} \qquad (27.3)$$

A proton consists of (uud) and a neutron of (udd) valence quarks. The required charges may now be read off directly from Table 16.1. The result is

$$\mathscr{A}_{^2\mathrm{H}} = -\frac{Gq^2}{2\pi\alpha\sqrt{2}} \frac{2}{5} \qquad (27.4)$$

The SLAC results are shown in Fig. 27.1. The simple result in Eq. (27.4) is also indicated. It gives a very nice first explanation of the data. A much more sophisticated analysis of the parity-violating DIS process is presented in [Ca78].

As a second example, consider parity violation in elastic scattering from the nucleon. The single-nucleon matrix element of the weak neutral current in the standard model must have the general form[1]

$$\langle p' | \mathscr{J}_\mu^{(0)}(0) | p \rangle = \frac{i}{\Omega} \bar{u}(p')[F_1^{(0)}\gamma_\mu + F_2^{(0)}\sigma_{\mu\nu}q_\nu + F_A^{(0)}\gamma_5\gamma_\mu - iF_P^{(0)}\gamma_5 q_\mu]u(p)$$

(27.5)

The matrix element of the electromagnetic current has the form given in Eqs. (19.6) and (19.7). It is then simply an exercise in Dirac algebra to show that for relativistic electrons the parity-violating asymmetry for $N(\vec{e}, e)N$ is given by [Po87]

$$\mathscr{A}\left\{[(F_1^\gamma)^2 + q^2(F_2^\gamma)^2]\cos^2\frac{\theta}{2} + \frac{q^2}{2m^2}(G_M^\gamma)^2\sin^2\frac{\theta}{2}\right\} =$$

$$-\frac{Gq^2}{2\pi\alpha\sqrt{2}}\left\{[F_1^{(0)}F_1^\gamma + q^2 F_2^{(0)}F_2^\gamma]\cos^2\frac{\theta}{2} + \frac{q^2}{2m^2}G_M^{(0)}G_M^\gamma\sin^2\frac{\theta}{2}\right.$$

$$\left. -\frac{\sin\theta/2}{m}\sqrt{q^2\cos^2\frac{\theta}{2} + \vec{q}^2\sin^2\frac{\theta}{2}}\, G_M^\gamma(1 - 4\sin^2\theta_W)F_A^{(0)}\right\}$$

(27.6)

Here the Sachs form factors are defined by

$$G_M = F_1 + 2mF_2$$

$$G_E = F_1 - \frac{q^2}{2m}F_2$$

(27.7)

The discussion in chapter 26 implies that within the framework of QCD in the nuclear domain of equal mass (u, d) quarks (which implies strong isospin invariance), the form factors appearing in this expression must have the form

$$F_{1,2}^{(0)} = F_{1,2}^V - 2\sin^2\theta_W F_{1,2}^\gamma$$

$$F_A^{(0)} = F_A^V$$

(27.8)

Here $F_{1,2}^{S,V}$ are obtained from electron scattering and F_A^V from charged current semi-leptonic weak interactions.

In the extended domain of (u, d, s, c) quarks and strong isospin invariance, Eq. (26.67) implies there is an additional isoscalar term in the

---

[1] Hermiticity of the current again implies that the form factors in this expression are real.

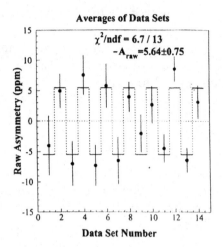

Fig. 27.2. Average value of raw asymmetry (difference/sum) observed with longitudinally polarized electron beam on a proton target for each data set. Odd data sets have the half-wave plate inserted in the laser beam (at the injector) and are expected to have the opposite asymmetry. Note the scale is *parts per million* (ppm). From the HAPPEX experiment at CEBAF [An99].

weak neutral current, so each of the form factors will have an additional isoscalar contribution

$$F_i^{(0)} \rightarrow F_i^{(0)} + \delta F_i^S \tag{27.9}$$

A parity-violation experiment to determine the distribution of weak neutral charge in the proton has been carried out at CEBAF. Figure 27.2 shows the measured asymmetry when nothing but the incident photon polarization is reversed at the injector on a macroscopic time scale using a half-wave plate. In the nuclear domain with only the light u and d quarks and their antiquarks, the weak neutral charge distribution should be identical to that of the electromagnetic charge. Any difference must arise from s (heavy) quarks. No difference is found, a result which has profound implications for our understanding of the structure of matter.

In more detail, a measurement of the parity-violating electroweak asymmetry in the elastic scattering of polarized electrons from the proton is presented in [An99]. The kinematic point $[\langle\theta_{\text{lab}}\rangle = 12.3°$ and $\langle Q^2 \rangle = 0.48\,\text{GeV}^2\text{c}^{-2}]$ is chosen to provide sensitivity, at a level that is of theoretical interest, to the strange electric form factor G_E^S. The result, $\mathscr{A} = -14.5 \pm 2.2\,\text{ppm}$, is consistent with the electroweak standard model and no additional contributions from the strange quarks. In particular, the measurement implies $G_E^S + 0.39 G_M^S = 0.023 \pm 0.034(\text{stat}) \pm 0.022\,(\text{syst}) \pm 0.026\,(\delta G_E^n)$, where the last uncertainty arises from the estimated uncertainty in the neutron electric form factor.

28

Excitation of nucleon resonances

One of the primary goals of electron scattering experiments is to understand the internal structure of the nucleon, both its static and dynamic properties. Ultimately, electron scattering data will provide benchmarks against which the theoretical predictions of QCD can be compared.

Elastic scattering from the nucleon has been discussed in chapter 22. There are no discrete bound states of the nucleon as there are in nuclei, and thus excited states of the nucleon show up as resonances in particle production processes. This is analogous to the situation with giant resonances in nuclei which lie above particle emission threshold. Nucleon resonances are characterized by strong interaction widths, a typical value for which is given by the time it takes a light signal to travel a pion Compton wavelength, or $\Gamma \approx \hbar c/(\hbar/m_\pi c) \approx m_\pi c^2 \approx 135 \, \mathrm{MeV}$.

The first inelastic process on the nucleon occurs with the production of the lightest hadron, the pion. The coincidence cross section for the reaction $\mathrm{N}(\mathrm{e}, \mathrm{e}' \pi)\mathrm{N}$ follows immediately from the general analysis in chapter 13. The angular distribution in the C-M system for arbitrary nucleon helicities is given by Eq. (13.68). If the nucleon target is unpolarized and its final polarization unobserved, the angular distribution reduces to that given in Eqs. (13.71) and (F.9). The analysis of pion electroproduction starting from the covariant, gauge invariant S-matrix and reducing it to the contribution of multipoles leading to states of definite J^π in the final π–N system is presented in detail in appendix H. Such a decomposition forms the basis for current phenomenological analyses of coincident electron scattering experiments aimed at extracting properties of nucleon resonances. Existing pion electroproduction data is presented in [Br82, Br83, Fo83] and discussed further in [Bu94]. The reader is referred to these references for previous applications.

To get some idea of the quality of the data that is now becoming

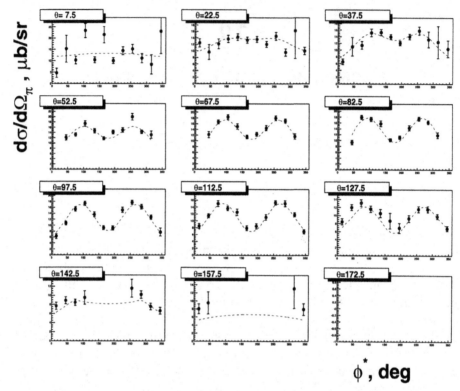

Fig. 28.1.   Preliminary angular distribution data $d\sigma/d\Omega_q$ in μb/sr for process $p(e,e'\pi^+)n$ on the first nucleon resonance from Hall B collaboration at TJNAF [Eg01]. Here $k^2 = 0.40\,(\text{GeV}/c)^2$, $W = 1.230\,\text{GeV}/c^2$, $\Delta k^2 = 0.100\,(\text{GeV}/c)^2$, and $\Delta W = 0.020\,\text{GeV}/c^2$. The plots are vs. $\phi^\star = \pi/2 - \phi_q$ for various θ_q. The author is grateful to H. Egiyan for preparing this figure.

available on the coincident electropion production process, we show in Fig. 28.1 some of the very first results for the process $p(e,e'\pi^+)n$ from the Hall B collaboration at TJNAF [Eg01].

QCD-inspired models of the internal structure of the nucleon give rise to a rich structure of dynamic excitations. One now has a quark-based picture of the underlying structure similar to that of the periodic table of the elements in atomic physics, or the shell model in nuclear physics [Bh88, Wa95]. The M.I.T. bag models confinement and asymptotic freedom with three massless quarks moving in a vacuum bubble [Ch74, Ch74a, De75, Ja76]. The constituent quark model has three non-relativistic quarks with masses $m_q \approx M/3$ moving in a confining potential, for example, a harmonic oscillator [Is77, Is80, Is81, Is85]. Electron scattering coincidence studies of reactions proceeding through these resonances $N(e,e')N^\star \to$

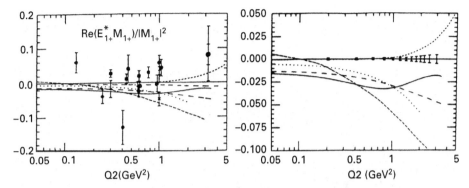

Fig. 28.2. (left) Existing world's data on $\mathrm{Re}\,(E_{1+}^{*}M_{1+})/|M_{1+}|^2$ at the $\Delta(1232)$ as of CEBAF PR89-037 [Bu89, Wa93]. Here $k \equiv Q$.

Fig. 28.3. (right) Projected range and error bars on $\mathrm{Re}\,(E_{1+}^{*}M_{1+})/|M_{1+}|^2$ at the $\Delta(1232)$ in CEBAF PR89-037 [Bu89, Wa93]. Here $k \equiv Q$.

N(e, e′ X)N promise to teach us much about the internal dynamics of the nucleon [Bu94].[1]

The quark model prediction for electron excitation of the first excited state of the nucleon, the $\Delta(1232)$, has been examined in chapter 24. It is clear from Fig. 12.8 that this first excited state is seen experimentally as a nice, isolated resonance.

The electric quadrupole transition amplitude E_{1+} to the $\Delta(1232)$ with $(J^\pi, T) = (\frac{3}{2}^+, \frac{3}{2})$ is particularly interesting. Quark bag models of the nucleon, with a one-gluon exchange interaction, indicate that the bag may deform — similar to the deformation of the deuteron arising from the tensor force. As with even–even deformed nuclei, the nucleon can have no quadrupole moment in its ground state, so the most direct evidence for such an intrinsic deformation would show up in this transition amplitude. In the quark model, the transition amplitude to the $P_{33}(1232)$ is predominantly a spin-flip magnetic dipole M_{1+}. The E_{1+} is, in fact, observed to be small, and it is currently only very poorly known. This is illustrated in Fig. 28.2, which shows the existing world's data on $\mathrm{Re}\,(E_{1+}^{*}M_{1+})/|M_{1+}|^2$ at the $\Delta(1232)$ as of the proposal CEBAF PR 89-037 [Bu89]. Figure 28.3 shows the projected range and error bars in that proposal [Bu89]. Note, in particular, the expansion of the vertical scale in this second figure. At TJ-NAF (CEBAF), the internal dynamics of the nucleon will be studied with unrivaled precision. These measurements will provide *deep insight into the dynamical consequences of QCD*. The accurate new data will continue to

---

[1] See this review article [Bu94] for an extensive list of further references on electron excitation of the nucleon and the quark model.

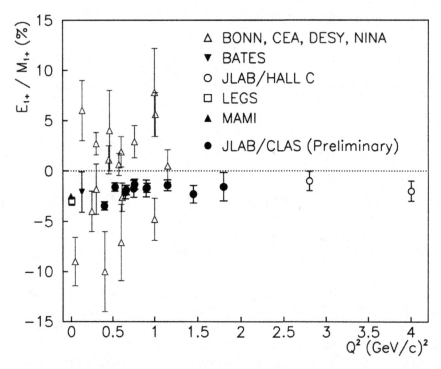

Fig. 28.4. Preliminary results obtained for E_{1+} by the Hall B collaboration at TJNAF from the reaction $p(e, e'\pi^0)p$ [HB01]. The author is grateful to V. Burkert and C. Smith for the preparation of this figure.

provide *benchmark tests for theoretical quark-model and QCD descriptions of the nucleon — the basic building block of matter.*

Figure 28.4 shows actual data on this ratio obtained from an analysis of the process $p(e, e'\pi^0)p$ by the Hall B collaboration at TJNAF [HB01]. Note the quality of these results.

It is evident from Fig. 12.8 and the Particle Data Book that the higher nucleon resonances are many, broad, and overlapping. It will be a challenge to isolate the individual resonance contributions, particularly when there is a substantial background contribution as is evident from Fig. 12.8. A second challenge is to have a completely relativistic description of the quark bound-state structure of the nucleon; this is essential when one goes to momentum transfers $k^2 \gg m_q^2$.[2]

As first shown in a non-relativistic static model by Chew and Low [Ch56a], and subsequently generalized to the relativistic case [Ch57, Fr60], the $\Delta(1232)$ can be alternatively obtained as a dynamic resonance in a pion–nucleon field theory (QHD). Here, instead of starting at short

---

[2] Relativistic corrections to the constituent quark model are examined in [Ca86, Ca87].

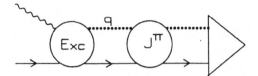

Fig. 28.5.   Electroexcitation of the first nucleon resonance.

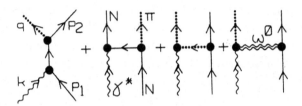

Fig. 28.6.   Generalized Feynman amplitude used as the excitation mechanism in the $|\pi N\rangle$ channel for the $\Delta(1232)$ [Pr69, Wa72]. It is assumed that $F_\pi \approx F_1^V$. [For the last graph, which plays a minor role for the $\Delta(1232)$, $g_{\omega\pi\gamma}$ is obtained from $\omega \to \pi + \gamma$, and $g_{\omega\pi\gamma}g_{\omega NN}$ from an overall fit to the inelastic resonance spectra; it is assumed that $F_{\omega\pi\gamma} \approx F_2^V/F_2^V(0)$.]

distances with an asymptotically free quark model, one starts at large distances with a pion–nucleon description of the structure of the nucleon. Electron excitation of this resonance can then be viewed as an excitation process into the proper π–N channel followed by a dynamic final-state enhancement that builds up the resonance, as illustrated in Fig. 28.5.

As a model for N(e, e′)Δ consider the following [Wa68, Pr69, Wa72]

$$a(W,k^2) = \frac{a^{\mathrm{lhs}}(W,k^2)}{D(W)}$$

$$D(W) = \exp\left\{ -\frac{1}{\pi} \int_{W_0}^{\infty} \frac{\delta(W')\,dW'}{W' - W - i\varepsilon} \right\} \qquad (28.1)$$

Here $a^{\mathrm{lhs}}(W,k^2)$ is the appropriate multipole projection of a set of Feynman graphs thought to play an important role in the excitation of the resonance and $D(W)$ is a final-state enhancement factor. The sum of excitation graphs is treated as a generalized Feynman amplitude in that renormalized coupling constants and electromagnetic form factors $F(k^2)$ are used at the vertices; the justification for this procedure is that this amplitude has the correct left-hand singularity structure arising from the pole terms in a dispersion treatment of this process [Fu58]. The graphs used in the present calculation [Wa68, Pr69, Wa72] are shown in Fig. 28.6. The multipole projections are obtained through the analysis in appendix H.

The excitation amplitude arising from the first three graphs is constructed below.

Division of numerator and denominator in Eq. (28.1) by $D(M_s)$ changes nothing and

$$\frac{D(W)}{D(M_s)} = \exp\left\{-\frac{(W - M_s)}{\pi}\int_{W_0}^{\infty}\frac{\delta(W')\,dW'}{(W' - M_s)(W' - W - i\varepsilon)}\right\} \quad (28.2)$$

This relation provides additional convergence. If one assumes that $a(W,k^2) \equiv a^{\text{lhs}}(W,k^2)$ at a given point, as in [Ch56a] where in the static limit one has a simple pole at $W = M$, then the quantity M_s is determined.[3]

This is a very simple model; but, it has several features to recommend it:

1. It has the correct analytic properties since $a^{\text{lhs}}(W,k^2)$ has the correct left-hand singularities and $D(W)$ has the right-hand unitarity cut;

2. It has the correct threshold behavior in both $|\mathbf{k}^*|$ and $|\mathbf{q}|$ [Bj66];

3. It satisfies Watson's theorem $a = |a|e^{i\delta}$ on the physical cut [Wa52]; here δ is the strong interaction π–N phase shift (see appendix H);

4. It is an approximate solution to the integral equation of Omnès [Om59];

5. The electroproduction amplitude resonates at the same W_R as elastic scattering;

6. The calculation is completely relativistic;

7. The current is conserved;

8. The k^2 dependence is explicit.

Let us elaborate on some of these points. The problem of constructing an analytic function $a(W,k^2)$ with a specified set of left-hand singularities in W given by $a^{\text{lhs}}(W,k^2)$, where $a^{\text{lhs}}(W,k^2)$ is real on the physical real axis and where the overall amplitude obeys Watson's theorem there, was formulated by Omnès as an integral equation [Om59]

$$a(W,k^2) = a^{\text{lhs}}(W,k^2) + \frac{1}{\pi}\int_{W_0}^{\infty}\frac{e^{-i\delta(W')}\sin\delta(W')\,a(W',k^2)}{W' - W - i\varepsilon}dW' \quad (28.3)$$

---

[3] The calculation shown uses $M_s = 0.95M$ and $\text{Re}\,\delta(W)$ everywhere in the integral.

The solution to this integral equation for $W_0 \leq W \leq \infty$ was also given by Omnès [Om59]

$$a(W,k^2) = e^{i\delta(W)} \left[a^{\text{lhs}}(W,k^2) \cos \delta(W) \right.$$
$$\left. + e^{\rho(W)} \frac{\mathscr{P}}{\pi} \int_{W_0}^{\infty} \frac{a^{\text{lhs}}(\xi,k^2) \sin \delta(\xi) e^{-\rho(\xi)}}{\xi - W} d\xi \right] \quad (28.4)$$

In this expression $\mathscr{P}$ is the Cauchy principal value and

$$\rho(W) = \frac{\mathscr{P}}{\pi} \int_{W_0}^{\infty} \frac{\delta(\zeta)\,d\zeta}{\zeta - W} \quad (28.5)$$

Assume now that $a^{\text{lhs}}(W,k^2)$ varies only slowly over the region where $\sin \delta(W) \neq 0$ on the physical cut, and factor it out of the integral. It then follows that

$$a(W,k^2) \approx a^{\text{lhs}}(W,k^2)\,\chi(W)$$
$$\chi(W) = \psi(W) \exp\left\{ \frac{1}{\pi} \int_{W_0}^{\infty} \delta(W')\,dW' \frac{1}{W' - W - i\varepsilon} \right\}$$
$$\psi(W) = \exp\left[-\frac{1}{\pi} \int_{W_0}^{\infty} \frac{\delta(W')\,dW'}{W' - W - i\varepsilon} \right]$$
$$+ \frac{1}{\pi} \int_{W_0}^{\infty} \frac{\sin \delta(\xi)\,d\xi}{\xi - W - i\varepsilon} \exp\left[-\frac{\mathscr{P}}{\pi} \int_{W_0}^{\infty} \frac{\delta(\zeta)\,d\zeta}{\zeta - \xi} \right] \quad (28.6)$$

The evident analytic properties of $\psi(W)$, and the observation that $\psi \to 1$ as $|W| \to \infty$, allow one to write an unsubtracted dispersion relation for $\psi(W) - 1$. A simple calculation shows that on the right-hand physical cut the discontinuity of this function vanishes, hence one concludes that $\psi(W) \equiv 1$!

It follows that

$$a(W,k^2) = \frac{a^{\text{lhs}}(W,k^2)}{D(W)}$$
$$D(W) = \exp\left[-\frac{1}{\pi} \int_{W_0}^{\infty} \frac{\delta(W')\,dW'}{W' - W - i\varepsilon} \right] \quad (28.7)$$

This is just Eq. 28.1. Here $D(W)$ serves as a final-state enhancement factor, and this final-state enhancement factor satisfies Watson's theorem

$$D(W) = |D(W)| e^{-i\delta(W)} \qquad ; W \geq W_0 \quad (28.8)$$

$D(W)$ is purely imaginary at a resonance in elastic scattering where $\delta(W_R) = \pi/2$. A Taylor series around the resonance then gives

$$D(W) \approx (W - W_R) \left[\frac{d\,\text{Re}\,D(W)}{dW} \right]_{W=W_R} + i\,\text{Im}\,D(W_R) \quad (28.9)$$

The electroproduction amplitude in this channel then resonates at the same W_R and has a Breit–Wigner form.

Finally, consider the Feynman diagrams in Fig. 28.6 as the excitation mechanism for the production of the low-lying nucleon resonances in Fig. 12.8. Treat these as generalized Feynman amplitudes using renormalized coupling constants and physical electromagnetic form factors $F(k^2)$ at the vertices; again, the justification for this procedure is that these terms give the correct pole contributions in the dispersion relations for the electroproduction amplitudes [Fu58]. The contribution of the nucleon and pion pole terms takes the form [Wa95]

$$\left(\frac{4\pi W}{M}\right)\mathscr{I}_\lambda^{\text{pole}}\varepsilon_\lambda = -g_\pi \bar{u}(p_2)\left\{\tau_\alpha M_\lambda^{(0)} + \delta_{\alpha 3} M_\lambda^{(+)} + \frac{1}{2}[\tau_\alpha,\tau_3]M_\lambda^{(-)}\right\}u(p_1)\varepsilon_\lambda$$

$$M_\lambda^{(i)} = \gamma_5 \frac{1}{i(\not{p}_1 + \not{k}) + M}[F_1^{(i)}\gamma_\lambda - F_2^{(i)}\sigma_{\lambda\rho}k_\rho]$$

$$+s_i[F_1^{(i)}\gamma_\lambda - F_2^{(i)}\sigma_{\lambda\rho}k_\rho]\frac{1}{i(\not{p}_2 - \not{k}) + M}\gamma_5$$

$$-if_i\gamma_5\frac{(2q-k)_\lambda}{(q-k)^2 + \mu^2}F_\pi \tag{28.10}$$

Here the spinor normalization is $\bar{u}u = 1$, the Feynman notation $\not{a} = a_\mu\gamma_\mu$ is employed, and the form factors are given by

$$\begin{aligned}2F^{(0)} &= F^S &\quad;\ s_0 &= +1 &\quad;\ f_0 &= 0 \\ 2F^{(+)} &= F^V &\quad;\ s_+ &= +1 &\quad;\ f_+ &= 0 \\ 2F^{(-)} &= F^V &\quad;\ s_- &= -1 &\quad;\ f_- &= 1\end{aligned} \tag{28.11}$$

If one assumes that $F_\pi(k^2) \approx F_1^V(k^2)$ in the region of interest, then the replacement $\varepsilon_\lambda \to k_\lambda$ gives zero; hence one concludes that this current is explicitly conserved. Multipoles can be projected from this amplitude through the procedures in appendix H.

The model presented here [Wa68, Pr69, Wa72] is a simple synthesis and summary of a great deal of theoretical work on $N(e, e')\Delta(1232)$ within a hadronic framework [Fu58, De61, Vi67, Za66, Ad68]. The result is shown as the theoretical curve in Fig. 12.9. Note that this QHD picture of the cross section to the first excited state of the nucleon holds out to $k^2 \approx 4\,\text{GeV}^2 = 100\,\text{fm}^{-2}$. The individual helicity amplitudes, which provide a much more detailed test of the picture, are compared with early experiments in [Pr70]. A coupled channel extension of this model exists that describes the inelastic form factors in the higher resonance regions in Fig. 12.8 [Pr69, Wa72].

Precise coincidence studies and measurement of all the amplitudes for all the excited states of the nucleon out to high k^2 will further challenge our understanding of the internal dynamics of the nucleon. Of major importance is the synthesis of a relativistic quark description of the internal dynamics of the nucleon with a meson field theory description of the dynamics of its external structure.[4] Ultimately, electron excitation of the nucleon will provide benchmark tests of *ab initio* calculations of QCD, perhaps through lattice gauge theory [Wi74].

---

[4] Various hybrid bag models are examined in [Bh88, Wa95].

Part 5
Future directions

29

TJNAF (CEBAF)

A top priority for the field of nuclear physics in the U.S. since the late 1970's, the Thomas Jefferson National Accelerator Facility (TJNAF) was approved for construction by Congress in 1987. This project was originally called CEBAF, the Continuous Electron Beam Accelerator Facility. It came into operation in Newport News, Virginia, in 1994. The first physics results were reported at the Particles and Nuclei International Conference (PANIC) held at the College of William and Mary in Williamsburg in 1996. The experimental program at TJNAF is now fully underway, and one can look forward to a steady output of significant experimental results providing insight into the structure of hadronic matter well into the 21st century.

Some of the new experimental results from TJNAF have already been referred to in this book, and we discuss more of the anticipated program in the next chapter. In order to fully understand the future opportunities this facility provides, we present a brief overview of the existing accelerator and major experimental equipment.

There is no single feature that makes TJNAF unique; each of the characteristics has been achieved previously at one location or another. Rather, it is the combination of properties that makes TJNAF (CEBAF) the world's most powerful microscope for looking at the nucleus. A schematic of the accelerator complex at TJNAF is shown in Fig. 29.1.

The accelerator itself is in the form of a racetrack 10 m underground. The basic accelerating structure is a superconducting niobium cavity fed with microwave power at a frequency of $\nu = 1497\,\mathrm{MHz}$. A longitudinal electric field in the cavities accelerates 0.5 mm-long packets of electrons down the linac, which is composed of self-contained cryomodules each containing four cavity pairs. After exiting from the injector, the electrons are moving at close to the velocity of light $c = 2.998 \times 10^{10}\,\mathrm{cm\,s^{-1}}$. The wavelength determining the longitudinal cavity dimension is then $\lambda = c/\nu = 20.0\,\mathrm{cm}$,

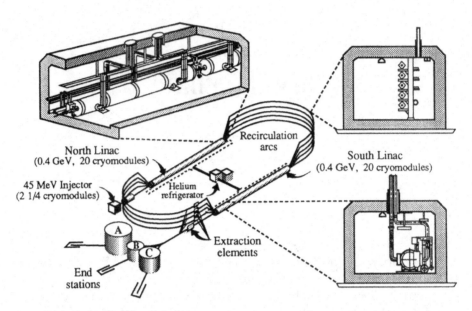

Fig. 29.1. Schematic of the accelerator complex at TJNAF [Le93].

and the period of the field oscillation is $\tau = 1/\nu = 0.668$ ns. The linac accelerates the electrons and increases the energy by 0.4 GeV on the first pass. The electron beam is then extracted from the first linac, passed through its own magnetic return arc, and re-injected into a second linac which adds an additional 0.4 GeV to the energy. The electron packets are again separated, magnetically returned, and re-injected into the first linac. It is a remarkable property of special relativity that even though the electrons may have different *energy*, they are moving with essentially the same *velocity*, namely the velocity of light c. The returned electron bunch can thus be placed spatially on top of a new bunch and accelerated with it. As designed, this process is repeated 5 times, until the electrons reach an energy of 4 GeV.[1] A precise electronic timing system keeps track of the location of each bunch, and on the final pass, the electrons are deflected out of the beam by an RF separator. They are then steered magnetically to one of three independent end stations. The electrons can, of course, be extracted after fewer circuits, to provide a beam of lower energy. In standard operation, every third bunch is steered to a given end station. The time between electron bunches is then $3\tau = 2.00$ ns during which time the electrons go 60.0 cm. At full current, 200 μA can be placed on target in each end station. Since the electron bunches come continuously, instead of in the well-separated macropulses of a room-temperature linac which

---

[1] The field gradient in the CEBAF cavities significantly exceeds design specification, and an accelerator upgrade to 6 GeV has been carried out.

requires recovery time to dissipate the generated heat, one says the beam is continuous; the *duty factor* of the machine is df = 100%. An electron beam polarized at the photo-cathode source can be readily transported through the accelerator and delivered on target. Polarizations as high as $\approx 75\%$ have now been obtained.

Although the cavities are superconducting, there are RF losses in the walls. The accelerating structures are immersed in superfluid ^{4}He at $\approx 2°$ K that carries off the generated heat. TJNAF is, in fact, the world's largest superfluid ^{4}He facility. Focusing magnetic elements keep the beam in line and well inside the machine aperture. The exiting beam has a remarkably low emittance, which means that it exits the machine as a fine pencil and stays that way for however far one wants to transport it on the site. The intrinsic energy resolution in the beam is $\delta E/E \approx 10^{-4}$ which implies that at 4 GeV one can resolve $\delta E \approx 400$ keV in the target, well suited to the energy scale of nuclear physics. Since not much happens at low temperatures, the beam is remarkably stable once the machine is up and running.

The experimental areas at TJNAF consist of three round halls of watertank construction that are independently fed by beam, and in which experiments can be carried out in parallel. In electron scattering experiments, it is always necessary to have one magnetic spectrometer to detect the scattered electron and define the virtual quantum of electromagnetic radiation interacting with the target. Since one of the great advantages of TJNAF is the 100% duty factor that allows coincidence experiments, at least one additional detector is required.

In Hall C, as illustrated in Fig. 29.2, there is a high momentum spectrometer (HMS) constructed from three superconducting quadrupoles and a superconducting dipole (QQQD). In simplest optical terms, the quadrupoles act as focusing lenses and the dipole as a dispersive prism with which momentum measurements are made. The final particle is observed in the detector hut with an appropriate detector package. The HMS is capable of detecting scattered electrons with momenta up to 6 GeV c^{-1}. The momentum resolution is moderate with $\delta p/p \approx 0.05$–0.1%. The solid angle acceptance is sizable with $\Delta\Omega \approx 6.7$ msr and the angular coverage for the scattered electron is $12°$–$90°$. Together with the HMS in Hall C, there is a short-orbit spectrometer (SOS) capable of detecting decaying secondaries with a maximum central momentum of 1.5 GeV c^{-1} and with $\delta p/p \approx 0.1\%$, $\Delta\Omega \approx 9$ msr, and angular range $12°$–$165°$.

In the initial commissioning phase, during the running of experiment E91-13, a brief test of the high momentum and short-orbit spectrometer pair for kaon detection was performed by the Hall C collaboration [HC96] (Fig. 29.3). In this experiment an electron collides with a proton, the nucleus of the element hydrogen, producing an electron and a

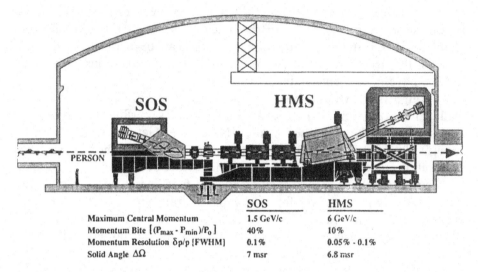

	SOS	HMS
Maximum Central Momentum	1.5 GeV/c	6 GeV/c
Momentum Bite $[(P_{max} - P_{min})/P_0]$	40%	10%
Momentum Resolution $\delta p/p$ [FWHM]	0.1%	0.05% - 0.1%
Solid Angle $\Delta\Omega$	7 msr	6.8 msr

Fig. 29.2.   Schematic of major detectors in Hall C at TJNAF [Do93a].

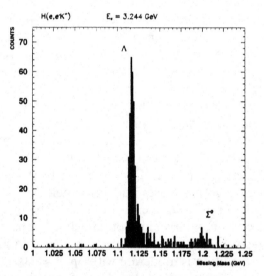

Fig. 29.3.   A brief test of the system for kaon detection in ^{1}H(e, e$'$ K$^+$) by the Hall C Collaboration during E91-13 at CEBAF [HC96, Wa97].

K^+ meson which are detected in coincidence, along with a variety of other particles which are not detected. In our notation this reaction is ^{1}H(e, e$'$ K$^+$). Quarks of positive and negative strangeness are created in pairs in this high-energy reaction, with the K^+ meson being the signature (and the carrier) of the positive strangeness quark. The missing mass (the collective masses of the unobserved particles) spectrum for this reaction,

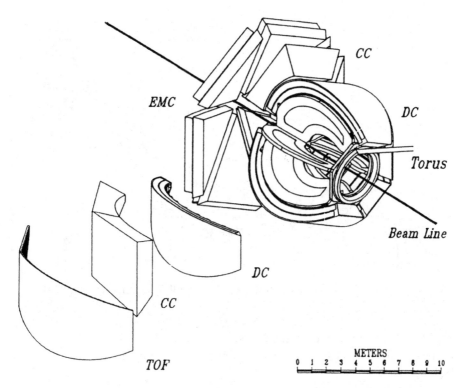

Fig. 29.4. Schematic of CLAS detector in Hall B at TJNAF [Do93a].

shown in Fig. 29.3, is indicative of the effectiveness of the system for kaon electroproduction studies that provide unique access to this process which implants strangeness (through the presence of the strange quarks) into the nucleus. Here the remaining Λ and Σ^0 hyperons produced in this reaction, which now carry negative strangeness, are clearly identified.

The CEBAF Large Acceptance Spectrometer (CLAS) is located in Hall B. This is a device designed to provide maximum coverage for particles emitted in coincidence with the scattered electron. A schematic of this device is shown in Fig. 29.4. Six thin superconducting current coils provide a toroidal field which surrounds the beam axis. Then, like segments of an orange, detector packages are inserted between the coils. The detectors contain drift chambers, Cerenkov counters, and time-of-flight scintillators. An electromagnetic calorimeter of segmented lead glass mounted in the forward direction allows one to detect the scattered electron. While the momentum resolution for the final electron is modest $\delta p/p \approx 1\%$, the detector is able to handle luminosities as high as $10^{34}\,\mathrm{cm}^{-2}\,\mathrm{s}^{-1}$ and has outstanding particle identification capability for

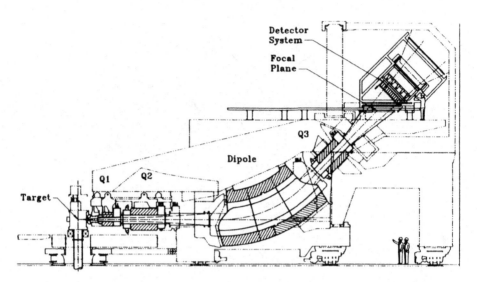

Fig. 29.5.   Schematic of HRS detectors in Hall A at TJNAF [Do93a].

single and multiparticle coincidences. An example of some of the initial data from the CLAS detector has already been given in Figs. 28.1 and 28.4.

In Hall A there is a pair of identical high resolution spectrometers (HRS) with properties matched to the outstanding quality of the TJNAF beam itself. The HRS is shown schematically in Fig. 29.5. The HRS is of a QQDQ design, with all magnetic elements superconducting. The momentum range is $0.3\text{-}4.0\,\mathrm{GeV}\,\mathrm{c}^{-1}$. The resolution is $\delta p/p \approx 10^{-4}$ and the solid angle coverage is a significant 7 msr. One spectrometer has a detector package for electrons, and the second for hadrons. A polarimeter exists which can be mounted in the hadron arm. The angular coverage of the electron spectrometer is $12.5°\text{-}165°$ and that of the hadron arm is $12.5°\text{-}130°$.

Polarization transfer has been discussed in appendix D. The polarization transfer experiment $^1\mathrm{H}(\vec{e}, e\,\vec{p})$ measures the product of the magnetic and electric form factors of the proton [Ar81]. Since the magnetic form factor is well known, this interference term allows an accurate determination of G_{Ep}. Figure 29.6 shows one of the first significant experimental results from TJNAF. This is a measurement of $G_{\mathrm{Ep}}/G_{\mathrm{Mp}}$ [Jo00]. The quality of the data is truly superb. The simplest interpretation of this data is that since the charge form factor falls off faster with Q^2 than the magnetic form factor, the charge density in the proton has a greater spatial extent than the magnetization density. This experiment provides fundamental information on the internal structure of the nucleon.

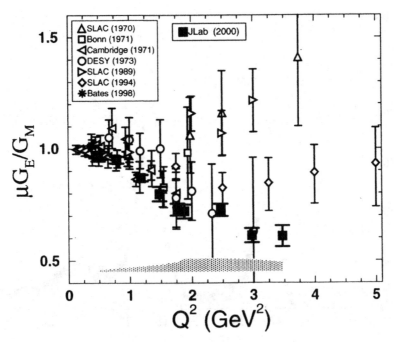

Fig. 29.6. Experimental data on the ratio G_{Ep}/G_{Mp} obtained from polarization transfer measurement $^1H(\vec{e},e\,\vec{p})$ at TJNAF [Jo00]. The author is grateful to C. Perdrisat and M. Jones for the preparation of this figure.

As we have seen in chapter 21, the reaction whereby a polarized electron incident on a nucleus produces a scattered electron and a polarized proton allows one to study how the nucleon spin propagates out from the nuclear interior. In this way one has a direct test of relativistic models of nuclear structure that describe the spin dependence of the nuclear shell model. One of the initial Hall A missing-mass spectra for this reaction on a $^{16}_{8}O$ target producing $^{15}_{7}N$ is shown in Fig. 29.7.

One of the first published experimental papers from TJNAF presents the results from CEBAF E89-12 shown in Fig. 29.8 [Bo98], along with results of the previous SLAC experiments NE8 and NE17, in addition to other measurements at lower photon (γ-ray) energy, E_γ. In the CEBAF experiment the electron beam strikes a target producing a beam of γ-rays which are used for the investigation of the quark structure of nucleons and nuclei. In E89-12 the photon strikes a deuterium nucleus, ^2_1H which consists of a bound proton and neutron, causing it to dissociate into a proton and a neutron. The SLAC experiment presented some evidence that this dissociation gave indications that QCD effects (the presence of sub-structure in the nucleons) played a role, because the observed reaction obeyed what are called simple constituent quark-

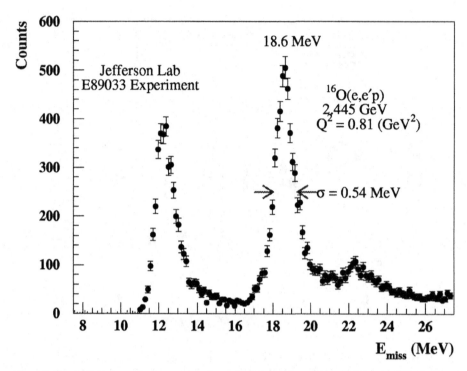

Fig. 29.7. Missing-mass spectrum in coincidence reaction $^{16}_{8}O(e, e'\, p)^{15}_{7}N^*$ taken at CEBAF at an incident electron energy of 2.445 GeV and four-momentum transfer squared of 0.81 $(GeV/c)^2$. The first peak is the $(1p_{1/2})^{-1}_{\pi}$ proton hole in $^{16}_{8}O$ and the second the $(1p_{3/2})^{-1}_{\pi}$. Note the overall energy resolution in $^{15}_{7}N$ of $\Delta E/E = 0.54/2445 = 2.2 \times 10^{-4}$ [Ma00a]. The author would like to thank C. Perdrisat and K. Wijesooriya for the preparation of this figure.

counting rules at high energy. This behavior is signaled by the fact that the appropriately energy weighted cross section "scales". So that, in this case, the product of the eleventh power of the square of the total energy in the center-of-momentum frame and the cross section becomes constant (see Fig. 29.8). The new CEBAF data exhibit a flat scaling behavior consistent with this rule, in photon energy approximately 1 to 4 GeV at a reaction angle of 90° (between the incident photon beam and the detected proton) for the $^{2}_{1}H(\gamma, p)n$ reaction. Furthermore, the new data also suggest that there is an onset of the scaling behavior above an energy of 3 GeV at a reaction angle of 37°. The results are consistent with an onset of scaling occurring at a transverse momentum of 1 GeV c^{-1}. The new data support the picture where six constituent quarks in the deuteron (each nucleon contains three constituent quarks)

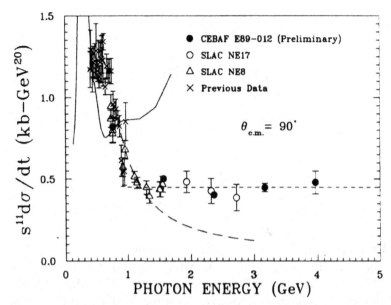

Fig. 29.8. The product of $s\,(square\ of\ total\ C - M\ energy)^{11}$ times cross section at 90° in $^2_1\mathrm{H}(\gamma,\mathrm{p})\mathrm{n}$ as a function of photon energy in measurements in E89-12 at CEBAF [Bo98]. The short dashed curve is the predicted flat scaling behavior from a simple constituent quark counting rule. Also shown are data from the previous NE8 and NE17 experiments at SLAC, as well as data from lower energy.

organize a concerted response involving the exchange of gluons among themselves.[2]

The author served as Scientific Director of CEBAF from 1986 to 1992 when the initial scientific program and design for the initial complement of equipment were established.[3] The reader is referred to an article which the author wrote for the publication *Physics News* in 1996 when the initial experimental results from that Laboratory appeared [Wa97]. Current information about the facilities and program at TJNAF can always be found on its website [TJ00].

---

[2] At least five gluon exchanges are required to reorganize the six quarks into two high-momentum outgoing nucleons. At very large s, the quark propagators each scale as $1/s$. The square of the amplitude, and conversion from solid angle to four-momentum transfer, then gives $d\sigma/dt \propto 1/s^{11}$.

[3] John Domingo, Associate Director for Physics, led the equipment design and construction effort (see [Do93a]).

30
Other facilities

The previous and subsequent chapters go into detail on TJNAF (CE-BAF) because that is the project in which the author was most deeply involved and about which he is most knowledgeable. Many other accelerator laboratories have played, and continue to play, an important role in electron scattering studies of nuclei and nucleons. Worth highlighting from the early years are the Nuclear Physics Laboratory at the University of Illinois, where the betatron provided a tool to do the very first study of nuclear structure with electrons [Ly51, Il87], and the High Energy Physics Laboratory at Stanford (HEPL), where Hofstadter carried out his pioneering work on charge and magnetization densities [Ho56, Ho63]. Many other important facilities sprang from the work at HEPL, including those at Amsterdam, Darmstadt, Mainz, Saskatchewan, Tohuko, and the Saclay Laboratory, which played a particularly important role in the development of the field. The Stanford Linear Accelerator (SLAC), under Wolfgang Panofsky's inspired leadership, found its roots in HEPL, as did TJNAF. A prototype of the CEBAF superconducting accelerator was first constructed at HEPL. An excellent discussion of the early years of electron scattering is to be found in [Il87].

It is the Bates Laboratory at M.I.T., where a variety of precision experiments truly demonstrated the power of electron scattering to study the nucleus, and the Stanford Linear Accelerator Center (SLAC), where high-energy experiments demonstrated the pointlike, asymptotically free, substructure of the nucleon and examined its weak neutral current, that are responsible for the role that electron scattering plays in nuclear and particle physics in the U.S. today.

The principal centers today for nuclear structure studies with electrons are TJNAF in Newport News, the Bates Laboratory at M.I.T. in Boston, and the Mainz Microtron (MAMI), in Germany. High energy studies, which probe the very short-range structure of nucleons are carried out principally at SLAC, in Stanford, and at the Deutches Electronen-Synchrotron (DESY) in Hamburg. Electromagnetic studies with very high energy muons are carried out at CERN in Geneva. The European community has an ongoing effort to design and fund a high-current electron accelerator to study physics in an energy regime intermediate between the few- and multi-GeV machines.

This book attempts to lay out the basic motivation, analysis, and goals of electron scattering studies of nuclei and nucleons. It is impossible in a work of this length to go into detail on all the existing facilities and programs. In fact, up-to-date information is always better, and more easily, found on the websites for the laboratories [TJ00, Ba00, Ma00, SL00, DE00]. It may be of some use, however, to provide a brief overview as guide to the four other principal current electron scattering centers: Bates, Mainz, SLAC, and DESY.

The Bates Linear Accelerator Center is a university-based facility for nuclear physics which is operated by the Massachusetts Institute of Technology for the Department of Energy as a National User Facility. Construction funding started in 1967. The accelerator is a room-temperature linac with a single-pass energy of 515 MeV at an average current of 100 μA with a 1% df. A single recirculation was subsequently added which brought the maximum unloaded energy to 1060 MeV, with a maximum average current of 40 μA. The energy spread in the beam for 80% current is 0.3%. From the beginning, the Energy Loss (dispersion matching) Spectrometer System (ELSSY) produced data of unprecedented resolution.[1] The best resolution achieved, for an extended period and limited by target thickness, is $\delta E/E = 4 \times 10^{-5}$. This was on $^{154}_{64}$Gd [He83, Tu00]. We have already demonstrated the type of nuclear information that can be obtained with this resolution in Fig. 12.7 and the accompanying discussion. The author considers that the work on the charge distribution in deformed nuclei at Bates, with ELSSY, was predominant in convincing the nuclear physics community in the U.S. of the power of electron scattering for nuclear physics. A 180° scattering facility at Bates allowed one to isolate the transverse contributions to the cross section and study magnetization distributions in elastic and inelastic scattering (Figs. 12.4, 12.6 and accompanying discussion).[2] A major recent addition to the detectors at Bates

---

[1] William Bertozzi and Stanley Kowalski led the effort to design and construct ELSSY. Note that the dispersion matching technique, whereby the incident beam is dispersed and the various components of the beam followed through the scattering, allows one to obtain scattering resolution orders of magnitude better than that in the primary beam itself.

[2] A magnet in front of the target bends the beam through a small angle θ; the backscattered beam is then bent through an *additional* angle θ, which removes it from the incident beam.

is an array of out-of-plane spectrometers (OOPS), which allows one to separate all the contributions in inclusive coincidence scattering (chapter 13). The polarized beam capability at Bates led to the ground breaking parity-violation experiment discussed in chapter 16, and to the study of the strange quark contribution to the intrinsic magnetism of the proton in the SAMPLE experiment. A major effort at Bates involved a parallel study of the three-body systems $^3_2$He, $^3_1$H. The latter involved an extensive radioactive target effort.[3] These studies resulted in data of the type shown in Fig. 23.13.

A storage ring was subsequently constructed in the South Hall at Bates. This gives rise to a continuous external beam capability for coincidence experiments through slow spill of the stored beam. More importantly, with a large circulating current of 200–300 mA, it allows experiments on very thin, polarized, *internal* targets. A major new detector, the Large Acceptance Spectrometer Toroid (BLAST) is under construction at Bates to take advantage of this opportunity and to measure spin-dependent electron scattering from polarized nuclei.

Even with the low duty factor, heroic, pioneering coincidence studies have been carried out with the Bates linac.

The Mainz Microtron MAMI is an electron accelerator which delivers a c.w. beam (100% df) of 100 μA at a maximum energy 855 MeV. Built under the guidance of H. Herminghaus and coming into operation in 1991, it consists of three cascaded racetrack microtrons with a 3.5 MeV injector linac. The last stage delivers beam from 180 to 855 MeV in 15 MeV steps with excellent emittance and stability.[4] The energy spread in the beam is 50 KeV for an impressive energy resolution of 6×10^{-5}. There is a polarized source.

At Mainz, the electron scattering Hall A contains three spectrometers, rotating about a common pivot, with resolution $\delta p/p = 10^{-4}$. The maximum momenta are 735, 870, and 551 MeV c^{-1}, with solid angle acceptances of 28, 5.6, and 28 msr, respectively. The angular coverages are $18°$–$160°$, $7°$–$62°$, and $18°$–$160°$, and all have angular resolution at the target of less than 3 mrad. The second spectrometer has the capability of going out-of-plane. This is an exceptional set of spectrometers, and with the accelerator capability, MAMI provides a superb facility for doing nuclear structure studies that provide a lower-energy complement to the work that will done at TJNAF. There is also a tagged photon facility at MAMI.

---

[3] Because electron scattering cross sections are small, radioactive targets are ordinarily not a serious problem at electron scattering facilities, and one can re-enter the experimental areas relatively quickly while conducting experiments.

[4] Mainz is currently adding a fourth stage to the microtron which will take the maximum energy to 1.5 GeV.

Current experiments at Mainz include nuclear (e, e'p) measurements, with polarizations, and (e, e'π) studies on the nucleon. There are also ongoing experiments on the form factor of the neutron and on parity violation.

The Stanford Linear Accelerator Center (SLAC) is located near the Stanford University campus. The 10,000-feet-long (2-mile) accelerator was originally designed to operate at a peak energy of 22.2 MeV with an RF frequency of 2856 MHz. It was designed with a (macroscopic) pulse length is 2.5 μsec and peak current of 25–50 mA, with an average current of 15–30 μA (df $\sim 10^{-3}$) [Ba65]. It met the design characteristics beautifully during the first year of operation, achieving an energy of 20.16 MeV with 43 mA peak current for a 1.6 μs pulse length [Lo67]. The linear accelerator has been continually upgraded over the years so that today the machine can achieve an energy of 48 GeV with 6×10^{11} particles in a 370-ns-long beam pulse (260 mA peak current) [De99].

End Station A at SLAC, designed for electron scattering experiments, was originally equipped with a complement of three spectrometers of maximum momenta 1.6 GeV c^{-1}, 8 GeV c^{-1}, and 20 GeV c^{-1}, respectively. They rotated about a common pivot and covered an angular region matched to their maximum accepted electron momenta ($25°$–$165°$, $12°$–$100°$, and, $0°$–$20°$). The spectrometers had solid angle acceptances of $4.1 \times 10^{-3}, 10^{-3}$, and 10^{-4} sr, respectively. The original resolution of the 1.6 GeV c^{-1} spectrometer was 0.08%, and of the higher energy spectrometers, $\sim 0.15\%$ [Pa70, Ki75].

The contribution of the SLAC deep-inelastic scattering experiments to the understanding of the quark–parton structure of the nucleon has been extensively discussed in this book. Many spectrometers have been assembled and disassembled in End Station A over the years for particular experiments. The 8 GeV c^{-1} spectrometer, in particular, has proven to be a workhorse over this period.

The physics contribution of SLAC over the years from colliding beam (e$^+$–e$^-$) experiments, where the annihilation creates a pure, virtual *time-like* quantum of electromagnetic radiation with definite quantum numbers, is well-known; nevertheless, important electron scattering experiments on hadronic targets continue to be done in End Station A up to the highest machine energies.

As one example, the measurement of the spin structure function in deep inelastic scattering was discussed in chapter 12. Polarized-beam,[5] polarized-target experiments to provide precision measurements of this quantity for both the proton and neutron are currently underway at SLAC using the highest energy of the accelerator. To illustrate the quality

---

[5] SLAC has played a key role in the successful effort to obtain high beam polarizations.

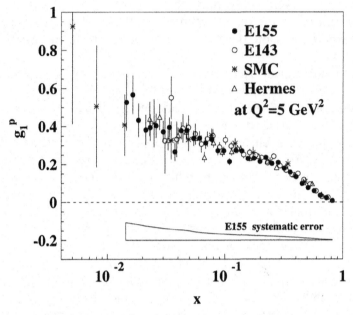

Fig. 30.1. Measurement of $g_1^p(x)$ from experiments E143 and E155 in End Station A at SLAC, together with some other results. The data are evolved to $Q^2 = 5\,\mathrm{GeV}^2\,\mathrm{c}^{-2}$ [An00]. The author is grateful to K. Griffioen and G. Mitchell for providing this figure.

of the currently available data, Fig. 30.1 shows results for $g_1^p(x)$ from experiments E143 and E155, evolved to $Q^2 = 5\,\mathrm{GeV}^2\,\mathrm{c}^{-2}$ [An00].[6]

Deep inelastic electron scattering (DIS) experiments are also carried out at the HERA collider at the Deutsches Electronen Synchroton (DESY) in Hamburg. The HERA collider can store electrons (positrons) of up to 30 GeV and protons of up to 820 GeV in two rings of 6.3 km circumference. The C-M energy is 314 GeV and the maximum achieved luminosity is $1.4\times 10^{31}\,\mathrm{cm}^{-2}\,\mathrm{s}^{-1}$. HERA has four interaction regions. The general purpose detectors H1 and ZEUS study the interactions between electron (positron) and proton colliding beams. The HERMES collaboration measures the spin structure of nucleons by the interaction of the polarized electron (positron) beam with polarized nucleons (nuclei) of a gas-jet target [Wo97].

The proton structure function $F_2(x, Q^2)$ has been measured at HERA over a wide range of x and Q^2, with values of Q^2 as high as 5000 $\mathrm{GeV}^2\,\mathrm{c}^{-2}$ and x as low as 10^{-5}. The most striking feature of the HERA data is the rapid rise of F_2 as $x \to 0$ which is seen to persist down to Q^2 values as small as $1.5\,\mathrm{GeV}^2\,\mathrm{c}^{-2}$. The Alterelli–Parisi (DGLAP)

---

[6] The Alterelli–Parisi QCD evolution equations relate the DIS structure functions at different Q^2 [Al77]; this is discussed in [Ro90, Wa95].

evolution equations describe the evolution of the parton densities with Q^2 [Al77, Ro90, St93, Wa95]. In order to solve these equations one must provide the parton densities as a function of x at some reference scale Q_0^2. With the assumption of a Regge behavior at very small x, perturbative QCD then implies that F_2 grows faster than any power of $\ln(1/x)$ as $x \to 0$ [Wo97]

$$F_2(x, Q^2) \approx C_0 \left\{ \frac{33 - 2n_f}{576\pi^2 \ln(1/x) \ln[\alpha_s(Q_0^2)/\alpha_s(Q^2)]} \right\}^{1/4}$$
$$\times \exp \sqrt{\frac{144 \ln(1/x)}{33 - 2n_f} \ln[\alpha_s(Q_0^2)/\alpha_s(Q^2)]} \qquad (30.1)$$

Here α_s is the strong coupling constant and n_f is the number of quark flavors. Improved evolution schemes which can give a faster rise for small x are under current investigation [Mu97, Wo97].

31

Future directions

We have seen several examples of existing experimental results from TJ-NAF (CEBAF), and have discussed their implications for nuclear and particle physics. In the author's opinion, the best way to get a feel for the quality and impact of the future CEBAF physics program is to show anticipated error bars, kinematic range, and event modeling in a few selected examples. While reluctant to show anticipated data because so much work lies ahead in actually carrying out the experiments, such a significant effort has already gone into modeling the detectors, magnetics, acceptances, efficiencies, electronics, and event rates for the real experiments that the author feels justified in presenting this material; it is taken from the proposals.[1] The experimental program is dynamic and constantly evolving. Where data now exist, they more than satisfy the expectations. The following discussion only represents one snapshot in time. It is based on talks the author gave on the CEBAF scientific program, when the experimental program was still one of anticipation [Wa93, Wa94].

As one example, Fig. 31.1 shows the anticipated errors on the charge form factor of the proton G_{Ep} (relative to the dipole fit) from the polarization transfer measurement $^1H(\vec{e}, e\,\vec{p})$ at CEBAF as anticipated in PR 89-014 [Pe89]. This polarization transfer experiment measures the product of the magnetic and electric form factors of the proton [Ar81]. Since the magnetic form factor is well known, this interference term allows an accurate determination of G_{Ep}. To get a feel for the validity of such projections, Fig. 29.6 shows subsequent actual data on the measurement of G_{Ep}/G_{Mp} at TJNAF [Jo00]. The data are indeed superb.

Figure 31.2 shows the anticipated error bars on the determination of G_{En} from two experiments: a polarization transfer measurement $^2_1H(\vec{e}, e'\,\vec{n})$ in CEBAF PR 89-005 [Ma89a] similar to that discussed above; and a

---

[1] The proposals are available in the library at TJNAF.

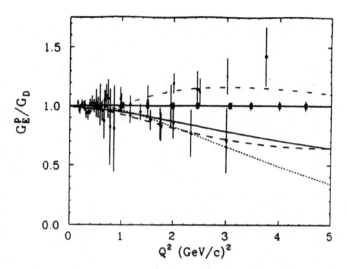

Fig. 31.1. Projected error bars in G_{Ep} in polarization transfer measurement $^1H(\vec{e}, e\,\vec{p})$ at CEBAF. From PR 89-014 [Pe89, Wa93]. Here $k \equiv Q$.

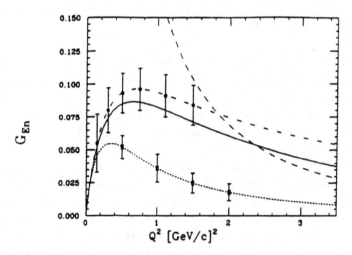

Fig. 31.2. Projected error bars on G_{En} from polarization transfer measurement $^2_1H(\vec{e}, e'\,\vec{n})$ in CEBAF PR 89-005 (upper); and polarized target experiment $^2_1\vec{H}(\vec{e}, e'\,n)$ in CEBAF PR 89-018 [Ma89a, Da89, Wa93]. Here $k \equiv Q$.

coincidence measurement with a polarized target $^2_1\vec{H}(\vec{e}, e'\,n)$ in CEBAF PR 89-018 which also determines G_{En} through an interference term [Da89].[2] Since the measurement of G_{En} ultimately involves nuclear physics (there are as yet no free neutron targets), it is important to have complementary

---

[2] The error bars are relative to the different theoretical estimates.

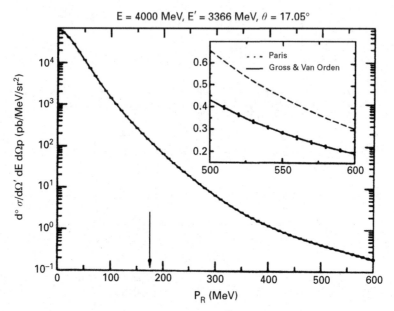

Fig. 31.3.   Projected range and error bars in $^2_1$H(e, e′ p) from CEBAF PR 89-028 [Fi89, Wa93].

determinations. Both of the charge distributions G_{En} and G_{Ep} directly reflect the internal structure of the baryon; the theoretical description of the accurate measurements of these charge distributions will continue to provide a benchmark challenge to quark models and QCD.

Consider next the nuclear coincidence reaction $^2_1$H($\vec{e}$, e′ $\vec{p}$) to be measured in CEBAF PR 89-028 [Fi89]. This polarization transfer experiment explores the spin structure of the deuteron in unrivaled detail; it also provides an important calibration for the measurement of G_{En} by a similar procedure. In the course of this experiment, the momentum distribution in the deuteron will be determined at the same kinematics. Plotted in Fig. 31.3 are the anticipated range and error bars in the determination of the basic nuclear coincidence cross section $^2_1$H(e, e′ p) to be measured in PR 89-028 [Fi89]. The arrow indicates the extent of existing data, and the inset demonstrates that the experiment will distinguish between different models; one calculation shown uses a good two-nucleon potential, the other a relativistic boson-exchange description.[3] Elastic charge scattering essentially measures the Fourier transform of the spatial density (square of the wave function); the (e, e′ p) reaction essentially measures the Fourier transform of the wave function (whose square is the momentum density)

---

[3] The calculation is for illustration; it assumes plane waves in the final state and neglects exchange currents.

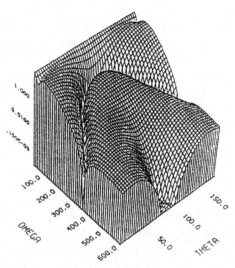

Fig. 31.4. Model calculation of in-plane Coulomb response surface for the reaction $^{18}_{10}$Ne(e,e′ 2p)$^{16}_{8}$O(g.s.) at $(\varepsilon_1, \theta_e) = (1\,\text{GeV}, 20°)$; see text [Da92, Wa93].

—these are complementary quantities, and by measuring both one can examine the structure of this fundamental two-nucleon bound state in unprecedented detail.

Consider the results of a very simple model calculation, meant only to provide some guidance for explorations into new territory. In principle, the most direct way to examine short-range correlations is to study two-nucleon emission with extreme kinematics. Figure 31.4 shows a preliminary analysis by John Dawson of the in-plane Coulomb response for the triple coincidence $^{18}_{10}$Ne(e,e′ 2p)$^{16}_{8}$O(g.s.) [Da92] — this reaction is forbidden in a single-particle model. Here the initial wave function is the correlated relative $^1S_0$ state obtained by solving the Bethe–Goldstone equation with a two-nucleon potential for the interacting $(\pi 1d_{5/2})^2$ pair in the presence of the $^{16}_{8}$O core.[4] The total energy and C-M momentum of the pair are $(\omega, \mathbf{P} = \mathbf{p}_1 + \mathbf{p}_2 = \boldsymbol{\kappa})$ and θ is the angle between the relative momentum of the pair $2\mathbf{p} = \mathbf{p}_1 - \mathbf{p}_2$ and $\boldsymbol{\kappa}$. Plane wave final states are used in this initial calculation. Note the characteristic diffraction minimum as ω is increased and characteristic angular distribution of the 2-proton final state. In the present approximation, this surface measures the sum of the Fourier transforms of the two-nucleon correlation function with respect to $\mathbf{p} \pm \boldsymbol{\kappa}/2$. This calculation was motivated by a presentation of William Hersman at PAC5 (Fifth Program Advisory Committee Meeting) at CEBAF, in which he showed a similar model surface for the basic nuclear two-proton

---

[4] This calculation and wave function are given in [Fe71].

coincidence experiment $^3_2$He(e, e′2p) that will be studied in CEBAF PR 89-031 [He89]. This experiment will map out the two-proton wave function in this three-nucleon system — an unprecedented measurement, fundamental to our understanding of nuclear physics.[5]

Consider next pion production and the internal dynamics of the nucleon. In CEBAF PR 89-037 [Bu89], precision angular distributions will be measured on the first nucleon resonance at $W = 1232\,\text{MeV}$ with varying k^2 for the reactions ^{1}H(e, e′ p)π^0, ^{1}H(e, e′ π^+)n, and $^2_1$H(e, e′ π^-)pp. The contributing multipoles can then be extracted from these angular correlation measurements. The resonant target transition is $(1/2^+, 1/2) \rightarrow (3/2^+, 3/2)$.

As discussed previously, the electric quadrupole transition E_{1+} is particularly interesting. Quark bag models of the nucleon, with a one-gluon exchange interaction, indicate that the bag may deform — similar to the deformation of the deuteron arising from the tensor force. As with even–even deformed nuclei, the nucleon can have no quadrupole moment in its ground state, so the most direct evidence for this deformation would show up is in this transition amplitude. In the quark model, the above transition to the $P_{33}(1232)$ is predominantly spin-flip magnetic dipole M_{1+}. The E_{1+} is, in fact, observed to be small, and it is only very poorly known; this is illustrated in Fig. 28.2, which shows the existing world's data on $\text{Re}\,(E^*_{1+}M_{1+})/|M_{1+}|^2$ at the $\Delta(1232)$ at the time of CEBAF PR 89-037 [Bu89]; Fig. 28.3 shows the projected range and error bars in that proposal. Note, in particular, the expansion of the vertical scale in this second figure. The subsequent actual experimental results for this quantity have been shown previously in Fig. 28.4, more than meeting expectations.

At CEBAF, the internal dynamics of the nucleon will be studied with unrivaled precision. These measurements will provide deep insight into the dynamical consequences of QCD. The accurate new data will continue to provide benchmark tests for theoretical quark-model and QCD descriptions of the nucleon — the basic building block of matter.

A simulation of the CLAS detector output for observation of meson production through the reaction ^{1}H(γ, p)X is shown in Fig. 31.5 from PR 91-008 [Ri91]. Here the tagging of the photon and the measurement of the proton determine the missing mass of X, and well-defined peaks are seen for the two-body reactions producing $(\pi^0, \eta, \omega, \eta')$ at $E_\gamma = 1.7\,\text{GeV}$. The π^0 production has already been referred to. The production of η with isospin $T = 0$ provides a selective mechanism to study the $T = 1/2$ nucleon resonances. CEBAF PR 89-039 [Dy89] utilizes the fact that the $S_{11}(1535)$ resonance has a large branching ratio into the η channel to selectively study the behavior of this state with high precision out to large k^2. This

---

[5] More detailed calculations of the process (e, e′ 2N) on nuclei are described in [Ry96, Ry97, Ry00].

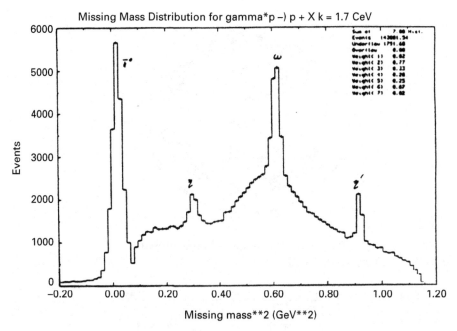

Fig. 31.5. CLAS simulation of missing-mass determination of meson production through the reaction ^{1}H(γ, p)X at $E_\gamma = 1.7$GeV. The abscissa is in GeV^2. From CEBAF PR 91-008 [Ri91, Wa93].

state is particularly interesting because its inelastic form factor appears to fall anomalously slowly. Both η production, and the production of the $T = 0$ ω meson studied in PR 91-024 [Fu91], can be used to selectively search for nucleon resonances that couple only very weakly to pions. The η and η' signals also provide the opportunity to study the structure of these mesons themselves in PR 91-008 [Ri91].

An important feature of coincident electron scattering is that the baryon levels in the $S = -1$ sector can *also* be accessed with the (e, e′ K$^+$) reaction. In fact, PR 89-024 will look at the resulting photon transitions between the low-lying levels in this sector — a lovely extension of traditional nuclear γ spectroscopy [Mu89]. A CLAS simulation of the reaction ^{1}H(γ, K$^+$)Λ and subsequent decay $\Lambda \to p + \pi^-$ from PR 89-004 is shown in Fig. 31.6 [Sc89]. The signature is very clear and this elementary process can be studied in unprecedented detail, as can the self-analyzing polarization of the Λ. The extension to $^2_1$H in PR 89-045 provides a neutron target and allows one to examine the two-baryon final-state interaction [Me89]. An examination of the hyperon production mechanism in a series of nuclei will be carried out in PR 91-014 [Hy91]. Figure 31.7 shows the projected rates and error bars in PR 91-016 for the production of the lightest bound hypernucleus through $^4_2$He(e, e′ K$^+$)$^4_\Lambda$H [Ze91]. The bound state is clearly

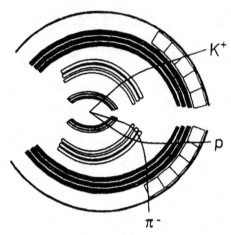

Fig. 31.6. CLAS simulation of ^{1}H$(\gamma, K^+)\Lambda$ and subsequent decay $\Lambda \to p + \pi^-$ from PR 89-004 [Sc89, Wa93].

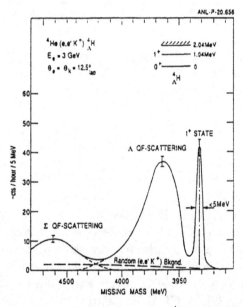

Fig. 31.7. Projected rates and error bars for $^4_2$He$(e, e'K^+)^4_\Lambda$H in PR 91-016 [Ze91, Wa93].

identified in this figure; the projected transition is almost entirely to the spin-flip 1^+. This experiment forms the prototype for the production of hypernuclei through the $(e, e' K^+)$ reaction at CEBAF — accessing a whole new dimension of nuclear structure.

Let us return to the subject of parity violation. The nuclear domain consists of (u, d) quarks and their antiquarks. Consider elastic scattering of polarized electrons from a $(0^+, 0)$ nucleus, for example $^{12}_{6}C(\vec{e}, e)$. The nuclear quantum numbers serve as a filter, and the standard model states that for such a transition in this sector the weak neutral current and electromagnetic current are strictly proportional

$$\mathscr{J}^{(0)}_\mu \doteq -2\sin^2\theta_W J^\gamma_\mu \tag{31.1}$$

The predicted parity-violating asymmetry $\mathscr{A} = (d\sigma_\uparrow - d\sigma_\downarrow)/(d\sigma_\uparrow + d\sigma_\downarrow)$ is then

$$\mathscr{A}_{12_C} = \frac{Gq^2}{\pi\alpha\sqrt{2}}\sin^2\theta_W \tag{31.2}$$

It is important to note that this result depends only on the existence of isospin symmetry; it holds *to all orders* in the strong interactions (QCD). As we have seen, this quantity has been measured in a tour de force experiment at Bates at $q = 150\,\mathrm{MeV}$ with the result that [So90] [6]

$$\begin{aligned}\mathscr{A}_{12_C}\, P_e\, &=\, 0.688\times 10^{-6} & &\text{; theory} \\ &=\, 0.60\pm 0.14\pm 0.02\times 10^{-6} & &\text{; experiment}\end{aligned} \tag{31.3}$$

This experiment serves as a demonstration of feasibility for the next generation of electron scattering parity-violation experiments.

Now consider the *extended domain* of (u, d, s, c) quarks and their anti-quarks. The standard model then has an additional isoscalar term in the weak neutral current

$$\delta\mathscr{J}^{(0)}_\mu = \frac{i}{2}[\bar{c}\gamma_\mu(1+\gamma_5)c - \bar{s}\gamma_\mu(1+\gamma_5)s] \tag{31.4}$$

The asymmetry for elastic scattering of polarized electrons on a $(0^+, 0)$ nucleus such as $^4_2$He then takes the form

$$\mathscr{A}_{^4\mathrm{He}} = \frac{Gq^2}{\pi\alpha\sqrt{2}}\sin^2\theta_W\left[1 - \frac{\delta F^{(0)}(q^2)}{2\sin^2\theta_W F^\gamma_0(q^2)}\right] \tag{31.5}$$

The additional weak neutral current form factor comes from the vector current in Eq.(31.4) — expected to arise predominantly from the much lighter strange quarks. Hence one has a direct measure of the strangeness current in nuclei. The total strangeness of the nucleus must vanish in the strong and electromagnetic sector, and hence $\delta F^{(0)}(0) = 0$; however, just as with the electromagnetic charge in the neutron, there can be a strangeness density, which is determined in this experiment.

Approval exists for the experimental measurements of the asymmetry in $^4_2$He$(\vec{e}, e)$ in CEBAF PR 91-004 [Be91], and the asymmetry for a similar

---

[6] The first error is statistical.

elastic scattering measurement on the nucleon itself ^{1}H($\vec{e}$, e) in CEBAF PR 91-010 [Fi91].[7] The measurement of the distribution of weak neutral current through ($\vec{e}$, e) and ($\vec{e}$, e′) will be one of the most important results at CEBAF. The beautiful experimental results that now exist on this latter experiment, and their deep implication for the structure of nuclei and nucleons, have already been presented in chapter 27.

In *summary*, let us try to pull all this material together with a statement of the nuclear physics goals of electron scattering studies: first, quite generally, one wants to examine the limits of the traditional, non-relativistic many-body description of the nucleus based on baryons interacting through static potentials fitted to two-body scattering and bound-state data. The nuclear shell model, for example, provides a remarkably successful description of the strongly interacting quantum mechanical nuclear many-body system. Just how far does that description hold, and when does it break down?

The degrees of freedom of the shell model are the nucleons, protons and neutrons. We know from electron scattering that additional sub-nucleonic hadronic degrees of freedom, mesons and isobars, come into play when one examines the nucleus at shorter and shorter distance scales. What is the role of these additional degrees of freedom? The only consistent description we have of a relativistic, interacting, hadronic many-body system is through a relativistic quantum field theory based on a local lagrangian density constructed from the hadronic degrees of freedom. What are the limits of a relativistic, hadronic field theory description of the nuclear system?

At shorter distances still, electron scattering first taught us that quark–gluon degrees of freedom are the relevant ones. At what distance scales are we forced to make the transition from a baryon–meson to a quark–gluon description of the nucleus? The constituent quark model provides a remarkably successful description of the interior structure of the hadrons themselves; however, it is still a model, and just as with the nuclear shell model, one wants to determine where this picture breaks down.

At a more fundamental level, one has a relativistic quantum field theory of the strong interactions, quantum chromodynamics (QCD) based on the strong color interactions of quark and gluons. This is the true relativistic, strongly coupled, nuclear many-body system. As with any theory, the experimental implications of QCD must continually be explored. Electron scattering data will provide the most direct benchmarks against which to test the experimental implications of QCD.

The standard model provides a marvelously successful unified description of the weak and electromagnetic interactions. The experimental im-

---

[7] Here the quantum numbers $\frac{1}{2}^+\frac{1}{2}$ allow other elastic form factors.

plications of the standard model must similarly continue to be explored. Electron scattering provides a tool for examining the weak neutral current distribution in nuclei, which, taken in conjunction with the study of the electromagnetic current distribution effectively doubles the power of electron scattering.

At very short-distance *particle physics* scales, one examines the quark distributions in the nucleon, including those contributing to its spin. At a more basic level, deep-inelastic electron scattering provides an unrivaled tool to examine the short-distance behavior of the relativistic quantum field theory describing the strong interactions, QCD.

Finally, at all levels, we are interested in exploring the *phenomena* manifest by the remarkable, strongly-coupled, quantum-mechanical, nuclear many-body system.

Appendix A
Long-wavelength reduction

This appendix is concerned with the long-wavelength reduction of the electromagnetic multipole operators. The analysis follows closely the arguments developed in [Bl52] (see also [de66]). Consider first the transverse electric and magnetic multipoles, which govern real photon transitions.[1]

The use of the relations $1/\hbar c = 5.07 \times 10^{10}\,\mathrm{cm}^{-1}\,\mathrm{MeV}^{-1}$ and $R \approx 1.2\,A^{1/3} \times 10^{-13}\,\mathrm{cm}$ allows one to write for real photons

$$kR \approx 6.1 \times 10^{-3}[E_\gamma(\mathrm{MeV})A^{1/3}] \tag{A.1}$$

Evidently $kR \ll 1$ for photons of a few MeV. In this case, the spherical Bessel functions can be expanded as[2]

$$j_J(kx) \to \frac{(kx)^J}{(2J+1)!!} \qquad ; kx \to 0 \tag{A.2}$$

One also needs from [Ed74]

$$\mathbf{L}Y_{lm} = \frac{1}{i}(\mathbf{r} \times \nabla)Y_{lm} = \sqrt{l(l+1)}\,\mathscr{Y}_{ll1}^m \tag{A.3}$$

With this relation, the multipole operators in Eqs. (9.16) take the form

$$\hat{T}_{JM}^{\mathrm{el}} = \frac{1}{k\sqrt{J(J+1)}} \int d^3x\, \Big\{ [\nabla \times \mathbf{L}j_J(kx)Y_{JM}] \cdot \hat{\mathbf{J}}_c(\mathbf{x})$$
$$+ k^2[\mathbf{L}j_J(kx)Y_{JM}] \cdot \hat{\boldsymbol{\mu}}(\mathbf{x}) \Big\}$$

$$\hat{T}_{JM}^{\mathrm{mag}} = \frac{1}{\sqrt{J(J+1)}} \int d^3x\, \Big\{ [\nabla \times \mathbf{L}j_J(kx)Y_{JM}] \cdot \hat{\boldsymbol{\mu}}(\mathbf{x})$$
$$+ [\mathbf{L}j_J(kx)Y_{JM}] \cdot \hat{\mathbf{J}}_c(\mathbf{x}) \Big\} \tag{A.4}$$

---

[1] Recall $\mathbf{x} \equiv \mathbf{r}$ and $x \equiv |\mathbf{x}| \equiv r$ in all these discussions.
[2] One has to get all the derivatives off the Bessel functions before they can be expanded — that is the point of the following exercise.

288

These expressions can now be manipulated in the following manner:

1. The differential orbital angular momentum operator **L** in Eq. (A.3) commutes with any function of the radial coordinate $[\mathbf{L}, f(r)] = 0$, and it is hermitian; thus it can be partially integrated in the last two terms on the r.h.s. in the above to get it over to the right [with a sign (-1)].

2. The divergence theorem in Eqs. (9.13) and (9.14) can be used on the first two terms on the r.h.s. of the above to get the curl to the right.

3. One can then get **L** to the right in these terms using the first argument [again with a (-1)]. This leads to two types of terms: first

$$\mathbf{L} \cdot \mathbf{v} = \frac{1}{i}(\mathbf{r} \times \nabla) \cdot \mathbf{v} = \frac{1}{i}(\nabla \times \mathbf{v}) \cdot \mathbf{r} = -\frac{1}{i}\nabla \cdot (\mathbf{r} \times \mathbf{v}) \qquad (A.5)$$

and second

$$\mathbf{L} \cdot (\nabla \times \mathbf{v}) = \frac{1}{i}(\mathbf{r} \times \nabla) \cdot (\nabla \times \mathbf{v}) = \frac{1}{i}[\nabla \times (\nabla \times \mathbf{v})] \cdot \mathbf{r}$$

$$= -\frac{1}{i}\nabla \cdot [\mathbf{r} \times (\nabla \times \mathbf{v})] \qquad (A.6)$$

Here the relation $\nabla \times \mathbf{r} = 0$ has been used in obtaining these equations.

4. Since all derivatives are now off the spherical Bessel functions and on the source terms, the Bessel functions may be expanded in the long-wavelength limit according to Eq. (A.2).

5. One next invokes the general vector identity

$$\int x^J Y_{JM} \nabla \cdot [\mathbf{r} \times (\nabla \times \mathbf{v})] \, d^3x = (J+1) \int x^J Y_{JM} \nabla \cdot \mathbf{v} \, d^3x \qquad (A.7)$$

This identity holds as long as the source terms $\mathbf{v}(\mathbf{x})$ vanish outside the nucleus.

With these steps the magnetic multipoles take the form

$$\hat{T}_{JM}^{\mathrm{mag}} \approx \frac{1}{i}\frac{k^J}{(2J+1)!!}\sqrt{\frac{J+1}{J}} \int d^3x \, x^J Y_{JM} \left\{ \nabla \cdot \hat{\boldsymbol{\mu}}(\mathbf{x}) + \frac{1}{J+1}\nabla \cdot [\mathbf{r} \times \hat{\mathbf{J}}_c(\mathbf{x})] \right\} \qquad (A.8)$$

Partial integration of this result then gives for the long-wavelength limit of the transverse magnetic multipoles

$$\hat{T}_{JM}^{\mathrm{mag}} \approx -\frac{1}{i}\frac{k^J}{(2J+1)!!}\sqrt{\frac{J+1}{J}} \int d^3x \, [\hat{\boldsymbol{\mu}}(\mathbf{x}) + \frac{1}{J+1}\mathbf{r} \times \hat{\mathbf{J}}_c(\mathbf{x})] \cdot \nabla x^J Y_{JM} \qquad (A.9)$$

Similarly, the electric multipole operators take the form

$$\hat{T}_{JM}^{\mathrm{el}} \approx \frac{1}{i}\frac{k^{J-1}}{(2J+1)!!}\sqrt{\frac{J+1}{J}} \int d^3x \left\{ \nabla \cdot \hat{\mathbf{J}}_c(\mathbf{x}) + \frac{k^2}{J+1}\nabla \cdot [\mathbf{r} \times \hat{\boldsymbol{\mu}}(\mathbf{x})] \right\} x^J Y_{JM} \qquad (A.10)$$

Now use the *continuity equation* on the first term

$$\nabla \cdot \hat{\mathbf{J}}_c(\mathbf{x}) = \nabla \cdot \hat{\mathbf{J}}(\mathbf{x}) = -\frac{1}{c}\frac{\partial \hat{\rho}}{\partial t} = -\frac{i}{\hbar c}[\hat{H}, \hat{\rho}] \qquad (A.11)$$

The matrix element of this relation yields

$$\langle f|[\hat{H}, \hat{\rho}]|i\rangle = (E_f - E_i)\langle f|\hat{\rho}|i\rangle = -\hbar k c \langle f|\hat{\rho}|i\rangle \qquad (A.12)$$

Thus, in the matrix element, one can replace[3] $\nabla \cdot \hat{\mathbf{J}}_c(\mathbf{x}) \rightarrow ik\hat{\rho}(\mathbf{x})$. Thus, for photon emission the long-wavelength limit of the transverse electric multipoles takes the form

$$\hat{T}^{el}_{JM} \approx \frac{k^J}{(2J+1)!!}\sqrt{\frac{J+1}{J}}\int d^3x\left\{x^J Y_{JM}\hat{\rho}(\mathbf{x}) - \frac{ik}{J+1}\hat{\boldsymbol{\mu}}(\mathbf{x})\cdot[\mathbf{r}\times\nabla x^J Y_{JM}]\right\}$$
$$(A.13)$$

The first term in Eq. (A.13) is just the JMth multipole of the charge density. The second term goes as $\hbar k c/mc^2 \ll 1$ and hence the contribution of this term is very small compared to that of the first term for real photons.[4]

Make a model where the nucleus is composed of individual nucleons, and where only the leading terms to order $1/m$ are retained in the current, that is, the terms in $\mathbf{p}(i)$ and $\boldsymbol{\sigma}(i)$ [see Eqs. (9.17) and (9.20)]. The $J = 1$ transverse magnetic dipole operator for $k \rightarrow 0$ then takes the form

$$\hat{T}^{mag}_{1M} \approx \frac{i\sqrt{2}}{3}\frac{\hbar k}{2mc}\sqrt{\frac{3}{4\pi}}\left\{\sum_{i=1}^{Z}\mathbf{l}(i) + \sum_{i=1}^{A}\lambda_i\boldsymbol{\sigma}(i)\right\}_{1M} \qquad (A.14)$$

This is the familiar magnetic dipole operator to within a numerical factor and power of k. Here the nucleon magnetic moments in nuclear magnetons are given by $\lambda_p = 2.793$ for the proton and $\lambda_n = -1.913$ for the neutron.

Static Moments. It is useful to make the connection between these general results for the electromagnetic nuclear moments and the static nuclear moments measured in time-independent electric and magnetic fields.

Consider first the static *electric* moments of the nucleus. Suppose one places a static charge distribution $\rho(\mathbf{r})$ in an *external* electrostatic potential $\Phi_{el}(\mathbf{r})$ where the external electric field is given by $\mathbf{E} = -\nabla\Phi_{el}(\mathbf{r})$ (see Fig. A.1). A relevant example is a nucleus in the field of the atomic electrons. The interaction energy is given by

$$U = e_p \int \rho(\mathbf{r})\Phi_{el}(\mathbf{r})\, d^3r \qquad (A.15)$$

---

[3] Note this is for photon *emission*; for photon *absorption* one has the opposite sign for this term.

[4] This term can become large in electron scattering where, as we shall see, the appropriate ratio is $\hbar qc/mc^2$ with q the momentum transfer.

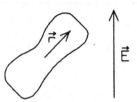

Fig. A.1.   Static electric nuclear moments.

The external field satisfies Laplace's equation since it is source-free over the nucleus

$$\nabla^2 \Phi_{el}(\mathbf{r}) = 0 \tag{A.16}$$

It is also finite there. Thus the external field in the region of the nucleus can be expanded in terms of the acceptable solutions to Laplace's equation

$$\Phi_{el}(\mathbf{r}) = \sum_{lm} a_{lm} r^l Y_{lm}(\Omega_r) \tag{A.17}$$

The numerical coefficients a_{lm} can be related to various derivatives of the field at the origin. Substitution of Eq. (A.17) into Eq. (A.15) yields

$$U = e_p \sum_{lm} a_{lm} \mathcal{M}_{lm}^{el} \tag{A.18}$$

Here the multipole moments of the charge density are defined by

$$\mathcal{M}_{lm}^{el} = \int x^l Y_{lm}(\Omega_x) \rho(\mathbf{x}) \, d^3 x \tag{A.19}$$

These are exactly the same expressions, to within a numerical factor and powers of k, as the first term in the transverse electric multipole operators in Eq. (A.13).[5] Note that the nuclear quadrupole moment is conventionally defined by

$$Q = \int (3z^2 - r^2) \rho(\mathbf{x}) \, d^3 x \tag{A.20}$$

which differs by a numerical constant from $\mathcal{M}_{20}^{el}$.

Consider next the nuclear *magnetic* moments. Take the ground-state expectation value that gives $\langle \partial \rho(\mathbf{x})/\partial t \rangle = (i/\hbar) \langle [\hat{H}, \hat{\rho}] \rangle = 0$. This implies

$$\nabla \cdot \langle \hat{\mathbf{J}}(\mathbf{x}) \rangle \;\; = \;\; \nabla \cdot \langle \hat{\mathbf{J}}_c(\mathbf{x}) \rangle = 0 \tag{A.21}$$

Here the general decomposition of current has been invoked

$$\hat{\mathbf{J}} = \hat{\mathbf{J}}_c + \nabla \times \hat{\boldsymbol{\mu}} \tag{A.22}$$

---

[5] The charge multipole *operators* are defined in terms of the charge density *operator*.

Since the divergence of the last quantity in Eq. (A.21) vanishes everywhere, it can be expressed as the curl of another vector $\mathbf{M}(\mathbf{x})$

$$\langle \hat{\mathbf{J}}_c(\mathbf{x}) \rangle = \nabla \times \mathbf{M}(\mathbf{x}) \tag{A.23}$$

One can assume that the additional magnetization $\mathbf{M}(\mathbf{x})$ vanishes outside the nucleus, for suppose it does not. Then since its curl vanishes outside the nucleus by Eq. (A.23), it can be written as $\mathbf{M}(\mathbf{x}) = \nabla \chi(\mathbf{x})$ in this region. Now choose a new magnetization $\mathbf{M}'(\mathbf{x}) = \mathbf{M}(\mathbf{x}) - \nabla \chi(\mathbf{x})$. This new magnetization has the same curl everywhere, and now, indeed, vanishes outside the nucleus.

The expectation value of the interaction hamiltonian for the nucleus in an external magnetic field now takes the form

$$\langle \hat{H}_{\text{int}} \rangle = -e_p \int [\nabla \times \mathbf{M}(\mathbf{x})] \cdot \mathbf{A}^{\text{ext}}(\mathbf{x}) \, d^3x - e_p \int \boldsymbol{\mu}(\mathbf{x}) \cdot \mathbf{B}^{\text{ext}}(\mathbf{x}) \, d^3x \tag{A.24}$$

Here $\boldsymbol{\mu} \equiv \langle \hat{\boldsymbol{\mu}} \rangle$. The use of Eqs. (9.13) and (9.14) permits this expression to be rewritten as

$$\langle \hat{H}_{\text{int}} \rangle = -e_p \int [\mathbf{M}(\mathbf{x}) + \boldsymbol{\mu}(\mathbf{x})] \cdot \mathbf{B}^{\text{ext}}(\mathbf{x}) \, d^3x \tag{A.25}$$

Since $\mathbf{B}^{\text{ext}}(\mathbf{x})$ is an external magnetic field with no sources over the nucleus, it satisfies Maxwell's equations there

$$\nabla \cdot \mathbf{B}^{\text{ext}} = \nabla \times \mathbf{B}^{\text{ext}} = 0 \tag{A.26}$$

Thus one can write in the region of interest

$$\begin{aligned} \mathbf{B}^{\text{ext}} &= -\nabla \Phi_{\text{mag}} \\ \nabla^2 \Phi_{\text{mag}} &= 0 \end{aligned} \tag{A.27}$$

One can now proceed with exactly the same arguments used on the electric moments. The energy of interaction is given by

$$\begin{aligned} \langle \hat{H}_{\text{int}} \rangle &= e_p \int [\mathbf{M}(\mathbf{x}) + \boldsymbol{\mu}(\mathbf{x})] \cdot \nabla \Phi_{\text{mag}}(\mathbf{x}) \, d^3x \\ &= -e_p \int \Phi_{\text{mag}} \nabla \cdot (\mathbf{M} + \boldsymbol{\mu}) \, d^3x \end{aligned} \tag{A.28}$$

The divergence in the last equation evidently plays the role of the "magnetic charge." Thus, just as before, when the general solution to Laplace's equation is substituted for the magnetic potential Φ_{mag}, all one needs are the magnetic charge multipoles given by

$$\begin{aligned} \mathscr{M}_{lm}^{\text{mag}} &= -\int x^l Y_{lm}(\Omega_x) \nabla \cdot (\mathbf{M} + \boldsymbol{\mu}) \, d^3x \tag{A.29} \\ &= -\int x^l Y_{lm}(\Omega_x) \nabla \cdot [\frac{1}{l+1} \mathbf{r} \times (\nabla \times \mathbf{M}) + \boldsymbol{\mu}] \, d^3x \end{aligned}$$

The second equality follows with the aid of the identity in Eq. (A.7). A partial integration, and the restoration to operator form yields the final result for the relevant static magnetic multipole operators

$$\hat{\mathcal{M}}_{lm}^{\text{mag}} = \int d^3x \left[\hat{\boldsymbol{\mu}}(\mathbf{x}) + \frac{1}{l+1} \mathbf{r} \times \hat{\mathbf{J}}_c(\mathbf{x}) \right] \cdot \nabla x^l Y_{lm} \qquad (A.30)$$

This is recognized to be, within a numerical factor and powers of k, the long-wavelength limit of the transverse magnetic multipole operator in Eq. (A.9).

Appendix B
Center of mass (C-M) motion

The center-of-mass (C-M) motion can, in fact, be handled correctly in the usual non-relativistic many-body problem. We follow the approach of [Fo69]. Introduce the usual C-M and internal coordinates as indicated in Fig. B.1.

$$\mathbf{X} \equiv \frac{1}{A}\sum_{i=1}^{A}\mathbf{x}_i$$
$$\mathbf{x}_i' \equiv \mathbf{x}_i - \mathbf{X} \qquad ; i = 1, 2, \ldots, A \qquad \text{(B.1)}$$

It follows that [Fo69]

$$\sum_{i=1}^{A}\mathbf{x}_i' = 0 \qquad\qquad \text{(B.2)}$$

$$d^3x_1\, d^3x_2 \cdots d^3x_A = d^3(AX)\, d^3x_1'\, d^3x_2' \cdots d^3x_A'\, \delta^{(3)}\left(\sum_{i=1}^{A}\mathbf{x}_i'\right)$$

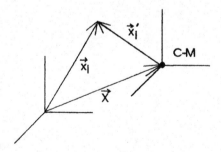

Fig. B.1.   C-M and internal coordinates; $i = 1, 2, \ldots, A$ labels the particles.

294

The rewriting of the volume element in the second equation is particularly useful. The target wave function can be written quite generally as

$$\Psi_i(\mathbf{x}_1, \ldots, \mathbf{x}_A) = \frac{1}{\sqrt{A^3\Omega}} e^{i\mathbf{p}\cdot\mathbf{X}} \psi_i(\mathbf{x}'_1, \ldots, \mathbf{x}'_A) \tag{B.3}$$

Now the nuclear charge density operator, for example, is given as

$$\hat{\rho}(\mathbf{x}) = \sum_{i=1}^{Z} \delta^{(3)}(\mathbf{x} - \mathbf{x}_i) \tag{B.4}$$

Its Fourier transform is written in terms of C-M and internal coordinates as

$$\int e^{-i\mathbf{q}\cdot\mathbf{x}} \hat{\rho}(\mathbf{x}) d^3x = e^{-i\mathbf{q}\cdot\mathbf{X}} \left(\sum_{i=1}^{Z} e^{-i\mathbf{q}\cdot\mathbf{x}'_i} \right) \tag{B.5}$$

The integral over the C-M coordinate can be done in the big box of volume Ω with p.b.c., and the result is

$$\langle f | \int e^{-i\mathbf{q}\cdot\mathbf{x}} \hat{\rho}(\mathbf{x}) d^3x | i \rangle = \delta_{\mathbf{p},\mathbf{p}'+\mathbf{q}} \langle \psi_f | \int e^{-i\mathbf{q}\cdot\mathbf{x}} \hat{\rho}(\mathbf{x}) d^3x | \psi_i \rangle \tag{B.6}$$

The remaining matrix element is now written in internal coordinates in the C-M system.

$$\langle \psi_f | \int e^{-i\mathbf{q}\cdot\mathbf{x}} \hat{\rho}(\mathbf{x}) d^3x | \psi_i \rangle = \int d^3x'_1 \cdots d^3x'_A \, \delta^{(3)} \left(\sum_{i=1}^{A} \mathbf{x}'_i \right)$$

$$\times \psi_f^*(\mathbf{x}'_1, \ldots, \mathbf{x}'_A) \left(\sum_{i=1}^{Z} e^{-i\mathbf{q}\cdot\mathbf{x}'_i} \right) \psi_i(\mathbf{x}'_1, \ldots, \mathbf{x}'_A) \tag{B.7}$$

Now use

$$\delta^2_{\mathbf{p},\mathbf{p}'+\mathbf{q}} = \delta_{\mathbf{p},\mathbf{p}'+\mathbf{q}}$$

$$\sum_f \delta_{\mathbf{p},\mathbf{p}'+\mathbf{q}} = \sum_f{}' \sum_{\mathbf{p}'} \delta_{\mathbf{p},\mathbf{p}'+\mathbf{q}} = \sum_f{}' \tag{B.8}$$

Here $\sum_f{}'$ goes over all internal quantum numbers. This allows one to write the sum over final states of the square of the matrix element as

$$\sum_f \left| \langle f | \int e^{-i\mathbf{q}\cdot\mathbf{x}} \hat{\rho}(\mathbf{x}) d^3x | i \rangle \right|^2 = \sum_f{}' \left| \langle \psi_f | \int e^{-i\mathbf{q}\cdot\mathbf{x}} \hat{\rho}(\mathbf{x}) d^3x | \psi_i \rangle \right|^2 \tag{B.9}$$

Now the analysis proceeds as in the text.

In summary, assume the current density has the form

$$\hat{J}_v(\mathbf{x}) = \sum_{i=1}^{A} [j_v(\mathbf{x}_i)\delta^{(3)}(\mathbf{x} - \mathbf{x}_i)] \tag{B.10}$$

Its Fourier transform can then be written

$$\int e^{-i\mathbf{q}\cdot\mathbf{x}}\hat{J}_v(\mathbf{x})\,d^3x = \sum_{i=1}^{A}[j_v(\mathbf{x}_i)e^{-i\mathbf{q}\cdot\mathbf{x}_i}] \tag{B.11}$$

Assume further that this expression can be written in terms of C-M and internal coordinates as

$$\int e^{-i\mathbf{q}\cdot\mathbf{x}}\hat{J}_v(\mathbf{x})\,d^3x = e^{-i\mathbf{q}\cdot\mathbf{X}}\sum_{i=1}^{A}[j_v(\mathbf{x}_i')e^{-i\mathbf{q}\cdot\mathbf{x}_i'}] \tag{B.12}$$

This holds true in the following cases:

- It is true for the charge density operator [see Eq. (B.5)];
- It is true for the spin current density [see Eq. (9.17)];
- It is true for the *transverse part* of the convection current density.

We give a proof of this third case. Consider the transverse part of the current defined by (here $\lambda = \pm 1$)

$$\hat{\mathbf{J}}(\mathbf{x}) \cdot \mathbf{e}_{\mathbf{q}\lambda}^{\dagger} = \sum_{i=1}^{Z} \left[\frac{\mathbf{p}(i)}{m}, \delta^{(3)}(\mathbf{x} - \mathbf{x}_i)\right]_{\text{sym}} \cdot \mathbf{e}_{\mathbf{q}\lambda}^{\dagger} \tag{B.13}$$

Since $m\dot{\mathbf{x}}_i = m\dot{\mathbf{X}} + m\dot{\mathbf{x}}_i'$, it follows that

$$\mathbf{p}(i) = \frac{1}{A}\mathbf{p} + \mathbf{p}'(i) \tag{B.14}$$

Now note that the transverse part of the convection current from the C-M, when the target is initially at rest, satisfies

$$\frac{1}{2A}(\mathbf{p} + \mathbf{p}') \cdot \mathbf{e}_{\mathbf{q}\lambda}^{\dagger} = -\frac{1}{2A}\mathbf{q} \cdot \mathbf{e}_{\mathbf{q}\lambda}^{\dagger} = 0 \tag{B.15}$$

Hence the C-M momentum does not contribute, and one can rewrite Eq. (B.13) as

$$\hat{\mathbf{J}}(\mathbf{x}) \cdot \mathbf{e}_{\mathbf{q}\lambda}^{\dagger} = \sum_{i=1}^{Z} \left[\frac{\mathbf{p}'(i)}{m}, \delta^{(3)}(\mathbf{x} - \mathbf{x}_i)\right]_{\text{sym}} \cdot \mathbf{e}_{\mathbf{q}\lambda}^{\dagger} \tag{B.16}$$

Thus the stated result is established.

In *conclusion*, it follows that

$$\overline{\sum_i \sum_f} |\langle f| \int e^{-i\mathbf{q}\cdot\mathbf{x}} \hat{\rho}(\mathbf{x}) d^3x|i\rangle|^2 \quad = \tag{B.17}$$

$$\overline{\sum_i \sum_f}{}' |\langle \psi_f| \int e^{-i\mathbf{q}\cdot\mathbf{x}} \hat{\rho}(\mathbf{x}) d^3x|\psi_i\rangle|^2$$

$$\overline{\sum_i \sum_f} \sum_{\lambda=\pm 1} |\langle f| \int e^{-i\mathbf{q}\cdot\mathbf{x}} \hat{\mathbf{J}}(\mathbf{x}) \cdot \mathbf{e}^{\dagger}_{\mathbf{q}\lambda} d^3x|i\rangle|^2 \quad =$$

$$\overline{\sum_i \sum_f}{}' \sum_{\lambda=\pm 1} |\langle \psi_f| \int e^{-i\mathbf{q}\cdot\mathbf{x}} \hat{\mathbf{J}}(\mathbf{x}) \cdot \mathbf{e}^{\dagger}_{\mathbf{q}\lambda} d^3x|\psi_i\rangle|^2$$

All of the subsequent analysis proceeds exactly as in the text.

A few comments are relevant. These are *exact* relations within non-relativistic quantum mechanics. The matrix elements are computed in the internal space according to Eq. (B.7); however, there is an A-body constraint $\delta^{(3)}(\sum_{i=1}^{A} \mathbf{x}'_i)$ in them. One usually does not deal correctly with this A-body constraint in calculations involving one or more valence particles, but there are models, such as the harmonic oscillator model, where it is possible to do so.

Within the framework of many particles in a harmonic oscillator potential, the center-of-mass motion can be taken into account by writing [El55, Ta58]

$$f(\kappa) = f_{\mathrm{CM}}(\kappa) F_{\mathrm{SM}}(\kappa) \tag{B.18}$$

Here $F_{\mathrm{SM}}(\kappa)$ is calculated with *A independent* nucleons in a harmonic oscillator shell-model potential, and $f(\kappa)$ is the transition form factor calculated with an intrinsic wave function with coordinates measured with respect to the center-of-mass; this is clearly what one is after. The C-M correction factor is

$$f_{\mathrm{CM}}(\kappa) = \exp\left(\frac{y}{A}\right)$$

$$y \equiv \left(\frac{\kappa b_{\mathrm{osc}}}{2}\right)^2$$

$$\hbar\omega_{\mathrm{osc}} = \frac{\hbar^2}{m b_{\mathrm{osc}}^2} \tag{B.19}$$

Note that the correction factor goes as $1/A$ where A is the number of nucleons. In calculations, this additional factor can always be conveniently lumped, together with the single-nucleon form factor of chapter 19, into an effective Mott cross section

$$\bar{\sigma}_{\mathrm{M}} \equiv f_{\mathrm{SN}}^2(\kappa) f_{\mathrm{CM}}^2(\kappa) \sigma_{\mathrm{Mott}} \tag{B.20}$$

Unless stated otherwise, this expression is used in all the traditional nuclear physics calculations carried out in this text.

We proceed to demonstrate the result in Eq. (B.19) in the case of four nucleons in the $1s$ state of the three-dimensional simple harmonic oscillator.[1] The independent-particle wave function in this case is

$$\Psi_{\text{SM}} \sim \exp\left(-\frac{1}{2b_{\text{osc}}^2}\sum_{i=1}^{A}\mathbf{r}_i^2\right) \tag{B.21}$$

The norm is discussed below. Introduce C-M and relative coordinates according to

$$\mathbf{R} \equiv \frac{1}{A}\sum_{i=1}^{A}\mathbf{r}_i$$
$$\mathbf{r}_i' \equiv \mathbf{r}_i - \mathbf{R} \tag{B.22}$$

Use the simple, crucial identity

$$\sum_{i=1}^{A}\mathbf{r}_i^2 = \sum_{i=1}^{A}\mathbf{r}_i'^2 + A\mathbf{R}^2 \tag{B.23}$$

Hence

$$\Psi_{\text{SM}} \sim \exp\left(-\frac{A\mathbf{R}^2}{2\,b_{\text{osc}}^2}\right)\psi_{\text{int}}(\mathbf{r}_i') \tag{B.24}$$

Now compute the charge form factor

$$F_{\text{SM}}(\kappa) \equiv \frac{1}{Z}\langle\Psi_{\text{SM}}|\sum_{i=1}^{Z}e^{-i\mathbf{q}\cdot\mathbf{r}_i}|\Psi_{\text{SM}}\rangle \tag{B.25}$$

$$\sim \frac{1}{Z}\int d^3R\, e^{-i\mathbf{q}\cdot\mathbf{R}}\exp\left(-\frac{A\mathbf{R}^2}{b_{\text{osc}}^2}\right)\langle\psi_{\text{int}}|\sum_{i=1}^{Z}e^{-i\mathbf{q}\cdot\mathbf{r}_i'}|\psi_{\text{int}}\rangle$$

The Fourier transform of the gaussian is immediately performed to give

$$F_{\text{SM}}(\kappa) = \exp\left(-\frac{b_{\text{osc}}^2\mathbf{q}^2}{4A}\right)\frac{1}{Z}\langle\psi_{\text{int}}|\sum_{i=1}^{Z}e^{-i\mathbf{q}\cdot\mathbf{r}_i'}|\psi_{\text{int}}\rangle \tag{B.26}$$

One can now check the normalization. Set $\kappa \equiv |\mathbf{q}| = 0$, and since both the shell-model and internal wave function are normalized, the overall factor is correct. This result is now solved for the true internal form factor

$$F_{\text{SM}}(\kappa) = \exp\left(-\frac{y}{A}\right)f_{\text{int}}(\kappa)$$
$$f_{\text{int}}(\kappa) = f_{\text{CM}}(\kappa)F_{\text{SM}}(\kappa) \tag{B.27}$$

---

[1] For the extension, see [de66].

This is the stated result. Note that $f_{int}(\kappa)$ is now calculated with true internal wave functions, the true internal volume element (see above discussion), and with the constraint $\sum_{i=1}^{A} \mathbf{r}'_i = 0$ thereby incorporated. The physics of this result is the following. The independent-particle model includes motion of the center-of-mass. This smears out the charge (probability) density. The true internal density is more compact, and hence its form factor falls off more slowly with κ. With many nucleons, the C-M motion does not smear out the density as much. Of course, this discussion is still all within the framework of the harmonic oscillator shell model. The extension to other forms of the potential, and especially to the fully relativistic case, is still an open problem.

Appendix C
Weizsäcker–Williams approximation

A useful approximation for the electron scattering cross section at low q^2 follows from the results in chapter 11; it is due to Weizsäcker and Williams (WWA). This approximation gives the dominant part of the inelastic cross section whenever the electron is undetected as it passes through matter, for then one has to sum over all possible momentum transfers, and the WWA cross section increases as $1/q^2$ for small q^2. Furthermore, the form of the WWA result derived below provides a stepping stone into the renormalization group evolution equations for quantum field theory [Al77], as discussed, for example, in [Ro90, Wa95].

We relate the electron scattering process in Fig. 11.1 to the corresponding *real photon* process illustrated in Fig. 11.4. This will allow us to express the electron scattering cross section as $q^2 \to 0$ in terms of a cross section measured in photoabsorption. In the course of the analysis, we will be able to identify a *probability of finding a photon* in the field of the electron. The classical basis for the WWA is described, for example, in [Ja62]. The Coulomb field of a relativistic electron Lorentz contracts and becomes predominantly transverse; the electron current produces a transverse magnetic field of comparable magnitude (Fig. C.1). This transverse field configuration is equivalent to a collection of *real* photons with a certain, specified momentum distribution.

The QED analysis here follows [Dr64, Wa84]. Recall the structure of the response tensor in Eqs. (11.20) and (11.27) for a target of mass m

$$
\begin{aligned}
W_{\mu\nu} &= (2\pi)^3 \overline{\sum_i} \sum_f \delta^{(4)}(q + p' - p)\langle p|J_\nu(0)|p'\rangle\langle p'|J_\mu(0)|p\rangle(\Omega E) \\
&= W_1(q^2, q \cdot p)\left(\delta_{\mu\nu} - \frac{q_\mu q_\nu}{q^2}\right) \\
&\quad + W_2(q^2, q \cdot p)\frac{1}{m^2}\left(p_\mu - \frac{p \cdot q}{q^2}q_\mu\right)\left(p_\nu - \frac{p \cdot q}{q^2}q_\nu\right) \quad (C.1)
\end{aligned}
$$

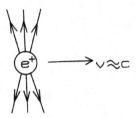

Fig. C.1. Lorentz contracted electric field of relativistic electron; basis for Weizsäcker–Williams approximation.

The (unpolarized) cross section for real photon processes follows directly from this response tensor. The relationship is derived in chapter 11, and the photoabsorption cross section is given by Eq. (11.39)

$$
\sigma_\gamma = \frac{(2\pi)^2\alpha}{\sqrt{(k\cdot p)^2}}\frac{1}{2}W_{\mu\mu}
$$

$$
= \frac{(2\pi)^2\alpha}{\sqrt{(k\cdot p)^2}}W_1(0, -k\cdot p) \qquad (C.2)
$$

The first line follows from the covariant polarization sum, and the second from a change to incoming photon momentum. Note that the real photon limit ($q^2 \to 0$) of Eq. (C.1) is perfectly *finite*; there are no singularities of the r.h.s. in this limit. Hence one establishes the following relations as $q^2 \to 0$ (chapter 11)

$$
W_2(q^2, q\cdot p) = O(q^2) \qquad\qquad ; q^2 \to 0
$$

$$
W_1(q^2, q\cdot p) = \frac{(p\cdot q)^2}{m^2q^2}W_2(q^2, q\cdot p) \qquad (C.3)
$$

These equations can be inverted to give for $q^2 \to 0$

$$
W_1 \doteq \frac{\sqrt{(q\cdot p)^2}}{(2\pi)^2\alpha}\sigma_\gamma\left(\frac{q\cdot p}{m}\right)
$$

$$
W_2 \doteq \frac{m^2q^2}{(p\cdot q)^2}W_1 \qquad (C.4)
$$

The electron scattering cross section can be written in terms of the variables in Fig. 11.1 as (chapter 11)

$$
d\sigma_e = \frac{4\alpha^2}{q^4}\frac{d^3k_2}{2\varepsilon_2}\frac{1}{\sqrt{(k_1\cdot p)^2}}\left\{q^2W_1 + \left[\frac{2(k_1\cdot p)(k_2\cdot p)}{m^2} - \frac{1}{2}q^2\right]W_2\right\} \quad (C.5)
$$

The overall dependence of $1/q^4$ coming from the square of the virtual photon propagator implies that in the integrated cross section, most of the contribution arises from the region where $q^2 \to 0$. In this case, one can replace the structure functions by their limiting forms in Eqs. (C.4)[1]

$$d\sigma_e \doteq \frac{4\alpha^2}{q^4} \frac{d^3k_2}{2\varepsilon_2} \frac{\sqrt{(q \cdot p)^2}}{\sqrt{(k_1 \cdot p)^2}} \frac{1}{(2\pi)^2\alpha} \sigma_\gamma \left(\frac{q \cdot p}{m} \right)$$

$$\times \left\{ q^2 + \frac{m^2 q^2}{(p \cdot q)^2} \left[\frac{2(k_1 \cdot p)(k_2 \cdot p)}{m^2} - \frac{1}{2}q^2 \right] \right\} \qquad \text{(C.6)}$$

This expression is Lorentz invariant. It is exact in the limit $q^2 \to 0$; at finite, but small q^2, it forms the Weizsäcker–Williams approximation. Equation (C.6) is the principal result of this appendix.

Let us, however, further develop this expression by using some kinematics. From Fig. 11.1 one has in the lab frame

$$q \cdot p = m(\varepsilon_1 - \varepsilon_2) = m\omega$$
$$k_1 \cdot p = -m\varepsilon_1 \qquad \text{(C.7)}$$

Also, the expression in brackets in Eq. (C.6) can be rewritten as

$$\{\cdots\} = q^2 + \frac{4\varepsilon_1\varepsilon_2 \sin^2 \theta/2}{\omega^2} \left[2\varepsilon_1\varepsilon_2 - 2\varepsilon_1\varepsilon_2 \sin^2 \frac{\theta}{2} \right]$$

$$= q^2 + \frac{8\varepsilon_1^2\varepsilon_2^2 \sin^2 \theta/2 \cos^2 \theta/2}{\omega^2} = q^2 + \frac{2\varepsilon_1^2\varepsilon_2^2 \sin^2 \theta}{\omega^2} \qquad \text{(C.8)}$$

Hence the result in Eq. (C.6) becomes

$$d\sigma_e \doteq \frac{8\alpha^2}{q^4} \frac{d^3k_2}{2\varepsilon_2} \frac{\omega}{\varepsilon_1} \frac{1}{(2\pi)^2\alpha} \sigma_\gamma(\omega) \left[\frac{\varepsilon_1^2\varepsilon_2^2 \sin^2 \theta}{\omega^2} + \frac{1}{2}q^2 \right] \qquad \text{(C.9)}$$

Now change variables using

$$\omega = \varepsilon_1 - \varepsilon_2$$
$$q^2 = 2\varepsilon_1\varepsilon_2(1 - \cos \theta) \qquad \text{(C.10)}$$

Hence (after an immediate integration over $d\phi$)

$$\frac{d^3k_2}{2\varepsilon_2} = \frac{\varepsilon_2\varepsilon_2 d\omega}{2\varepsilon_2} 2\pi \frac{dq^2}{2\varepsilon_1\varepsilon_2} = \frac{\pi}{2\varepsilon_1} d\omega dq^2 \qquad \text{(C.11)}$$

The limit $q^2 \to 0$ is achieved at finite ε_2 by going to small angles where $\theta \to 0$. In this case one has

$$\varepsilon_1^2\varepsilon_2^2 \sin^2 \theta \doteq \varepsilon_1^2\varepsilon_2^2\theta^2 \doteq q^2\varepsilon_1\varepsilon_2 \qquad \text{(C.12)}$$

---

[1] Here the symbol $\doteq$ implies an approximate relation that is exact in the limit $q^2 \to 0$.

Hence

$$d\sigma_e \ \dot{=} \ \frac{4\pi\alpha}{q^2}\frac{d\omega}{\varepsilon_1}\frac{\omega}{\varepsilon_1}\frac{1}{(2\pi)^2}\left[\frac{\varepsilon_1\varepsilon_2}{\omega^2}+\frac{1}{2}\right]dq^2\sigma_\gamma(\omega) \qquad (C.13)$$

Now introduce the *momentum fraction* of the virtual photon

$$\frac{\omega}{\varepsilon_1} \equiv z \qquad\qquad \frac{\varepsilon_2}{\varepsilon_1} = 1 - z \qquad (C.14)$$

Also introduce the differential of the so-called *resolution* in the electron scattering process defined here by

$$d\tau \ \equiv \ d\ln\left(\frac{q^2}{q_0^2}\right) = \frac{dq^2}{q^2} \qquad (C.15)$$

The electron scattering cross section in Eq. (C.13) can then be rewritten as

$$d\sigma_e \ \dot{=} \ \frac{\alpha}{2\pi}d\tau\, z\, dz\left[\frac{2(1-z)}{z^2}+1\right]\sigma_\gamma(z) \qquad (C.16)$$

We are now in a position to provide a more detailed interpretation of this result [Al77]. The contribution from the accompanying photon field to the electron scattering cross section for a beam of N electrons can be written as the following product: [number of photons $d\gamma(z,\tau)dz$ viewed with resolution between τ and $\tau+d\tau$ carrying a momentum fraction between z and $z+dz$ of the beam]$\times$ (photoabsorption cross section at that z). The first factor can in turn be related to the probability that at that τ, a photon carrying momentum fraction z is produced by an electron; we define that differential probability by $(\alpha/2\pi)P_{\gamma\leftarrow e}(z)d\tau dz$. It follows that

$$\begin{aligned} Nd\sigma_e &\equiv \ [d\gamma(z,\tau)dz]\,\sigma_\gamma(z) \\ &\equiv \ \left[N\frac{\alpha}{2\pi}P_{\gamma\leftarrow e}(z)d\tau dz\right]\sigma_\gamma(z) \end{aligned} \qquad (C.17)$$

One is now in a position to identify the *splitting function* $P_{\gamma\leftarrow e}(z)$ which forms the heart of the analysis of the evolution equations of QED and QCD. A comparison of Eqs. (C.16) and (C.17) gives

$$P_{\gamma\leftarrow e}(z) = z\left[\frac{2(1-z)}{z^2}+1\right] = \frac{1}{z}[(z-1)^2+1] \qquad (C.18)$$

Note that the splitting function as calculated here is independent of τ. For the additional splitting functions in QED and QCD, see for example [Qu83, Wa95].

Appendix D

Polarization and spin-1/2 fermions

It is essential to know the technique for dealing with the spin and polarization of any spin-1/2 fermion, electron or nucleon, entering into a
scattering process. Consider first the case of a massless fermion, for example a relativistic electron. The positive energy, stationary state, momentum
eigenstate of the Dirac equation in this case satisfies[1]

$$
\begin{aligned}
\boldsymbol{\alpha} \cdot \mathbf{p}\,\psi &= E_p \psi \\
E_p &= |\mathbf{p}|
\end{aligned}
\tag{D.1}
$$

Introduce the Dirac matrix γ_5 with the properties

$$
\begin{aligned}
\gamma_5 &\equiv \gamma_1\gamma_2\gamma_3\gamma_4 \\
\gamma_5\gamma_\mu + \gamma_\mu\gamma_5 &= 0 \\
\gamma_5^2 &= 1
\end{aligned}
\tag{D.2}
$$

In the standard representation, γ_5 and $\gamma_5\boldsymbol{\alpha}$ take the form

$$
\gamma_5 = \begin{pmatrix} 0 & -1 \\ -1 & 0 \end{pmatrix} \quad ; \ \gamma_5\boldsymbol{\alpha} = \boldsymbol{\alpha}\gamma_5 = \begin{pmatrix} -\boldsymbol{\sigma} & 0 \\ 0 & -\boldsymbol{\sigma} \end{pmatrix} \equiv -\boldsymbol{\sigma}
\tag{D.3}
$$

Introduce the projection operators defined by

$$
P_\downarrow = \frac{1}{2}(1+\gamma_5) \qquad P_\uparrow = \frac{1}{2}(1-\gamma_5)
\tag{D.4}
$$

These satisfy

$$
\begin{aligned}
P_\downarrow^2 &= P_\downarrow \\
P_\uparrow^2 &= P_\uparrow \\
P_\downarrow P_\uparrow &= 0
\end{aligned}
\tag{D.5}
$$

---

[1] Recall $\hbar = c = 1$ here. The reader can extend the arguments to the negative energy
solutions.

Define

$$\psi_\downarrow = P_\downarrow \psi \qquad\qquad \psi_\uparrow = P_\uparrow \psi \qquad\qquad (D.6)$$

Now multiply Eq. (D.1) on the left by, for example, $P_\downarrow$. Since γ_5 and $\boldsymbol{\alpha}$ commute, this gives

$$\boldsymbol{\alpha} \cdot \mathbf{p}\, \psi_\downarrow = |\mathbf{p}|\, \psi_\downarrow \qquad\qquad (D.7)$$

Multiply this equation on the left by γ_5 and make use of the above relations

$$-\boldsymbol{\sigma} \cdot \mathbf{p}\, \psi_\downarrow = |\mathbf{p}|\, \psi_\downarrow$$

$$\boldsymbol{\sigma} \cdot \left(\frac{\mathbf{p}}{|\mathbf{p}|}\right) \psi_\downarrow = -\psi_\downarrow \qquad\qquad (D.8)$$

One concludes that $P_\downarrow$ projects out of the Dirac spinors that part with *negative helicity*. $P_\uparrow$ does just the opposite.

One can now compute the cross section for massless fermions of any helicity by inserting either $P_\uparrow$ or $P_\downarrow$ before the appropriate Dirac spinors and then summing over *all* helicities. This converts the required expressions to traces, and only the appropriate helicity will contribute to the answer.

Suppose the fermion has a non-zero rest mass m. One can then go to the rest frame of the particle. In this frame, the four-vector $p_\mu = (\mathbf{0}, im)$. The Dirac spinors for a particle at rest reduce to simple Pauli spinors, and the spin operator in this frame is just $\boldsymbol{\sigma}/2$. The spin can be quantized along *any* convenient z-axis in this rest frame. Introduce a spin vector which points along this z-direction

$$\mathbf{S} \equiv \frac{\mathbf{z}}{|\mathbf{z}|} \qquad ; \text{ rest frame} \qquad\qquad (D.9)$$

Evidently

$$\boldsymbol{\sigma} \cdot \mathbf{S}\psi = \pm\psi \qquad\qquad (D.10)$$

One can readily construct projection operators for spin up or down along this *z*-axis in the rest frame

$$P_\uparrow = \frac{1}{2}(1 + \boldsymbol{\sigma} \cdot \mathbf{S}) \qquad\qquad P_\downarrow = \frac{1}{2}(1 - \boldsymbol{\sigma} \cdot \mathbf{S}) \qquad (D.11)$$

Now define a four-vector S_μ to be the result obtained by Lorentz transforming $S_\mu \equiv (\mathbf{S}, 0)$ from the rest frame. One evidently has the Lorentz invariant relations

$$p \cdot S = 0 \qquad\qquad S^2 = 1 \qquad\qquad (D.12)$$

The projection operators can be put into covariant form by using Eqs. (D.3), with the result that for positive energy spinors (for which, in the rest frame, $\beta\psi = \psi$)

$$
\begin{aligned}
P_\uparrow &= \frac{1}{2}(1 - \gamma_5\,\boldsymbol{\alpha}\cdot\mathbf{S}) \\
&= \frac{1}{2}(1 + i\gamma_5\,\boldsymbol{\gamma}\cdot\mathbf{S}\,\beta) \\
&= \frac{1}{2}(1 + i\gamma_5\,\gamma_\mu S_\mu)
\end{aligned}
\tag{D.13}
$$

This result can now be readily transformed from the rest frame to any other Lorentz frame. A similar result is obtained for $P_\downarrow$, and the reader can verify that Eqs. (D.5) are again satisfied.

These projection operators can now be inserted in front of the appropriate Dirac spinors and sums then taken over *all* spins, which converts spin sums to traces. Only the appropriate spin states will contribute. The result will be expressed in terms of Lorentz invariant expressions involving the four-vector S_μ which has a simple interpretation in the rest frame of the particle in terms of the direction of its spin.

Let us illustrate these developments with a simple exercise. Consider the scattering of longitudinally polarized, relativistic (massless) electrons from point Dirac nucleons with one-photon exchange. Let $h = \pm 1$ represent the helicity of the incident beam with $P_h = (1 - h\gamma_5)/2$. Calculate the polarization of the final nucleon defined by

$$
P_S \equiv \frac{N_\uparrow - N_\downarrow}{N_\uparrow + N_\downarrow} \equiv \frac{\mathscr{N}}{\mathscr{D}}
\tag{D.14}
$$

Here the arrows refer to the direction $\mathbf{S}$ in the rest frame. Since all common factors cancel in the ratio, one only needs to consider the Dirac traces obtained upon insertion of the appropriate projection operators. One needs the contraction of

$$
\begin{aligned}
\tilde{\eta}_{\mu\nu} &= \operatorname{trace}\{\gamma_\mu(1 - h\gamma_5)(-ik_\rho\gamma_\rho)\gamma_\nu(-ik'_\sigma\gamma_\sigma)\} \tag{D.15}\\
\tilde{W}_{\mu\nu} &= \operatorname{trace}\{\gamma_\mu(1 + i\gamma_5\gamma_\lambda S_\lambda)(M - ip_\alpha\gamma_\alpha)\gamma_\nu(M - ip'_\beta\gamma_\beta)\}
\end{aligned}
$$

At least four gamma matrices must be paired off with the γ_5 to get a non-zero result, and

$$
\operatorname{trace}\{\gamma_\mu\gamma_\nu\gamma_\rho\gamma_\sigma\gamma_5\} = 4\varepsilon_{\mu\nu\rho\sigma}
\tag{D.16}
$$

Hence

$$
\begin{aligned}
\tilde{\eta}_{\mu\nu} &= -4(k_\mu k'_\nu + k_\nu k'_\mu - \delta_{\mu\nu}\,k\cdot k') - 4h\,\varepsilon_{\mu\rho\nu\sigma}k_\rho k'_\sigma \\
\tilde{W}_{\mu\nu} &= 4M^2\delta_{\mu\nu} - 4(p_\mu p'_\nu + p_\nu p'_\mu - \delta_{\mu\nu}\,p\cdot p') \\
&\quad - 4M(\varepsilon_{\mu\lambda\nu\beta}S_\lambda p'_\beta + \varepsilon_{\mu\lambda\alpha\nu}S_\lambda p_\alpha)
\end{aligned}
\tag{D.17}
$$

In the contraction of the two tensors, both terms must either be even or odd in the interchange of μ and v to get a non-zero result. For the contraction of the antisymmetric terms use

$$\varepsilon_{\mu v \rho \sigma} \, \varepsilon_{\mu v \alpha \beta} \;\; = \;\; 2(\delta_{\rho\alpha}\delta_{\sigma\beta} - \delta_{\rho\beta}\delta_{\sigma\alpha})$$

$$\varepsilon_{\mu v \rho \sigma} \, \varepsilon_{\mu v \alpha \beta} \, a_\rho b_\sigma c_\alpha d_\beta \;\; = \;\; 2\, a \cdot c \, b \cdot d - 2\, a \cdot d \, b \cdot c \qquad \text{(D.18)}$$

Hence for the polarization P_S in Eq. (D.14) one has for massless electrons

$$\begin{aligned}
\mathcal{N} \;\; &\doteq \;\; +2hM(S \cdot k \, p' \cdot k' - S \cdot k' \, p' \cdot k - S \cdot k \, p \cdot k' + S \cdot k' \, p \cdot k) \\
&= \;\; -hM \, q^2 \, S \cdot (k + k') \\
\mathcal{D} \;\; &\doteq \;\; 2M^2 k \cdot k' + 2k \cdot p \, k' \cdot p' + 2k \cdot p' \, k' \cdot p \qquad \text{(D.19)}
\end{aligned}$$

In the second line $q \equiv k' - k = p - p'$ has been used. One only has a non-zero P_S in this case if h is non-zero and there is a polarization *transfer* (see [Ar81]).

Appendix E

Symmetry properties of matrix elements

In this appendix we derive symmetry properties of matrix elements of the electromagnetic multipole operators that follow from hermiticity of the current and time-reversal invariance of the strong and electromagnetic interactions [Pr65, Wa84].[1] The electromagnetic current is an observable and an hermitian operator

$$\hat{\mathbf{J}}(\mathbf{x})^{\dagger} = \hat{\mathbf{J}}(\mathbf{x})$$
$$\hat{\rho}(\mathbf{x})^{\dagger} = \hat{\rho}(\mathbf{x}) \tag{E.1}$$

The properties of the spherical and vector spherical harmonics under complex conjugation follow by inspection

$$Y_{JM}^{\star} = (-1)^{M} Y_{J,-M}$$
$$\mathscr{Y}_{JJ1}^{M\star} = (-1)^{1+M} \mathscr{Y}_{JJ1}^{-M} \tag{E.2}$$

The adjoints of the multipole operators then follow from their definition

$$\hat{\mathscr{T}}_{JM_J}(\kappa)^{\dagger} = (-1)^{M_J+\eta} \hat{\mathscr{T}}_{J,-M_J}(\kappa)$$
$$\eta \equiv 1 \qquad \quad ; \text{ current multipoles}$$
$$\equiv 0 \qquad \quad ; \text{ charge multipoles} \tag{E.3}$$

It is useful to include isospin in the analysis. Define spherical components of τ

$$\tau_{\pm 1} = \mp\frac{1}{\sqrt{2}}(\tau_1 \pm i\tau_2)$$
$$\tau_0 = \tau_3 \tag{E.4}$$

---

[1] Selection rules from parity invariance of these interactions are discussed in the text.

Now isolate the isospin dependence of a multipole operator in a factor

$$I_{TM_T} \equiv \frac{1}{2} \qquad ; T = 0$$

$$\equiv \frac{1}{2}\tau_{1,M_T} \qquad ; T = 1 \tag{E.5}$$

It follows that the multipole adjoints further satisfy

$$\hat{\mathscr{T}}^\dagger_{TM_T} = (-1)^{M_T}\hat{\mathscr{T}}_{T,-M_T} \tag{E.6}$$

A combination of these results gives the full adjoints of the multipole operators

$$\hat{\mathscr{T}}_{JM_J;TM_T}(\kappa)^\dagger = (-1)^{M_T+M_J+\eta}\,\hat{\mathscr{T}}_{J,-M_J;T,-M_T}(\kappa) \tag{E.7}$$

We shall now derive from this the following relation on a general reduced matrix element of a multipole operator

$$\langle J_f T_f \stackrel{\cdots}{\cdots} \hat{\mathscr{T}}_{J,T}(\kappa) \stackrel{\cdots}{\cdots} J_i T_i\rangle^\star = (-1)^{J_f-J_i+T_f-T_i+\eta}\langle J_i T_i \stackrel{\cdots}{\cdots} \hat{\mathscr{T}}_{J,T}(\kappa) \stackrel{\cdots}{\cdots} J_f T_f\rangle \tag{E.8}$$

Here the symbol $\stackrel{\cdots}{\cdots}$ indicates a reduced matrix element with respect to both angular momentum and isospin. The proof of this relation follows from the Wigner–Eckart theorem [Ed74]

$$\langle J_f M_f T_f \bar{M}_f|\hat{\mathscr{T}}_{JM_J;TM_T}|J_i M_i T_i \bar{M}_i\rangle = (-1)^{J_f-M_f} \begin{pmatrix} J_f & J & J_i \\ -M_f & M_J & M_i \end{pmatrix}$$

$$\times [J \rightleftharpoons T] \times \langle J_f T_f \stackrel{\cdots}{\cdots} \hat{\mathscr{T}}_{J,T} \stackrel{\cdots}{\cdots} J_i T_i\rangle \tag{E.9}$$

Now take the complex conjugate of this relation and use the definition of the adjoint $\langle f|\hat{\mathscr{T}}|i\rangle^\star = \langle i|\hat{\mathscr{T}}^\dagger|f\rangle$

$$(-1)^{J_f-M_f} \begin{pmatrix} J_f & J & J_i \\ -M_f & M_J & M_i \end{pmatrix} \times [J \rightleftharpoons T] \times \langle J_f T_f \stackrel{\cdots}{\cdots} \hat{\mathscr{T}}_{J,T} \stackrel{\cdots}{\cdots} J_i T_i\rangle^\star$$

$$= (-1)^{M_J+M_T+\eta}(-1)^{J_i-M_i} \begin{pmatrix} J_i & J & J_f \\ -M_i & -M_J & M_f \end{pmatrix}$$

$$\times [J \rightleftharpoons T] \times \langle J_i T_i \stackrel{\cdots}{\cdots} \hat{\mathscr{T}}_{J,T} \stackrel{\cdots}{\cdots} J_f T_f\rangle \tag{E.10}$$

Here the Wigner–Eckart theorem has been used once more on the last matrix element. Now use the properties of the 3-j symbols [Ed74] to rewrite the right hand side

$$\text{r.h.s} = (-1)^{J_f-J_i+T_f-T_i+\eta}(-1)^{J_f-M_f} \begin{pmatrix} J_f & J & J_i \\ -M_f & M_J & M_i \end{pmatrix}$$

$$\times [J \rightleftharpoons T] \times \langle J_i T_i \stackrel{\cdots}{\cdots} \hat{\mathscr{T}}_{J,T} \stackrel{\cdots}{\cdots} J_f T_f\rangle \tag{E.11}$$

Equation (E.8) has now been established.

Let us now investigate the restrictions imposed by time-reversal invariance. Recall that the time-reversal operator is *anti-unitary* and satisfies

$$\hat{T} i \hat{T}^{-1} = -i$$
$$\langle f|\hat{T}^{-1}|i\rangle = \langle Tf|i\rangle^\star \tag{E.12}$$

The properties of the electromagnetic current under time reversal follow from classical correspondence

$$\hat{T} \hat{\mathbf{J}}(\mathbf{x}) \hat{T}^{-1} = -\hat{\mathbf{J}}(\mathbf{x})$$
$$\hat{T} \hat{\rho}(\mathbf{x}) \hat{T}^{-1} = \hat{\rho}(\mathbf{x}) \tag{E.13}$$

Thus the multipole operators satisfy

$$\hat{T} \hat{\mathscr{T}}_{JM_J,TM_T} \hat{T}^{-1} = (-1)^{M_J} \hat{\mathscr{T}}_{J,-M_J;TM_T} \tag{E.14}$$

Note that the current only involves $M_T = 0$ and hence time reversal does not affect the isospin here. Our states are defined to transform according to [2]

$$\hat{T} |JM_J; TM_T\rangle = (-1)^{J+M_J} |J, -M_J; TM_T\rangle \tag{E.15}$$

Time-reversal invariance then says

$$\langle J_f M_f T_f \bar{M}_f | \hat{\mathscr{T}}_{JM_J;TM_T} | J_i M_i T_i \bar{M}_i \rangle$$
$$= \langle J_f M_f T_f \bar{M}_f | \hat{T}^{-1} \hat{T} \, \hat{\mathscr{T}}_{JM_J;TM_T} \, \hat{T}^{-1} \hat{T} | J_i M_i T_i \bar{M}_i \rangle$$
$$= (-1)^{J_i+M_i} (-1)^{J_f+M_f} (-1)^{M_J}$$
$$\times \langle J_f, -M_f T_f \bar{M}_f | \hat{\mathscr{T}}_{J,-M_J;TM_T} | J_i, -M_i T_i \bar{M}_i \rangle^\star \tag{E.16}$$

Now use the Wigner–Eckart theorem on both sides and the properties of the 3-j symbols [Ed74]

$$(-1)^{J_f-M_f} \begin{pmatrix} J_f & J & J_i \\ -M_f & M_J & M_i \end{pmatrix} \times [J \rightleftharpoons T]_{(1)} \times \langle J_f T_f \mathbin{\vdots\vdots} \hat{\mathscr{T}}_{J,T} \mathbin{\vdots\vdots} J_i T_i \rangle$$

$$= (-1)^{J_i+M_i} (-1)^{J_f+M_f} (-1)^{M_J} (-1)^{J_f+M_f} \begin{pmatrix} J_f & J & J_i \\ M_f & -M_J & -M_i \end{pmatrix}$$

$$\times [J \rightleftharpoons T]_{(1)} \times \langle J_f T_f \mathbin{\vdots\vdots} \hat{\mathscr{T}}_{J,T} \mathbin{\vdots\vdots} J_i T_i \rangle^\star \tag{E.17}$$

Since the isospin factors are identical, this relation implies

$$\langle J_f T_f \mathbin{\vdots\vdots} \hat{\mathscr{T}}_{J,T} \mathbin{\vdots\vdots} J_i T_i \rangle^\star = (-1)^J \langle J_f T_f \mathbin{\vdots\vdots} \hat{\mathscr{T}}_{J,T} \mathbin{\vdots\vdots} J_i T_i \rangle \tag{E.18}$$

---

[2] Note that this involves a *phase convention*.

Table E.1. Selection rules for multipole operators from parity and time reversal in elastic scattering; this quantity must be +1.

	$\hat{M}_{JM}(\kappa)$	$\hat{T}^{el}_{JM}(\kappa)$	$\hat{T}^{mag}_{JM}(\kappa)$
Parity	$(-1)^J$	$(-1)^J$	$(-1)^{J+1}$
Time Reversal	$(-1)^J$	$(-1)^{J+1}$	$(-1)^{J+1}$

A combination of Eq. (E.8) and Eq. (E.18) then leads to

$$\langle J_f T_f \:\vdots\: \hat{\mathscr{T}}_{J,T} \:\vdots\: J_i T_i \rangle = (-1)^{J+\eta+J_f-J_i+T_f-T_i} \langle J_i T_i \:\vdots\: \hat{\mathscr{T}}_{J,T} \:\vdots\: J_f T_f \rangle \quad \text{(E.19)}$$

This is the basic result of this appendix. It follows from the hermiticity of the current, time-reversal invariance of the strong and electromagnetic interactions, *and* a phase convention on the states. This relation allows one to turn around the matrix elements. If the initial and final states are identical, as is the case in elastic electron scattering, this relation leads to a *selection rule*. It states that

$$(-1)^{J+\eta} = 1 \qquad ; \text{ elastic scattering} \qquad \text{(E.20)}$$

Thus $J + \eta$ must be an even integer in elastic scattering. Hence only the even charge multipoles and odd current multipoles can contribute to elastic scattering. The selection rules for the various multipoles from both parity and time reversal in the case of elastic scattering are shown in Table E.1. For the charge and transverse magnetic multipoles, time-reversal and parity invariance lead to identical selection rules, that is, only charge multipoles with even J and transverse magnetic multipoles with odd J contribute to elastic electron scattering. For the transverse electric multipoles, parity implies J must be even while time reversal implies J must be odd. Hence invariance under *both* parity and time-reversal invariance implies there are no transverse electric multipoles in elastic electron scattering

$$\langle i| \hat{T}^{el}_{JM}(\kappa)|i \rangle = 0 \qquad ; \text{ parity and time reversal} \qquad \text{(E.21)}$$

Appendix F
Angular correlations

Consider the basic coincidence reaction

$$A(S_1^{\pi_1}) \, [e, e' \, X(S_X^{\pi_X})] \, A'(S_2^{\pi_2}) \tag{F.1}$$

The angular distribution of particle X in the C-M system can be analyzed in more detail using some basic results from [Ja59]. If particle X is massive, so that all helicity states are present, one can make a change of basis to L–S coupling states for the final two-particle system.

$$|JM\lambda_2\lambda_X\rangle = \sum_{LS} \langle J; LS|J; \lambda_2\lambda_X\rangle \, |JM; LS\rangle$$

$$\langle J; LS|J; \lambda_2\lambda_X\rangle = \sqrt{(2L+1)(2S+1)}(-1)^{S-S_2+S_X-L-2\lambda}$$

$$\times \begin{pmatrix} L & S & J \\ 0 & \lambda & -\lambda \end{pmatrix} \begin{pmatrix} S_2 & S_X & S \\ \lambda_2 & -\lambda_X & -\lambda \end{pmatrix} \tag{F.2}$$

Here $\lambda = \lambda_2 - \lambda_X$. This transformation reproduces the usual non-relativistic L–S coupling wave functions [Ja59]; however, it is also a completely *general* unitary transformation, for with some algebra [Wa84], one establishes the relations

$$\sum_{LS} \langle LS|\lambda_1\lambda_2\rangle \langle LS|\lambda_1'\lambda_2'\rangle = \delta_{\lambda_1\lambda_1'}\delta_{\lambda_2\lambda_2'}$$

$$\sum_{\lambda_1\lambda_2} \langle LS|\lambda_1\lambda_2\rangle \langle L'S'|\lambda_1\lambda_2\rangle = \delta_{LL'}\delta_{SS'} \tag{F.3}$$

Here J is suppressed. The transformation thus remains valid for arbitrary relativistic motion of the final two particles. The transformation in Eq. (F.2) is real, and the coefficients are independent of M just as in the proof of the Wigner–Eckart theorem.

The L–S basis states have two advantages. First, since they reduce to the usual non-relativistic L–S wave functions, one can use angular-momentum barrier arguments in this case to classify the contributions. Second, they produce *eigenstates of parity*, for again with some algebra [Wa84], one establishes the relation

$$P|JM;LS\rangle = \eta_2\eta_X(-1)^L|JM;LS\rangle \tag{F.4}$$

The change of basis in Eq. (F.2) can now be substituted in the expression for the bilinear product of current matrix elements appropriately summed and averaged over the final and initial helicities in Eq. (13.68). The result is, again after some algebra [Wa84]

$$\overline{(\mathscr{I}^{\lambda_k})^\star_{\lambda_f,\lambda_i}(\mathscr{I}^{\lambda'_k})_{\lambda_f,\lambda'_i}} = \frac{1}{4k^\star q}\frac{1}{2S_1+1}\sum_{\lambda_1}\sum_{J}\sum_{J'}\sum_{L}\sum_{L'}\sum_{S}\sum_{l}$$

$$\times(2J+1)(2J'+1)\sqrt{(2L+1)(2L'+1)}(-1)^{l+J+J'-S+\lambda_i}$$

$$\times\begin{pmatrix} L & l & L' \\ 0 & 0 & 0 \end{pmatrix}\begin{Bmatrix} J & J' & l \\ L' & L & S \end{Bmatrix}\sqrt{4\pi(2l+1)}Y_{l,\lambda'_k-\lambda_k}(\theta_q,\phi_q)$$

$$\times\begin{pmatrix} J & J' & l \\ \lambda_i & -\lambda'_i & \lambda_k-\lambda'_k \end{pmatrix}\langle LS|T^J|\lambda_1\lambda_k\rangle^\star\langle L'S|T^{J'}|\lambda_1\lambda'_k\rangle \tag{F.5}$$

Here $\lambda_i = \lambda_1-\lambda_k$ and $\lambda'_i = \lambda_1-\lambda'_k$, and a 6-j coefficient has been introduced [Ed74].

Transition amplitudes into states of definite parity can be defined by

$$c(LS;J;\lambda_1) \equiv \frac{\kappa^\star}{\omega^\star}\langle LS|T^J|\lambda_1,0\rangle$$

$$t(LS;J;\lambda_1) \equiv \langle LS|T^J|\lambda_1,+1\rangle \tag{F.6}$$

Recall these are functions of (W,k^2) and still contain all the dynamics. Parity invariance then implies that

$$\langle LS|T^J|\lambda_1,\lambda_k\rangle = \eta(-1)^{L+J-S_1}\langle LS|T^J|-\lambda_1,-\lambda_k\rangle \tag{F.7}$$

Again $\eta \equiv \eta_1\eta_2^\star\eta_X^\star$. This relation allows one to eliminate $\langle LS|T^J|-\lambda_1,-1\rangle$ and leads to the selection rule

$$c(LS;J;\lambda_1) = \eta(-1)^{L+J-S_1}c(LS;J;-\lambda_1) \tag{F.8}$$

Upon substitution of the appropriate values of λ_k, one can identify the coefficients appearing in Eqs. (13.71) as

$$A_l = \sum K_{JJ'}^l (LL'S\lambda_1)$$

$$\times \begin{pmatrix} J & J' & l \\ \lambda_1 & -\lambda_1 & 0 \end{pmatrix} c(LS;J;\lambda_1)^* c(L'S;J';\lambda_1) \tag{F.9}$$

$$B_l = -2\sum K_{JJ'}^l (LL'S\lambda_1)$$

$$\times \begin{pmatrix} J & J' & l \\ \lambda_1 - 1 & -\lambda_1 + 1 & 0 \end{pmatrix} t(LS;J;\lambda_1)^* t(L'S;J';\lambda_1)$$

$$C_l = \frac{-2}{\sqrt{l(l+1)}} \sum K_{JJ'}^l (LL'S\lambda_1)$$

$$\times \begin{pmatrix} J & J' & l \\ \lambda_1 & -\lambda_1 + 1 & -1 \end{pmatrix} \mathrm{Re}\, c(LS;J;\lambda_1)^* t(L'S;J';\lambda_1)$$

$$D_l = \frac{-1}{\sqrt{(l-1)l(l+1)(l+2)}} \sum K_{JJ'}^l (LL'S\lambda_1)(-1)^{L'+J'-S_1}$$

$$\times \begin{pmatrix} J & J' & l \\ \lambda_1 - 1 & -\lambda_1 - 1 & 2 \end{pmatrix} t(LS;J;\lambda_1)^* t(L'S;J';-\lambda_1)$$

Here $\sum \equiv \sum_J \sum_{J'} \sum_S \sum_L \sum_{L'} \sum_{\lambda_1}$ and the common summand factor is defined by

$$K_{JJ'}^l(LL'S\lambda_1) \equiv \frac{2l+1}{2S_1+1}(2J+1)(2J'+1)\sqrt{(2L+1)(2L'+1)}$$

$$\times (-1)^{J+J'+l-S+\lambda_1} \begin{Bmatrix} J & J' & l \\ L' & L & S \end{Bmatrix} \begin{pmatrix} L & L' & l \\ 0 & 0 & 0 \end{pmatrix} \tag{F.10}$$

Thus we have derived a general expression for the angular distribution in the C-M system for the coincidence reaction in Eq. (F.1). The derivation is completely relativistic, as long as particle X has non-zero rest mass so that all helicity amplitudes are present in the reaction.

For a 0^+ nuclear target, these angular correlation coefficients are discussed and tabulated in [Kl83, Wa84]. We give one other application here.

Consider pion electroproduction from the nucleon so that particle X is a pion and the initial and final target states are the nucleon with $J^\pi = 1/2^+$. For the pseudoscalar pion $S_X = 0$ and $\eta_X = -1$. For the nucleon $S_1 = S_2 = 1/2$ and $\eta_1 = \eta_2 = +1$. It follows from Eq. (F.2) that only one value of the total spin $S = 1/2$ enters the analysis, and this quantum number will subsequently be suppressed. The parity of the final π–N states follows from Eq. (F.4)

$$P|JM;L\rangle = (-1)^{L+1}|JM;L\rangle \tag{F.11}$$

There are now only two values of the initial nucleon helicity $\lambda_1 = \pm 1/2$, and the sum over this quantity can be immediately performed. Introduce the notation

$$c(LJ\lambda_1) \equiv \frac{k^\star}{\omega^\star} \langle L|T^J(W,k^2)|\lambda_1,0\rangle$$

$$t(LJ\lambda_1) \equiv \langle L|T^J(W,k^2)|\lambda_1,+1\rangle \tag{F.12}$$

Equations (F.9), which give the angular distributions in the C-M system through Eqs. (13.71) then reduce to the form

$$A_l = \sum K^l_{JJ'}(LL') \tag{F.13}$$
$$\times \begin{pmatrix} J & J' & l \\ 1/2 & -1/2 & 0 \end{pmatrix} c(LJ\tfrac{1}{2})^\star c(L'J'\tfrac{1}{2})$$

$$B_l = -\sum K^l_{JJ'}(LL')$$
$$\times \left[\begin{pmatrix} J & J' & l \\ -1/2 & 1/2 & 0 \end{pmatrix} t(LJ\tfrac{1}{2})^\star t(L'J'\tfrac{1}{2}) \right.$$
$$\left. - \begin{pmatrix} J & J' & l \\ -3/2 & 3/2 & 0 \end{pmatrix} t(LJ,-\tfrac{1}{2})^\star t(L'J',-\tfrac{1}{2}) \right]$$

$$C_l = \frac{-1}{\sqrt{l(l+1)}} \sum K^l_{JJ'}(LL') \operatorname{Re} c(LJ\tfrac{1}{2})^\star$$
$$\times \left[\begin{pmatrix} J & J' & l \\ 1/2 & 1/2 & -1 \end{pmatrix} t(L'J'\tfrac{1}{2}) \right.$$
$$\left. -\eta(-1)^{L+J-1/2} \begin{pmatrix} J & J' & l \\ -1/2 & 3/2 & -1 \end{pmatrix} t(L'J',-\tfrac{1}{2}) \right]$$

$$D_l = \frac{-1}{\sqrt{(l-1)l(l+1)(l+2)}} \sum K^l_{JJ'}(LL')(-1)^{L'+J'-1/2}$$
$$\times \begin{pmatrix} J & J' & l \\ -1/2 & -3/2 & 2 \end{pmatrix} \operatorname{Re} t(LJ\tfrac{1}{2})^\star t(L'J',-\tfrac{1}{2})$$

Here one is left with $\sum \equiv \sum_J \sum_{J'} \sum_L \sum_{L'}$, and the common summand is now

$$K^l_{JJ'}(LL') \equiv (2l+1)(2J+1)(2J'+1)\sqrt{(2L+1)(2L'+1)}$$
$$\times (-1)^{J+J'+l} \begin{Bmatrix} J & J' & l \\ L' & L & 1/2 \end{Bmatrix} \begin{pmatrix} L & L' & l \\ 0 & 0 & 0 \end{pmatrix} \tag{F.14}$$

Also

$$\eta \equiv \eta_1\eta_2\eta_X = -1 \tag{F.15}$$

As one application, suppose the pion electroproduction proceeds entirely through the first excited state of the nucleon with $J^\pi = 3/2^+$. In this case only one total angular momentum contributes so that $J = J'$. Furthermore, since $L = J \mp 1/2$ the positive parity picks out $L = L' = 1$ from Eq. (F.11). The summand can be evaluated with the aid of [Ed74] to give $K_{3/2,3/2}^2(11) = 8\sqrt{5}$, and further evaluation of the required 3-j symbols leads to the explicit angular distributions

$$\overline{|\mathscr{I}_c|^2} = \frac{1}{k^\star q}[1 + P_2(\cos\theta_q)]\left|c\left(1\frac{3}{2}\frac{1}{2}\right)\right|^2$$

$$\overline{|\mathscr{I}^{+1}|^2} + \overline{|\mathscr{I}^{-1}|^2} = \frac{1}{k^\star q}\left\{\left[\left|t\left(1\frac{3}{2}\frac{1}{2}\right)\right|^2 + \left|t\left(1\frac{3}{2},-\frac{1}{2}\right)\right|^2\right]\right.$$

$$\left.+ P_2(\cos\theta_q)\left[\left|t\left(1\frac{3}{2}\frac{1}{2}\right)\right|^2 - \left|t\left(1\frac{3}{2},-\frac{1}{2}\right)\right|^2\right]\right\}$$

$$\mathrm{Im}\,\overline{\mathscr{I}_c^\star(\mathscr{I}^{+1} + \mathscr{I}^{-1})} = \frac{1}{k^\star q}\sin\phi_q\,P_2^{(1)}(\cos\theta_q)$$

$$\times\left[-\frac{1}{\sqrt{3}}\mathrm{Re}\,c\left(1\frac{3}{2}\frac{1}{2}\right)^\star t\left(1\frac{3}{2},-\frac{1}{2}\right)\right]$$

$$\mathrm{Re}\,\overline{(\mathscr{I}^{+1})^\star(\mathscr{I}^{-1})} = \frac{1}{k^\star q}\cos 2\phi_q\,P_2^{(2)}(\cos\theta_q)$$

$$\times\left[\frac{1}{2\sqrt{3}}\mathrm{Re}\,t\left(1\frac{3}{2}\frac{1}{2}\right)^\star t\left(1\frac{3}{2},-\frac{1}{2}\right)\right] \qquad \text{(F.16)}$$

The integrals over the angle-dependent terms vanish when $\int d\Omega_q$ is performed, leaving just the angle-independent terms in the inclusive cross section.

Appendix G
Relativistic quasielastic scattering

If one scatters an electron from a nucleon at rest to a final state of discrete mass, then, as was shown in chapter 12, the Lorentz invariant response surfaces take the following form[1]

$$W_i(q^2, q \cdot p) = w_i(q^2) \frac{m^2}{p_0'} \delta(p_0 - p_0' - q_0) \qquad ; i = 1, 2 \qquad \text{(G.1)}$$

For a Dirac nucleon

$$
\begin{aligned}
w_1 &= \frac{q^2}{4m^2}(F_1 + 2mF_2)^2 \\
w_2 &= F_1^2 + \frac{q^2}{4m^2}(2mF_2)^2
\end{aligned}
\qquad \text{(G.2)}
$$

For elastic scattering from an isolated nucleon, it was shown in chapter 12 that

$$
\begin{aligned}
\int d\varepsilon_2 \, \delta(m - E' - q_0) &= \frac{E'}{m} r \\
r^{-1} &\equiv \left(1 + \frac{2\varepsilon_1 \sin^2 \theta/2}{m}\right)
\end{aligned}
\qquad \text{(G.3)}
$$

Hence the differential cross section for elastic scattering is given by

$$\frac{d\sigma}{d\Omega} = \sigma_M \left[w_2(q^2) + 2w_1(q^2)\tan^2 \frac{\theta}{2}\right] r \qquad \text{(G.4)}$$

This is the celebrated Rosenbluth cross section.

---

[1] In this section, momenta denote four-vectors so that $q^2 \equiv q_\mu^2$. We explicitly denote the three-vectors by $\mathbf{q}$, etc.

317

An alternative way to proceed is to rewrite the energy-conserving delta function in Eq. (G.1) as

$$
\begin{aligned}
\frac{m^2}{p_0'}\delta(p_0 - p_0' - q_0) &= 2m^2\,\delta[(p_0 - q_0)^2 - p_0'^2] \\
&= 2m^2\,\delta[(p - q)^2 - p'^2] \\
&= 2m^2\,\delta(2p\cdot q - q^2) \\
&= m\,\delta\left(v - \frac{q^2}{2m}\right) \\
&= \frac{m}{v}\,\delta(1 - x)
\end{aligned}
\tag{G.5}
$$

Here

$$
v \equiv \frac{p\cdot q}{m} \qquad ; \; x \equiv \frac{q^2}{2mv}
\tag{G.6}
$$

The quantity $v = \varepsilon_1 - \varepsilon_2$ is the electron energy loss in the lab, and x is the Bjorken scaling variable. Three-momentum conservation has been used in arriving at the second equality in Eq. (G.5) and the fact that this is elastic scattering so that $p^2 = p'^2 = -m^2$ in the third.

Note also that the combination

$$
\begin{aligned}
\frac{q^2}{4m^2}\delta\left(v - \frac{q^2}{2m}\right) &= \frac{1}{2m}\frac{q^2}{2mv}\delta\left(1 - \frac{q^2}{2mv}\right) \\
&= \frac{1}{2m}\delta(1 - x)
\end{aligned}
\tag{G.7}
$$

Hence for elastic scattering from an isolated nucleon, the response surfaces are given by

$$
\begin{aligned}
\frac{v}{m}W_2 &= \delta(1 - x)w_2(q^2) \\
\frac{2m}{m}W_1 &= \delta(1 - x)\bar{w}_1(q^2) \\
\bar{w}_1 &\equiv \frac{4m^2}{q^2}w_1(q^2) = (F_1 + 2mF_2)^2 \\
w_2 &= F_1^2 + \frac{q^2}{4m^2}(2mF_2)^2
\end{aligned}
\tag{G.8}
$$

If one now models the nucleus as a collection of non-interacting nucleons at rest, the nucleon cross sections can just be summed; equivalently, the structure functions take the form

$$
\begin{aligned}
\frac{v}{m}W_2^{(A)} &= \delta(1 - x)[Z\,w_2^p(q^2) + N w_2^n(q^2)] \\
2W_1^{(A)} &= \delta(1 - x)[Z\,\bar{w}_1^p(q^2) + N\bar{w}_1^n(q^2)]
\end{aligned}
\tag{G.9}
$$

This is the world's most naive model of the nucleus; however, it does have the following features to recommend it:

- It is completely covariant, assuming only that the nucleons are at rest in the lab frame and remain nucleons after the scattering. The nuclear response tensor has the correct Lorentz covariant structure;

- The nuclear current is conserved, and the structure of the nuclear response tensor reflects this fact;

- The nucleons can have *arbitrarily large* final four-momentum $p' = p - q$; the calculation still holds;

- When divided by the appropriate single-nucleon response functions, the nuclear response tensors exhibit *Bjorken scaling*, depending only on the variable v through the Bjorken scaling variable x appearing in the factor $\delta(1 - x)$.

It is a simple matter to generalize the above to the situation where the target nucleon is moving with momentum **p**. There are two changes that one has to consider:

1) From the definition of the initial flux as the number of particles crossing unit area transverse to the beam per unit time, one has

$$
\begin{aligned}
I_{\text{inc}} &= \frac{1}{\Omega} \mathbf{v}_{\text{rel}} \cdot \left(\frac{\mathbf{k}_1}{k_1}\right) \\
&= \frac{1}{\Omega} \left(\frac{\mathbf{k}_1}{k_1} - \frac{\mathbf{p}}{E}\right) \cdot \left(\frac{\mathbf{k}_1}{k_1}\right) \\
&= \frac{1}{\Omega} \frac{\sqrt{(p \cdot k_1)^2}}{E k_1}
\end{aligned}
\tag{G.10}
$$

This is exactly the same expression used previously in obtaining the invariant form of the cross section in Eq. (11.20). Hence it is appropriate to start from there.

2) Since the electron tensor is conserved, the terms in the nucleon tensor proportional to q_μ and q_ν can be discarded in the contraction of the two. The required replacements are therefore:

$$
\begin{aligned}
\eta_{\mu\nu} \delta_{\mu\nu} &\to \eta_{\mu\nu} \delta_{\mu\nu} \\
\eta_{\mu\nu} \frac{p_\mu p_\nu}{m^2} &\to \frac{1}{m^2} [2(p \cdot k_1)(p \cdot k_2) + (k_1 \cdot k_2)m^2] \\
&= \frac{1}{m^2} \left[2(p \cdot k_1)^2 + q^2(p \cdot k_1) - \frac{1}{2}q^2 m^2\right]
\end{aligned}
\tag{G.11}
$$

The first contraction, given entirely in electron variables, is unchanged. For a nucleon at rest in the lab frame with $p_\mu = (\mathbf{0}, im)$, the second contraction takes the previous form

$$\frac{1}{m^2}[2(p \cdot k_1)(p \cdot k_2) + (k_1 \cdot k_2)m^2] = 2\varepsilon_1\varepsilon_2 \cos^2 \frac{\theta}{2} \qquad (G.12)$$

For a moving nucleon, one simply evaluates Eqs. (G.11) for

$$p_\mu = (\mathbf{p}, iE) = (\mathbf{p}, i\sqrt{\mathbf{p}^2 + m^2}) \qquad (G.13)$$

The cross section for scattering a massless Dirac electron from a Dirac nucleon moving with initial momentum $\mathbf{p}$ in the lab is thus given by

$$\left(\frac{d^2\sigma}{d\varepsilon_2 d\Omega_2}\right)_{\text{mov nucl}} = \sigma_M \frac{m\varepsilon_1}{\sqrt{(k_1 \cdot p)^2}} 2m \, \delta(2p \cdot q - q^2) \left\{ 2w_1(q^2) \tan^2 \frac{\theta}{2} \right.$$

$$\left. + w_2(q^2)\frac{1}{2m^2\varepsilon_1\varepsilon_2 \cos^2 \theta/2} \left[2(p \cdot k_1)^2 + q^2(p \cdot k_1) - \frac{1}{2}q^2 m^2 \right] \right\} \qquad (G.14)$$

Now suppose the nucleus is modeled as a collection of non-interacting nucleons where there are $n(\mathbf{p}^2)\,d^3p$ nucleons moving with momentum between $\mathbf{p}$ and $\mathbf{p}+d\mathbf{p}$. This could, for example, be the momentum distribution for nucleons in an independent-particle shell model[2]

$$n^{(\alpha)}(\mathbf{p}^2) = \sum_i |\phi_i^{(\alpha)}(\mathbf{p})|^2 \qquad ; \alpha = p, n \qquad (G.15)$$

One can again just add the individual cross sections.

The third modification required for this case, in addition to the previous two, is as follows:

3) The expression for the energy-conserving delta function now takes the form

$$2m \int n(\mathbf{p}^2)\,d^2p_\perp\,dp_\| \,\delta(2p \cdot q - q^2) = \frac{2m}{2q} \int n(\mathbf{p}^2)\,d^2p_\perp\,dW \left(\frac{\partial p_\|}{\partial W}\right) \delta(W)$$

$$= \frac{m}{q} \int n(\mathbf{p}_\perp^2 + p_\|^2)\,d^2p_\perp \left(\frac{\partial p_\|}{\partial W}\right)$$

$$W \equiv p_\| + \frac{\omega}{q}(\mathbf{p}_\perp^2 + p_\|^2 + m^2)^{1/2} - \frac{q_\mu^2}{2q} = 0$$

$$\frac{\partial W}{\partial p_\|} = \frac{Eq + \omega p_\|}{Eq} \qquad (G.16)$$

The equation $W = 0$ determines $p_\|(\mathbf{p}_\perp^2, q, \omega)$ where now $q \equiv |\mathbf{q}|$ and $\omega = -q_0 = \varepsilon_1 - \varepsilon_2$.

---

[2] Closed shells are assumed and hence the distribution is a function of $\mathbf{p}^2$.

The resulting nuclear cross section is given by an incoherent sum

$$
\left(\frac{d^2\sigma}{d\varepsilon_2 d\Omega_2} \right)^{(A)} = \sigma_M \frac{m}{q} \sum_{\alpha=n,p} \int n^{(\alpha)} (\mathbf{p}_\perp^2 + p_\parallel^2) \, d^2 p_\perp
$$

$$
\times \left(\frac{Eq}{Eq + \omega p_\parallel} \right) \frac{m\varepsilon_1}{\sqrt{(k_1 \cdot p)^2}} \left\{ 2 w_1^{(\alpha)} (q^2) \tan^2 \frac{\theta}{2} \right.
$$

$$
\left. + w_2^{(\alpha)} (q^2) \frac{1}{2m^2 \varepsilon_1 \varepsilon_2 \cos^2 \theta/2} \left[2(p \cdot k_1)^2 + q^2 (p \cdot k_1) - \frac{1}{2} q^2 m^2 \right] \right\} \quad \text{(G.17)}
$$

Here $p_\parallel$ is again determined from $W = 0$.

These are exact results within this model. The nuclear current is again conserved, and the nucleon can be scattered through arbitrarily large (q, ω). While achieving these goals, it is important to note that the kinematics for electron scattering on a free nucleon have been employed, as well as the dispersion relation for a free initial nucleon in Eq. (G.13). Final-state interactions and modification of the initial nucleon spinors have been neglected.

To obtain some insight into this answer, specialize to the case where $|\mathbf{p}/E| = |(\mathbf{v}/c)_{\text{initial}}| \ll 1$. To leading order, the coefficients in the cross section reduce to those in Eq. (G.8), and the only change is to introduce a new quantity into the previous y-scaling analysis in Eq. (23.36)

$$
\tilde{y} \equiv \frac{m\omega}{q} - \frac{q_\mu^2}{2q} \quad \text{(G.18)}
$$

This is energy–momentum conservation to order $(v/c)_{\text{initial}}^2$. Note again, (q, ω) can be arbitrarily large as long as the nucleon remains a nucleon. y-scaling is discussed in much more detail in the review article [Da90], and also in [Do99].

Appendix H
Pion electroproduction

Much can be said about the amplitude for pion electroproduction from the nucleon, $N(e, e' \pi)N$, on general grounds. This is the first inelastic process one encounters in scattering electrons from protons or neutrons. The development in this appendix follows [Fu58, Wa68, Pr69, Wa84]. The pioneering work on the *photoproduction* process was carried out by CGLN [Ch57]. Other important early references on pion electroproduction include [De61, Za66, Vi67, Ad68, Pr70].

The kinematic situation is shown in Fig. 13.1; here particle X is now a pion. The laboratory cross section is given in terms of the covariant matrix elements of the current in Eq. (13.41) by Eq. (13.47). The angular distribution of the pions in the C–M system is given in terms of the helicity amplitudes by Eq. (13.68). With a transition to the L–S basis, and unobserved polarizations, the angular distribution takes the form in Eqs. (13.71, F.13). Here for the nucleon $J^\pi = 1/2^+$ and for pseudoscalar pions $\eta = \eta_1 \eta_2 \eta_X = -1$.

From Lorentz invariance, the S-matrix for the process $N(e, e' \pi)N$ in the one-photon-exchange approximation can be written as

$$S_{fi} = -\frac{(2\pi)^4}{\Omega} i\delta^{(4)}(k_1 + p_1 - k_2 - p_2 - q) \left(\frac{m^2}{2\omega_q E_1 E_2 \Omega^3} \right)^{1/2} T_{fi}$$

$$T_{fi} = 4\pi\alpha [i\bar{u}(k_2)\gamma_\mu u(k_1)] \frac{1}{k^2} J_\mu$$

$$J_\mu = \left(\frac{2\omega_q E_1 E_2 \Omega^3}{m^2} \right)^{1/2} \langle q p_2^{(-)} | J_\mu | p_1 \rangle \qquad \text{(H.1)}$$

Assume one has a theory for the pion–nucleon interaction with a set of Feynman diagrams and Feynman rules so that an expression for T_{fi} is at hand. Define the Møller potential by

$$\varepsilon_\mu \equiv \bar{u}(k_2)\gamma_\mu u(k_1)\frac{1}{k^2} \tag{H.2}$$

where, as before, $k \equiv k_1 - k_2$. The quantity $\varepsilon_\mu J_\mu$ is then a Lorentz scalar.[1] Conservation of the electromagnetic current states that the amplitude must vanish under the replacement $\varepsilon_\mu \to k_\mu$

$$k_\mu J_\mu = 0 \tag{H.3}$$

With the aid of the Dirac equation and current conservation, the transition amplitude can always be reduced to the following form

$$\varepsilon_\mu J_\mu = \bar{u}(p_2)\left[\sum_{i=1}^{6} a_i(W, \Delta^2, k^2)\,\varepsilon_\mu M_\mu^{(i)}\right]u(p_1) \tag{H.4}$$

The Dirac spinors for the nucleon are now normalized to $\bar{u}u = 1$. The four-momentum transfer to the nucleon used here, and mean four-momentum used below, are defined by

$$\Delta \equiv \frac{1}{2}(k - q)$$

$$P \equiv \frac{1}{2}(p_1 + p_2) \tag{H.5}$$

There are six independent kinematic invariants, and they can be taken to be [Fu58]

$$M_A = \frac{1}{2}i\gamma_5\left[\not{\varepsilon}\,\not{k} - \not{k}\,\not{\varepsilon}\right]$$

$$M_B = 2i\gamma_5\left[(P \cdot \varepsilon)(q \cdot k) - (P \cdot k)(q \cdot \varepsilon)\right]$$

$$M_C = \gamma_5\left[\not{\varepsilon}\,(q \cdot k) - \not{k}\,(q \cdot \varepsilon)\right]$$

$$M_D = 2\gamma_5\left[\not{\varepsilon}\,(P \cdot k) - \not{k}\,(P \cdot \varepsilon)\right] - im\gamma_5\left[\not{\varepsilon}\,\not{k} - \not{k}\,\not{\varepsilon}\right]$$

$$M_E = i\gamma_5\left[(k \cdot \varepsilon)(q \cdot k) - (q \cdot \varepsilon)k^2\right]$$

$$M_f = \gamma_5\left[\not{k}\,(k \cdot \varepsilon) - \not{\varepsilon}\,k^2\right] \tag{H.6}$$

Here the Feynman notation $\not{v} \equiv \gamma_\mu v_\mu$ is employed. Current conservation is evidently satisfied since the replacement $\varepsilon \to k$ causes each invariant to vanish identically.[2] Furthermore, in photoproduction, the last two invariants are absent since $k^2 = k \cdot \varepsilon = 0$ in that case [Ch57].

---

[1] Strictly speaking one must renormalize the electron wave functions with a factor $(E/m_e)^{1/2}$ so that $\bar{u}u = 1$ for this to be true [Bj65]; however, since all subsequent expressions in this appendix are linear in ε (and we know how to get the correct cross section) the overall normalization of ε here plays no role.

[2] Recall $\not{v}\not{v} = v^2$.

Without loss of generality, one can reduce the transition amplitude to an expression taken between two-component Pauli spinors by substituting the explicit form of the Dirac spinors introduced previously in Eq. (19.9) and now normalized to $\bar{u}u = 1$

$$u(\mathbf{p}, s) = \left(\frac{E_p + m}{2m}\right)^{1/2} \begin{pmatrix} \eta_s \\ \dfrac{\boldsymbol{\sigma} \cdot \mathbf{p}}{E_p + m} \eta_s \end{pmatrix} \tag{H.7}$$

Here $\eta_\uparrow = \begin{pmatrix} 1 \\ 0 \end{pmatrix}$ and $\eta_\downarrow = \begin{pmatrix} 0 \\ 1 \end{pmatrix}$ represent spin up and down along the z-axis, taken to be the direction of the incident nucleon in the C-M system as in Fig. 13.3. Substitution of Eq. (H.7) in Eq. (H.6) and explicit evaluation of the Dirac matrix products leads to the following equivalent, but still exact, expression for the spatial part of the transition matrix element expressed in term of Pauli matrices in the C-M system

$$\hat{\boldsymbol{\varepsilon}} \cdot \mathbf{J} = \eta_{s_2}^\dagger \left[\sum_{i=1}^{6} G_i(W, \Delta^2, k^2) m_i \right] \eta_{s_1}$$

$$m_1 = i\boldsymbol{\sigma} \cdot \hat{\boldsymbol{\varepsilon}}$$

$$m_2 = \boldsymbol{\sigma} \cdot \hat{\mathbf{q}} \left[\boldsymbol{\sigma} \cdot (\hat{\mathbf{k}} \times \hat{\boldsymbol{\varepsilon}}) \right]$$

$$m_3 = i\boldsymbol{\sigma} \cdot \hat{\mathbf{k}} (\hat{\mathbf{q}} \cdot \hat{\boldsymbol{\varepsilon}}) \qquad\qquad m_5 = i\boldsymbol{\sigma} \cdot \hat{\mathbf{q}} (\hat{\mathbf{k}} \cdot \hat{\boldsymbol{\varepsilon}})$$

$$m_4 = i\boldsymbol{\sigma} \cdot \hat{\mathbf{q}} (\hat{\mathbf{q}} \cdot \hat{\boldsymbol{\varepsilon}}) \qquad\qquad m_6 = i\boldsymbol{\sigma} \cdot \hat{\mathbf{k}} (\hat{\mathbf{k}} \cdot \hat{\boldsymbol{\varepsilon}}) \tag{H.8}$$

In this expression $\hat{\mathbf{v}}$ denotes a unit vector. The linear relations between the amplitudes a_i referred to as $\{A, B, \dots, E\}$ and the G_i is given by [Wa68, Wa84]

$$G_1 = \left[\frac{(E_1 + m)(E_2 + m)}{4m^2}\right]^{1/2} (W - m)$$

$$\times \left[A + (W - m)D - \frac{k \cdot q}{W - m}(C - D) + \frac{k^2}{W - m}F \right]$$

$$G_2 = \frac{|\mathbf{q}| k^*(W + m)}{[4m^2(E_1 + m)(E_2 + m)]^{1/2}}$$

$$\times \left[-A + (W + m)D - \frac{k \cdot q}{W + m}(C - D) + \frac{k^2}{W + m}F \right]$$

$$G_3 = |\mathbf{q}| k^*(W + m) \left[\frac{E_2 + m}{4m^2(E_1 + m)}\right]^{1/2}$$

$$\times \left[C - D + (W - m)B - \frac{k^2}{W + m}E \right]$$

$$G_4 = \mathbf{q}^2(W-m)\left[\frac{E_1+m}{4m^2(E_2+m)}\right]^{1/2}$$
$$\times\left[C-D-(W+m)B+\frac{k^2}{W-m}E\right]$$

$$G_5 = \frac{|\mathbf{q}|\,k^\star}{[4m^2(E_1+m)(E_2+m)]^{1/2}}$$
$$\times\{k_0\,[-A+(W+m)(D-F)-k\cdot q(B-E)]$$
$$-k\cdot q\,[C-D-(W+m)(B-E)]\}$$

$$G_6 = k^{\star 2}\left[\frac{E_2+m}{4m^2(E_1+m)}\right]^{1/2}$$
$$\times[-A+k\cdot q(E-B)-(W+m)F-(W-m)D] \qquad (\text{H.9})$$

The Coulomb matrix element can be obtained from these results by current conservation

$$\langle qp_2^{(-)}|\mathbf{J}\cdot\hat{\mathbf{k}}|p_1\rangle = \left(\frac{k_0}{k^\star}\right)\langle qp_2^{(-)}|\rho|p_1\rangle \qquad (\text{H.10})$$

If the Coulomb matrix element is evaluated directly, the result is

$$\left(\frac{2\omega_q E_1 E_2 \Omega^3}{m^2}\right)^{1/2}\langle qp_2^{(-)}|(-1)J_0\varepsilon_0|p_1\rangle = \eta_{s_2}^\dagger[m_7 G_7 + m_8 G_8]\eta_{s_1}$$

$$m_7 = -i\varepsilon_0\,\boldsymbol{\sigma}\cdot\hat{\mathbf{q}}$$
$$m_8 = -i\varepsilon_0\,\boldsymbol{\sigma}\cdot\hat{\mathbf{k}} \qquad (\text{H.11})$$

Equation (H.10) allows the identification

$$G_7 = \frac{k^\star}{k_0}[G_5+(\hat{\mathbf{k}}\cdot\hat{\mathbf{q}})\,G_4]$$

$$G_8 = \frac{k^\star}{k_0}[G_1+(\hat{\mathbf{k}}\cdot\hat{\mathbf{q}})\,G_3+G_6] \qquad (\text{H.12})$$

It is convenient to take out the same overall factor as in Eq. (13.41), and one defines new transition amplitudes by

$$\mathscr{I}_i \equiv \frac{m}{4\pi W}G_i \qquad i=1,\ldots,8 \qquad (\text{H.13})$$

It then follows from Eq. (H.8) that

$$\hat{\boldsymbol{\varepsilon}}\cdot\mathscr{I} = \eta_{s_2}^\dagger\left[\sum_{i=1}^6 \mathscr{I}_i(W,\Delta^2,k^2)m_i\right]\eta_{s_1} \qquad (\text{H.14})$$

To carry out a multipole analysis of the transition amplitude of the current, the covariant transition matrix element of the current is expanded according to Eq. (13.58)

$$
\eta^{\dagger}_{\lambda_2} \left(\sum_{i=1}^{6} m_i \mathscr{J}_i \right) \eta_{s_1} = \tag{H.15}
$$

$$
\frac{1}{(4k^\star q)^{1/2}} \sum_{J} (2J+1) \mathscr{D}^{J}_{\lambda_1-\lambda_k, \lambda_2}(-\phi_p, -\theta_p, \phi_p)^\star \, \langle \lambda_2 | T^{J}(W, k^2) | \lambda_1 \, \lambda_k \rangle
$$

Here $\lambda_1 (\equiv s_1)$ and λ_2 are the initial and final nucleon helicities, and λ_k is the virtual photon helicity. The C-M configuration is shown in Fig. 13.3. A little study shows that the Pauli spinor $\eta^{\star}_{\lambda_2}$ can be expressed in terms of the previous spinor $\eta^{\star}_{s_2}$ (representing spin up or down along the $-\hat{k}^\star$ axis) by the rotation

$$
\eta^{\dagger}_{\lambda_2} = \sum_{s_2} \mathscr{D}^{1/2}_{\lambda_2, s_2}(-\phi_p, \theta_p, \phi_p) \, \eta^{\dagger}_{s_2} \tag{H.16}
$$

Now one has the invariant amplitude expressed in terms of helicity amplitudes. This relation can be inverted using the orthonormality properties of the rotation matrices [Ed74]. Thus, given any invariant amplitude for pion electroproduction, one has all the equivalent helicity amplitudes.

Recall the transformation coefficients to the L–S basis, which provides eigenstates of parity. For the case of the π–N, the transformation in Eq. (F.2) takes the form (again S is suppressed)

$$
|JL\rangle = \sum_{\lambda_2} \sqrt{2L+1}(-1)^{1+L+\lambda_2+1/2} \begin{pmatrix} L & 1/2 & J \\ 0 & \lambda_2 & -\lambda_2 \end{pmatrix} |J\lambda_2\rangle \tag{H.17}
$$

Substitution of this expression in the definition of the transition amplitude in Eq. (F.6) gives

$$
c(LJ\tfrac{1}{2}) = \frac{k^\star}{\omega^\star} \sum_{\lambda_2} \sqrt{2L+1}(-1)^{1+L+\lambda_2+1/2} \begin{pmatrix} L & 1/2 & J \\ 0 & \lambda_2 & -\lambda_2 \end{pmatrix} \langle \lambda_2 | T^{J} | \tfrac{1}{2}, 0 \rangle
$$

$$
t(LJ\lambda_1) = \sum_{\lambda_2} \sqrt{2L+1}(-1)^{1+L+\lambda_2+1/2} \begin{pmatrix} L & 1/2 & J \\ 0 & \lambda_2 & -\lambda_2 \end{pmatrix} \langle \lambda_2 | T^{J} | \lambda_1, +1 \rangle
$$

$$
\tag{H.18}
$$

In the second relation $\lambda_1 = \pm 1/2$, and the sum in both relations goes over $\lambda_2 = \pm 1/2$. Thus, once the helicity amplitudes have been obtained, the transition amplitudes into eigenstates of parity follow immediately. The angular correlation coefficients are then given by Eq. (F.13). The transition multipole amplitudes into states of definite parity are sometimes more conventionally defined according to

$$c(LJ, \frac{1}{2}) = \pm (4k^\star q)^{1/2} \frac{1}{\sqrt{2}} \frac{k^\star}{\omega^\star} L_{l\pm}$$

$$t(LJ, \frac{1}{2}) = \pm (4k^\star q)^{1/2} \frac{1}{\sqrt{2}} T_{1/2}^{l\pm}$$

$$t(LJ, -\frac{1}{2}) = \pm (4k^\star q)^{1/2} \frac{1}{\sqrt{2}} T_{3/2}^{l\pm} \qquad (H.19)$$

Here $J = L \pm 1/2$ with $L \equiv l$.

Although we now have all one needs to obtain the general angular distribution in pion electroproduction, it is useful in comparing with current analyses [Bu94] to derive an equivalent expression directly from Eq. (H.14) by taking simple (two-component) traces. The cross section is given by Eq. (13.47) where the helicity unit vectors are defined in Eq. (13.43) with $\hat{\varepsilon}_0 = \hat{\varepsilon}_{k3}$. The result is readily shown to be

$$\overline{|\mathscr{J}_6|^2} = |\mathscr{J}_7|^2 + |\mathscr{J}_8|^2 + 2\mathrm{Re}\,\mathscr{J}_7^*\mathscr{J}_8 \cos\theta_q \qquad (H.20)$$

$$\overline{|\mathscr{J}^{+1}|^2} + \overline{|\mathscr{J}^{-1}|^2} = 2(|\mathscr{J}_1|^2 + |\mathscr{J}_2|^2 - 2\mathrm{Re}\,\mathscr{J}_1^*\mathscr{J}_2 \cos\theta_q) + \sin^2\theta_q$$
$$\times \left(|\mathscr{J}_3|^2 + |\mathscr{J}_4|^2 + 2\mathrm{Re}\,\mathscr{J}_1^*\mathscr{J}_4 + 2\mathrm{Re}\,\mathscr{J}_2^*\mathscr{J}_3 + 2\mathrm{Re}\,\mathscr{J}_3^*\mathscr{J}_4 \cos\theta_q\right)$$

$$\overline{\mathrm{Im}\,\mathscr{J}_6^*\,[\mathscr{J}^{+1} + \mathscr{J}^{-1}]} = -(1/\sqrt{2})\sin\phi_q \sin\theta_q\,[2\mathrm{Re}\,\mathscr{J}_1^*\mathscr{J}_7 + 2\mathrm{Re}\,\mathscr{J}_4^*\mathscr{J}_7$$
$$+ 2\mathrm{Re}\,\mathscr{J}_2^*\mathscr{J}_8 + 2\mathrm{Re}\,\mathscr{J}_3^*\mathscr{J}_8$$
$$+ \cos\theta_q(2\mathrm{Re}\,\mathscr{J}_3^*\mathscr{J}_7 + 2\mathrm{Re}\,\mathscr{J}_4^*\mathscr{J}_8)]$$

$$\overline{\mathrm{Re}\,(\mathscr{J}^{+1})^*\,(\mathscr{J}^{-1})} = -(1/2)\cos 2\phi_q \sin^2\theta_q$$
$$\times \left(|\mathscr{J}_3|^2 + |\mathscr{J}_4|^2 + 2\mathrm{Re}\,\mathscr{J}_1^*\mathscr{J}_4 + 2\mathrm{Re}\,\mathscr{J}_2^*\mathscr{J}_3 + 2\mathrm{Re}\,\mathscr{J}_3^*\mathscr{J}_4 \cos\theta_q\right)$$

The amplitudes $\mathscr{J}_i$ for $i = 1, \ldots, 4$ are expressed in terms of more familiar multipole amplitudes by

$$\mathscr{J}_1 = \sum_l \{[lM_{l+} + E_{l+}]P'_{l+1}(x) + [(l+1)M_{l-} + E_{l-}]P'_{l-1}(x)\}$$

$$\mathscr{J}_2 = \sum_l \{[(l+1)M_{l+} + lM_{l-}]P'_l(x)\}$$

$$\mathscr{J}_3 = \sum_l \{[E_{l+} - M_{l+}]P''_{l+1}(x) + [E_{l-} + M_{l-}]P''_{l-1}(x)\}$$

$$\mathscr{J}_4 = \sum_l \{[M_{l+} - E_{l+} - M_{l-} - E_{l-}]P''_l(x)\} \qquad (H.21)$$

Here $x = \cos\theta_q$ and $P'_l(x) = dP_l(x)/dx$. The notation $l\pm$ indicates that $J = l \pm 1/2$. For $k^2 \to 0$, that is the limit of photoproduction, these four equations reduce to those of CGLN [Ch57].[3] In pion electroproduction, the multipole amplitudes are still functions of both the energy in the C-M frame and the four-momentum transfer (W, k^2). In addition in electroproduction, there are the Coulomb multipoles

$$\mathscr{I}_7 = \frac{k^\star}{k_0}(\mathscr{I}_5 + x\mathscr{I}_4) = \sum_l \{[C_{l-} - C_{l+}]P_l'(x)\} \tag{H.22}$$

$$\mathscr{I}_8 = \frac{k^\star}{k_0}(\mathscr{I}_1 + x\mathscr{I}_3 + \mathscr{I}_6) = \sum_l \{[C_{l+}P_{l+1}'(x) - C_{l-}P_{l-1}'(x)]\} \tag{H.23}$$

Here $k_0 \equiv \omega^\star$.

These equations can be inverted to solve for the mutlipole amplitudes themselves. Define

$$\mathscr{I}_l^i(W, k^2) = \frac{1}{2}\int_{-1}^{1} P_l(x)\mathscr{I}^i(w, k^2, x)\, dx \tag{H.24}$$

Then use of the properties of the Legendre polynomials [Ed74] and a little algebra lead to

$$lE_{l-} = \mathscr{I}_l^1 - \mathscr{I}_{l-1}^2 + \frac{l+1}{2l+1}[\mathscr{I}_{l+1}^3 - \mathscr{I}_{l-1}^3] + \frac{l}{2l-1}[\mathscr{I}_l^4 - \mathscr{I}_{l-2}^4]$$

$$lM_{l-} = -\mathscr{I}_l^1 + \mathscr{I}_{l-1}^2 - \frac{1}{2l+1}[\mathscr{I}_{l+1}^3 - \mathscr{I}_{l-1}^3]$$

$$(l+1)E_{l+} = \mathscr{I}_l^1 - \mathscr{I}_{l+1}^2 - \frac{l}{2l+1}[\mathscr{I}_{l+1}^3 - \mathscr{I}_{l-1}^3] - \frac{l+1}{2l+3}[\mathscr{I}_{l+2}^4 - \mathscr{I}_l^4]$$

$$(l+1)M_{l+} = \mathscr{I}_l^1 - \mathscr{I}_{l+1}^2 + \frac{1}{2l+1}[\mathscr{I}_{l+1}^3 - \mathscr{I}_{l-1}^3]$$

$$C_{l+} = \mathscr{I}_{l+1}^7 + \mathscr{I}_l^8$$

$$C_{l-} = \mathscr{I}_{l-1}^7 + \mathscr{I}_l^8 \tag{H.25}$$

Finally, to close the loop, we give the relations between these multipoles and the helicity amplitudes into states of definite parity defined in Eqs. (H.19)

$$(l+1)M_{l+} = -\frac{i}{\sqrt{2}}[T_{1/2}^{l+} + \left(\frac{l+2}{l}\right)^{1/2} T_{3/2}^{l+}]$$

$$(l+1)E_{l+} = -\frac{i}{\sqrt{2}}[T_{1/2}^{l+} - \left(\frac{l}{l+2}\right)^{1/2} T_{3/2}^{l+}]$$

$$lM_{l-} = -\frac{i}{\sqrt{2}}[T_{1/2}^{l-} - \left(\frac{l-1}{l+1}\right)^{1/2} T_{3/2}^{l-}]$$

$$lE_{l-} = +\frac{i}{\sqrt{2}}[T_{1/2}^{l-} + \left(\frac{l+1}{l-1}\right)^{1/2} T_{3/2}^{l-}] \tag{H.26}$$

---

[3] Recall that $E_{1-} = M_{0+} \equiv 0$.

The longitudinal multipoles are defined in terms of the Coulomb multipoles with the aid of current conservation

$$C_{l\pm} \equiv \frac{k^\star}{k_0} i L_{l\pm} \equiv \frac{k^\star}{k_0} N_{l\pm} \tag{H.27}$$

If only the electron is detected in an *inclusive* experiment, one must integrate over the final pion direction. Only the terms (A_0, B_0) remain in Eq. (13.71), and the result is written, with the aid of Eqs. (H.19), as

$$\int \frac{d\Omega_q}{4\pi} \overline{|\mathscr{I}_{\mathscr{C}}|^2} = \sum_{J^\pi} \left(J + \frac{1}{2}\right) |C_{l\pm}|^2 \tag{H.28}$$

$$\int \frac{d\Omega_q}{4\pi} \left(\overline{|\mathscr{I}^{+1}|^2} + \overline{|\mathscr{I}^{-1}|^2}\right) = \sum_{J^\pi} \left(J + \frac{1}{2}\right) \left(|T_{3/2}^{l\pm}|^2 + |T_{1/2}^{l\pm}|^2\right)$$

Consider the role of isospin in pion electroproduction. Let $\alpha = 1, 2, 3$ be the hermitian components of isospin for the produced pion. Recall that the electromagnetic current has the isospin structure

$$J_\mu^\gamma = J_\mu^S + J_\mu^{V_3} \tag{H.29}$$

Isospin invariance of the strong interactions implies that the transition matrix of the current must then have the covariant form

$$T = T^{(+)}\delta_{\alpha 3} + T^{(-)}\frac{1}{2}[\tau_\alpha, \tau_3] + T^{(0)}\tau_\alpha \tag{H.30}$$

The transition amplitudes into states of given total isospin from a proton target then follow as

$$T(\frac{3}{2}, p) = \left(\frac{2}{3}\right)^{1/2} (T^+ - T^-)$$

$$T(\frac{1}{2}, p) = -\left(\frac{1}{3}\right)^{1/2} (T^+ + 2T^- + 3T^0) \tag{H.31}$$

The relations between the multipoles presented in this appendix are all derived in detail in [Wa84]. The reader now has enough background to proceed from any covariant, gauge-invariant expression for the S-matrix in pion electroproduction in the form of Eqs. (H.1) and Eq. (H.13) to individual multipole amplitudes. The coincident angular distribution is then given by Eqs. (13.71, F.13), or by Eqs. (H.20). Simultaneously, one has all the information needed for a general phenomenological analysis of pion electroproduction in terms of contributing multipoles [Bu94].

Finally, for the transition below the two-pion threshold into a π–N state with given (J^π, T), there is a theorem due to Watson that the *phase* of the electroproduction amplitude is given by the strong-interaction elastic scattering phase shift in this channel [Wa52]. To understand Watson's theorem, consider a 2-channel process where the first channel $a + b \rightleftharpoons a + b$ is elastic scattering through the strong interaction in a given partial wave, the transition amplitude is weak, say of $O(e)$ as in $\gamma + a \rightleftharpoons a + b$, and the scattering in the second channel $\gamma + a \rightleftharpoons \gamma + a$ is of $O(e^2)$. Time-reversal invariance implies that the S-matrix for this process must be symmetric and unitarity implies that $\mathscr{S}^\dagger \mathscr{S} = 1$ [Ja59]. To $O(e)$, the first condition implies that the S-matrix in this channel must have the form

$$\mathscr{S} = \begin{pmatrix} e^{2i\delta} & 2it \\ 2it & 1 \end{pmatrix} \tag{H.32}$$

Explicit evaluation of the unitarity condition for this 2×2 matrix then leads to the relation

$$t = |t| e^{i\delta} \tag{H.33}$$

Thus the phase of the weak transition amplitude is that of the strong-interaction phase shift. This is Watson's theorem.

Appendix I
Light-cone variables

In this appendix we introduce light-cone variables and discuss the response function in deep-inelastic electron scattering (DIS) when analyzed in terms of these quantities. The discussion follows closely that in [De73], which provides a much more extensive introduction to this topic.

Suppose that in coordinate space one has a four-vector $x_\mu = (x_1, x_2, x_3, ix_0) = (x, y, z, ict)$.[1] The light-cone variables are defined by

$$
\begin{aligned}
x_\pm &\equiv \frac{1}{\sqrt{2}}(z \pm ct) \\
\mathbf{x}_\perp &\equiv (x, y)
\end{aligned}
\tag{I.1}
$$

The situation is illustrated in Fig. I.1, where the new axes are defined by the lines $x_\mp = 0$. The square of the four-vector x_μ is evidently

$$
x^2 = x_\mu x_\mu = 2x_+ x_- + \mathbf{x}_\perp^2
\tag{I.2}
$$

In inclusive DIS we have two kinematic four-vectors $q_\mu = (k_2 - k_1)_\mu = (q_x, q_y, q_z, iq_0)$ and $p_\mu = (p_x, p_y, p_z, ip_0)$. We similarly define light-cone combinations

$$
\begin{aligned}
p_\pm &\equiv \frac{1}{\sqrt{2}}(p_z \pm p_0) & &; \; \mathbf{p}_\perp = (p_x, p_y) \\
q_\pm &\equiv \frac{1}{\sqrt{2}}(q_z \pm q_0) & &; \; \mathbf{q}_\perp = (q_x, q_y)
\end{aligned}
\tag{I.3}
$$

The scalar products are given by

$$
\begin{aligned}
m\nu = p \cdot q &= p_+ q_- + p_- q_+ + \mathbf{p}_\perp \cdot \mathbf{q}_\perp \\
q^2 &= 2q_+ q_- + \mathbf{q}_\perp^2
\end{aligned}
\tag{I.4}
$$

---

[1] We restore $\hbar$ and c in this appendix for clarity.

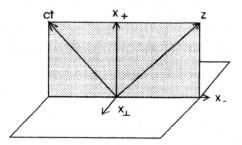

Fig. I.1.   Transformation to light-cone variables.

Assume the momentum transfer $\mathbf{q}$ defines the z-axis so that $\mathbf{q}_\perp = 0$. Further, assume for simplicity that $\mathbf{p}_\perp = 0$ (as is true, for example, in the lab). From the electron scattering kinematics, one has

$$
\begin{aligned}
q_z &= |\mathbf{k}_2 - \mathbf{k}_1| = (k_1^2 + k_2^2 - 2k_1 k_2 \cos\theta)^{1/2} \\
q_0 &= k_2 - k_1 \\
q_+ &= \frac{1}{\sqrt{2}}[(k_1^2 + k_2^2 - 2k_1 k_2 \cos\theta)^{1/2} - (k_1 - k_2)] \\
q_- &= \frac{1}{\sqrt{2}}[(k_1^2 + k_2^2 - 2k_1 k_2 \cos\theta)^{1/2} + (k_1 - k_2)]
\end{aligned}
\tag{I.5}
$$

Here we have written $|\mathbf{k}| \equiv k$.

The DIS limit is defined by $v \to \infty$, $q^2 \to \infty$, with constant $q^2/2mv = x_B$; it is evidently achieved by the following:

$$
\text{Fix } (q_+, p_\mu); \text{ and let } q_- \to \infty
\tag{I.6}
$$

In this case

$$
mv \to p_+ q_- \qquad ; q^2 \to 2q_+ q_- \qquad ; \frac{q^2}{2mv} \to \frac{q_+}{p_+}
\tag{I.7}
$$

To illustrate the arguments, we consider a very simplified, heuristic version of Eq. (14.18) where all indices and sums are suppressed

$$
w(p, q) \equiv \frac{1}{4\pi} \int e^{iq \cdot x} (p|[j(z), j(0)]|p) \, d^4 x
\tag{I.8}
$$

Now

$$
\begin{aligned}
d^4 x &= d^2 x_\perp dx_- dx_+ \\
q \cdot x &= q_+ x_- + q_- x_+
\end{aligned}
\tag{I.9}
$$

In the DIS limit of Eq. (I.6), the integrand in Eq. (I.8) oscillates very rapidly, and the resulting integral goes to zero, unless there is a finite contribution from the region where $x_+ \to 0$. If $x_+ \to 0$, then Eq. (I.2) implies that $x^2 \to \mathbf{x}_\perp^2$. This now represents a space-like separation of two points. The principle of *microscopic causality* states that the commutator of two hermitian observables (here the currents) must vanish for space-like separations since their measurements cannot interfere outside of the light cone. Hence the only contribution to the integral in Eq. (I.8) will come from the region where $(x_+, \mathbf{x}_\perp) \to 0$, which implies (for any finite x_-) that $x^2 \to 0$; this defines the light cone and illustrates the utility of the new variables. To obtain the asymptotic form of the response function in the DIS region, one is led to an investigation of the structure of the commutator of the two currents on the light cone.

Geometrically, the forward light cone is a cone around the ct axis that lies in the second quadrant in Fig. I.1. Both the x_+ and x_- axes lie in the surface of the cone. In the DIS limit, one is forced to the $x_+ = 0$ plane, which is tangent to the light cone along the negative x_- axis. Since by causality the commutator of the currents vanishes outside the light cone, the only contribution to the integral in Eq. (I.8) comes from the negative x_- axis in the DIS limit.

What kind of singularities exist on the light cone for the commutator of two hermitian operators in field theory? To get some insight, consider the very simple example of a free, massless, real (neutral), scalar field

$$\phi(x_\mu) = \frac{1}{\sqrt{\Omega}} \sum_k \left(\frac{\hbar}{2\omega_k} \right)^{1/2} \left(c_k e^{ik \cdot x} + c_k^\dagger e^{-ik \cdot x} \right) \tag{I.10}$$

It is one of the standard introductory exercises in field theory to show that the commutator of this field taken at two different space-time points is given by

$$[\phi(x_\mu), \phi(y_\mu)] = \frac{\hbar}{ic} \Delta(x_\mu - y_\mu)$$

$$\Delta(x_\mu) = \frac{i}{(2\pi)^3} \int d^4k \, \varepsilon(k_0) \, \delta(k^2) \, e^{ik \cdot x}$$

$$= \frac{1}{2\pi} \varepsilon(x_0) \, \delta(x^2) \tag{I.11}$$

Here $d^4k = d^3k \, dk_0$. The invariant commutator has a delta-function singularity on the light cone.

One can quite generally define

$$w(p, q) \equiv \frac{1}{i\pi} \int e^{iq \cdot x} G(p, x) D(x^2) \, d^4x \tag{I.12}$$

Assume now that the free-field singularities of the commutator have been isolated in $D(x^2)$ and that the function $G(p, x)$ contains the details of the currents and the states. From Lorentz invariance one must have $G(p, x) = G(p \cdot x, x^2)$. In the DIS limit one requires the singularities of the commutator on the light cone. In extracting the asymptotic limit, one can then replace the regular coefficient G by its value on the light cone

$$G(p \cdot x, x^2) \approx G(p \cdot x, 0) \equiv g(p \cdot x) \qquad ; \text{DIS} \qquad (I.13)$$

Introduce the Fourier transform of this function

$$g(\sigma) = \int e^{-i\alpha\sigma} F(\alpha) \, d\alpha \qquad (I.14)$$

Substitution into Eq. (I.12) then gives

$$w(p, q) \approx \frac{1}{i\pi} \int F(\alpha) \, d\alpha \int e^{-i(\alpha p - q) \cdot x} D(x^2) \, d^4x \qquad (I.15)$$

Again, for simplicity and illustration, suppose the light-cone singularity structure is that of Eq. (I.11). A four-dimensional Fourier transform then leads to

$$w(p, q) \approx 2 \int \varepsilon(\alpha p_0 - q_0) \, \delta[(\alpha p - q)^2] \, F(\alpha) d\alpha \qquad (I.16)$$

In the DIS limit $q_0 \to -\infty$ and with $p^2/q^2 \ll 1$,

$$w(p, q) \approx 2 \int \delta(2\alpha p \cdot q - q^2) F(\alpha) \, d\alpha$$

$$\approx \frac{1}{m\nu} F(x_B) \qquad (I.17)$$

One thus derives the scaling relation of the quark–parton model from the free-field singularities, and details of the structure, of the commutator of the currents on the light cone.

With local currents constructed out of bilinear combinations of quark fields, one first separates the points in the quark fields and introduces the notion of bilocal operators when evaluating the required current commutators [De73]. To quote from [De73], ... "The important lesson we learn ... is that the behavior of the structure function in the inelastic region is strictly related to the light-cone behavior of the commutator of the currents. The nature of the commutator singularity at $x_+ = 0$ determines the precise nature of the scaling, while the scaling function can be expressed as the Fourier transform of $g(\sigma)$, which in turn is related to the matrix element of a bilocal operator."

The idea of using the commutation relations of free-quark currents on the light cone to derive the DIS quark–parton results is due to Fritzsch and Gell-Mann [Fr71, De73]. It is Wilson's operator product expansion that provides a systematic way of looking at the short-distance behavior of a field theory [Wi69].

With the asymptotically-free theory QCD, one can justify the use of the free-field results at very short distances.[2] One can then proceed to calculate *corrections* to these free-field results. A useful way to proceed is to make use of the analysis in chapter 14 to rewrite the expression in Eq. (I.8). First, introduce a scattering amplitude analogous to that for forward virtual Compton scattering

$$a(p,q) \equiv \frac{i}{2\pi} \int e^{iq\cdot z} (p|P[j(z), j(0)]|p) \, d^4 z \qquad (I.18)$$

Here P denotes the time-ordered product

$$P[j(z), j(0)] \equiv j(z)j(0)\theta(z_0) + j(0)j(z)\theta(-z_0) \qquad (I.19)$$

This expression is immediately analyzed in terms of Feynman diagrams [Fe71]; the necessary Feynman rules for QCD are given in chapter 25. Insertion of a complete set of states and explicit evaluation of the integrals in Eq. (I.18), with the inclusion of an adiabatic damping factor for convergence in the time integrals, leads to the Low equation for the scattering amplitude

$$a(p,q) = \frac{1}{\pi} \sum_f (2\pi)^3 \left[\frac{\delta^{(3)}(\mathbf{q} + \mathbf{p}' - \mathbf{p})}{q_0 + p_0' - p_0 - i\eta} - \frac{\delta^{(3)}(\mathbf{q} - \mathbf{p}' + \mathbf{p})}{q_0 - p_0' + p_0 + i\eta} \right]$$
$$\times \langle p|j(0)|p' \rangle \langle p'|j(0)|p \rangle (\Omega E) \qquad (I.20)$$

Now take the imaginary part of this expression.[3] As in chapter 14, the second term does not contribute by the stability of the target, and

$$\text{Im}\, a(p,q) = \sum_f (2\pi)^3 \delta^{(4)}(q + p' - p)\langle p|j(0)|p'\rangle\langle p'|j(0)|p\rangle(\Omega E) \qquad (I.21)$$

The right side is recognized as the analog of Eq. (14.8) for the simplified response function in Eq. (I.8), and therefore

$$\text{Im}\, a(p,q) = w(p,q) \qquad (I.22)$$

Thus by taking the imaginary part of the scattering amplitude written in terms of Feynman diagrams, one can evaluate the response function in DIS.

The quark–parton result for the DIS response function in the impulse approximation in the $\mathbf{p} \to \infty$ frame is derived in chapter 12; it is evidently

---

[2] Indeed, the non-abelian gauge theory QCD was originally developed to do just that!
[3] More generally, take the *absorptive* part.

obtained by considering the imaginary part of the scattering diagram where the scattering takes place from a single non-interacting quark in the target (the so-called *handbag* diagram). The probability of finding such a quark in the target, $F(x_B)$, still depends on the strong-coupling aspects of the theory. From the above analysis, this result is equivalent to keeping the contribution of the singularities of the free-quark commutator on the light cone, with an amplitude $g(\sigma)$ again determined by the dynamics.

By considering additional Feynman diagrams, with radiative corrections, one can obtain perturbation-theory corrections to the response function of DIS. The evolution equations then allow one to obtain renormalization-group-improved results [Al77, Ch84, Ro90, Wa95].

The topics of operator product expansion, QCD radiative corrections, and evolution equations are explored in many texts (e.g. [Ch84]). In particular, the reader is referred to [Ro90] for an extensive discussion of the current theory of DIS scattering from the proton (with a summary of experimental results). Hopefully, the present text and this appendix will make that discussion more meaningful.

References

[Ab73] E. S. Abers and B. W. Lee, *Phys. Rep.* **9**, 1 (1973)

[Ad68] S. L. Adler, *Ann. of Phys.* **50**, 189 (1968)

[Ai89] I. J. R. Aitchison and A. J. G. Hey, *Gauge Theories in Particle Physics*, Adam Hilger, Bristol, England (1989)

[Al56] K. Alder, A. Bohr, T. Huus, B. Mottelson, and A. Winther, *Rev. Mod. Phys.* **28**, 432 (1956)

[Al77] G. Altarelli and G. Parisi, *Nucl. Phys.* **B126**, 298 (1977)

[Al88] W. M. Alberico, A. Molinari, T. W. Donnelly, E.L. Kronenberg, and J. W.Van Orden, *Phys. Rev.* **C38**, 1801 (1988)

[An99] K. A. Aniol, D.S. Armstrong, M. Baylac, E. Burtin, J. Calarco, G. D. Cates, C. Cavata, J.-P. Chen, E. Chudakov, D. Dale, C. W. de Jager, A. Deur, P. Djawotho, M. B. Epstein, S. Escoffier, L. Ewell, N. Falletto, J. M. Finn, K. Fissum, A. Fleck, B. Frois, J. Gao, F. Garibaldi, A. Gasparian, G. M. Gerstner, R. Gilman, A. Glamazdin, J. Gomez, V. Gorbenko, O. Hansen, F. Hersman, R. Holmes, M. Holtrop, B. Humensky, S. Incerti, J. Jardillier, M. K. Jones, J. Jorda, C. Jutier, W. Kahl, C. H. Kim, M. S. Kim, K. Kramer, K. S. Kumar, M. Kuss, J. KeRose, M. Leushner, D. Lhuillier, N. Liyanage, R. Lourie, R. Madey, D. J. Margaziotis, F. Marie, J. Martino, P. Mastromarino, K. McCormick, J. McIntyre, Z.-E. Meziani, R. Michaels, G. W. Miller, D. Neyret, C. Perdrisat, G. G. Petratos, R. Pomatsalyuk, J. S. Price, D. Prout, V. Punjabi, T. Pussieux, G. Quéméner, G. Rutledge, P. M. Rutt, A. Saha, P. A. Souder, M. Spradlin, R. Suleiman, J. Thompson, L. Todor, P. E. Ulmer, B. Vlahovic, K. Wijesooriya, R. Wilson, B. Wojtsekhowski, *Phys. Rev. Lett.* **82**, 1096 (1999)

[An00] P. L. Anthony, R. G. Arnold, T. Averett, H. R. Band, M. C. Berisso, H. Borel, P. E. Bosted, S. L. Blütmann, M. Buenerd, T. Chupp, S. Churchwell, G. R. Court, D. Crabb, D. Day, P. Decowski, P. DePietro, R. Erbacher, R. Erickson, A. Feltham, H. Fonvieille, E. Frlez, R. Gearhart,

337

V. Ghazikhanian, J. Gomez, K. A. Griffioen, C. Harris, M. A. Houlden, E. W. Hughes, C. E. Hyde-Wright, G. Igo, S. Incerti, J. Jensen, J. R. Johnson, P. M. King, Yu. G. Kolomensky, S. E. Kuhn, R. Lindgren, R. M. Lombard-Nelson, J. Marroncle, J. McCarthy, P. McKee, W. Meyer, G. S. Mitchell, J. Mitchell, M. Olson, S. Penttila, G. A. Peterson, G. G. Petratos, R. Pitthan, D. Pocanic, R. Prepost, C. Prescott, L. M. Qin, B. A. Raue, D. Reyna, L. S. Rochester, S. Rock, O. A. Rondon-Aramayo, F. Sabatie, I. Sick, T. Smith, L. Sorrell, F. Staley, S. St. Lorant, L. M. Stuart, Z. Szalata, Y. Terrien, A. Tobias, L. Todor, T. Toole, S. Trentalange, D. Walz, R. C. Welsh, F. R. Wesselmann, T. R. Wright, C. C. Young, M. Zeier, H. Zhu, and B. Zihlmann, *Phys. Lett.* **B493**, 19 (2000)

[Ar78]   R. G. Arnold, B. T. Chertok, S. Rock, W. P. Schütz, Z. M. Szalata, D. Day, J. S. McCarthy, F. Martin, B. A. Mecking, I. Sick, and G. Tamas, *Phys. Rev. Lett.* **40**, 1429 (1978)

[Ar81]   R. G. Arnold, C. E. Carlson, and F. Gross, *Phys. Rev.* **C23**, 363 (1981)

[Au83]   J. J. Aubert *et al.*, *Phys. Lett.* **123B**, 275 (1983)

[Au85]   S. Auffret, J.-M. Cavedon, J.-C. Clemens, B. Frois, D. Goutte, M. Huet, F. P. Juster, P. Leconte, J. Martino, Y. Mizuno, X. H. Phan, S. Platchkov, and I. Sick, *Phys. Rev. Lett.* **55**, 1362 (1985)

[Ba65]   J. Ballam, G. A. Loew, and R. B. Neal, SLAC-PUB-128 (1965)

[Ba73]   W. Bartel, F.-W. Büsser, W.-R. Dix, R. Felst, D. Harmes, H. Krehbiel, P. E. Kuhlmann, J. McElroy, J. Meyer, and G. Weber, *Nucl. Phys.* **B58**, 429 (1973)

[Ba00]   Website for Bates Laboratory: *http://mitbates.mit.edu/*

[Be66]   K. Berkelman, in *Proc. Int. Symp. on Electron and Photon Inter.* Hamburg, (1965), Deutsche, Phys. Gesell. cV. Hamburg, Germany (1966) Vol. II, p. 299

[Be90]   D. H. Beck, *Phys. Rev. Lett.* **64**, 268 (1990)

[Be98]   V. Bernard, H. W. Fearing, T. R. Hemmert, and U.-G. Meissner, *Nucl. Phys.* **A635** 121 (1998); **A642**, 563 (1998)

[Be91]   CEBAF PR 91-004, spks. E. Beise

[Bh88]   R. K. Bhaduri, *Models of the Nucleon*, Addison-Wesley Publ. Co., Reading, Massachusetts (1988)

[Bi89]   R. P. Bickerstaff and A. W. Thomas, *J. Phys.* **G15**, 1523 (1989)

[Bj60]   J. D. Bjorken, *unpublished*

[Bj65]   J. D. Bjorken and S. D. Drell, *Relativistic Quantum Mechanics*, McGraw-Hill Book Company, Inc., New York (1965)

[Bj65a]   J. D. Bjorken and S. D. Drell, *Relativistic Quantum Fields*, McGraw-Hill Book Company, Inc., New York (1965)

[Bj66]   J. D. Bjorken and J. D. Walecka, *Ann. of Phys.* **38**, 35 (1966)

[Bj69]   J. D. Bjorken, *Phys. Rev.* **179**, 1547 (1969)

[Bj69a] J. D. Bjorken and E. A. Paschos, *Phys. Rev.* **185**, 1975 (1969)

[Bl37]   F. Bloch and A. Nordsieck, *Phys. Rev.* **73**, 54 (1937)

[Bl52]   J. M. Blatt and V. F. Weisskopf, *Theoretical Nuclear Physics*, John Wiley and Sons, Inc., New York (1952)

[Bl68]   E. Bloom *et al.*, SLAC Group A, reported by W. K. Panofsky in *Int. Conf. on High Energy Phys., Vienna, 1968* CERN, Geneva (1968) p.23

[Bl91]   P. G. Blunden and E. J. Kim, *Nucl. Phys.* **A531**, 461 (1991)

[Bo69]   A. Bohr and B. R. Mottelson, *Nuclear Structure Vol. I, Single-Particle Motion* (1969) and *Vol. II, Nuclear Deformations* (1975), W. A. Benjamin, Inc., Reading, Massachusetts

[Bo98]   C.Bochna, B. P. Terburg, D. J. Abbott, A. Ahmidouch, C. S. Armstrong, J. Arrington, K. A. Assamagan, O. K. Baker, S. P. Barrow, D. P. Beatty, D. H. Beck, S. Y. Beedoe, E. J. Beise, J. E. Belz, P. E. Bosted, E. J. Brash, H. Breuer, R. V. Cadman, L. Cardman, R. D. Carlini, J. Cha, N. S. Chant, G. Collins, C. Cothran, W. J. Cummings, S. Danagoulian, F. A. Duncan, J. A. Dunne, D. Dutta, T. Eden, R. Ent, B. W. Filippone, T. A. Forest, H. T. Fortune, V. V. Frolov, H. Gao, D. F. Geesaman, R. Gilman, L. J. Gueye, K. K. Gustafsson, J.-O. Hansen, M. Harvey, W. Hinton, R. J. Holt, H. E. Jackson, C. E. Keppel, M. A. Khandaker, E. R. Kinney, A. Klein, D. M. Koltenuk, G. Kumbartzki, A. F. Lung, D. J. Mack, R. Madey, P. Markowitz, K. W. MacFarlane, R. D. McKeown, D. G. Meekins, Z.-E. Meziani, M. A. Miller, J. H. Mitchell, H. G. Mkrtchyan, R. M. Mohring, J. Napalitano, A. M. Nathan, G. Niculescu, I. Niculesco, T. G. O'Neill, B. R. Owen, S. F. Pate, D. H. Potterveld, J. W. Price, G. L. Rakness, R. Ransome, J. Reinhold, P. M. Rutt, C. W. Salgado, G. Savage, R. E. Segel, N. Simicevic, P. Stoler, R. Suleiman, L. Tang, D. van Westrum, W. F. Vulcan, S. Williamson, M. T. Witkowski, S. A. Wood, C. Yan, and B. Zeidman, *Phys. Rev. Lett.* **81**, 4576 (1998)

[Br71]   M. Breidenbach, Ph.D. thesis, M.I.T. (1971), *unpublished*

[Br82]   H. Breuker *et al.*, *Z. Phys.* **C13**, 113 (1982)

[Br83]   H. Breuker *et al.*, *Z. Phys.* **C17**, 121 (1983)

[Bu89]   CEBAF PR 89-037, spks. V. Burkert

[Bu94]   V. D. Burkert, *Perspectives in the Structure of Hadronic Systems*, eds. M. N. Harakeh *et al.*, Plenum Press, N.Y. (1994) p. 101

[Ca63]   N. Cabibbo, *Phys. Rev. Lett.* **10**, 531 (1963)

[Ca78]   R. N. Cahn and F. J. Gilman, *Phys. Rev.* **D17**, 1313 (1978)

[Ca80]   J. R. Calarco in *Proc. 1980 RCHP Int'l. Symposium on Highly Excited States in Nuclear Reactions*, eds. N. Ikegami and M. Muresaka, RCHP, Osaka (1980) p. 543

[Ca82] J. M. Cavedon, B. Frois, D. Goutte, M. Huet, Ph. Leconte, J. Martino, X-H. Phan, S. K. Platchkov, S. E. Williamson, W. Boeglin, I. Sick, P. de Witt-Huberts, L. S. Cardman, and C. N. Papanicolas, *Phys. Rev. Lett.* **49**, 986 (1982)

[Ca86] S. Capstick and N. Isgur, *Phys. Rev.* **D34**, 2809 (1986)

[Ca87] S. Capstick, *Phys. Rev.* **D36**, 2800 (1987)

[Ca90] C. Carlson, CEBAF Theory Study Group, CEBAF, Newport News, Virginia (1990), *unpublished*

[Ca91] J. Carlson, V. R. Pandharipande, and R. Schiavilla, in *Modern Topics in Electron Scattering*, eds. B. Frois and I. Sick, World Scientific, Singapore (1991) p. 177

[Ca94] J. Calarco, *Nucl. Phys.* **A569**, 363c (1994)

[CD90] Conceptual Design Report Basic Experimental Equipment, CEBAF (1990)

[Ce97] R. Cenni, T. W. Donnelly, and A. Molinari, *Phys. Rev.* **C56**, 276 (1997)

[Ch56] E. E. Chambers and R. Hofstadter, *Phys. Rev.* **103**, 1454 (1956)

[Ch56a] G. F. Chew and F. E. Low, *Phys. Rev.* **101**, 1570 (1956)

[Ch57] G. F. Chew, M. L. Goldberger, F. E. Low, and Y. Nambu, *Phys. Rev.* **106**, 1337 (1957); **106**, 1345 (1957)

[Ch58] G. F. Chew, R. Karplus, S. Gasiorowicz, and F. Zachariasen, *Phys. Rev.* **110**, 265 (1958)

[Ch69] B. T. Chertok, E. C. Jones, W. L. Bendel, and L. W. Fagg, *Phys. Rev. Lett.* **23**, 34 (1969)

[Ch74] A. Chodos, R. L. Jaffe, K. Johnson, C. B. Thorn, and V. Weisskopf, *Phys. Rev.* **D9**, 3471 (1974)

[Ch74a] A. Chodos, R. L. Jaffe, K. Johnson, and C. B. Thorn, *Phys. Rev.* **D10**, 2599 (1974)

[Ch84] T.-P. Cheng and L.-F. Li, *Gauge Theory of Elementary Particle Physics*, Clarendon Press, Oxford (1984)

[Co65] H. Collard, R. Hofstadter, E. B. Hughes, A. Johansson, M. R. Yearian, R. B. Day, and R. T. Wagner, *Phys. Rev.* **138**, B57 (1965)

[Co93] E. D. Cooper, S. Hama, B. C. Clark, and R. L. Mercer, *Phys. Rev.* **C47**, 297 (1993)

[Cu66] L. S. Cutler, *Ph.D. Thesis*, Stanford University (1966), *unpublished*

[Cu83] E. D. Cummins and P. H. Buchsbaum, *Weak Interactions of Leptons and Quarks*, Cambridge University Press, Cambridge, England (1983)

[Cz63] W. Czyż and K. Gottfried, *Ann. of Phys.* **21**, 47 (1963)

[Da51] R. H. Dalitz, *Proc. Roy. Soc.* **A206**, 509 (1951)

[Da89] CEBAF PR 89-018, spks. D. Day

[Da90] D. B. Day, J. S. McCarthy, T. W. Donnelly, and I. Sick, *Ann. Rev. Nucl. Part. Sci.* **40**, 357 (1990)

[Da92] J. F. Dawson, *private communication* (1992)

[De61] P. Dennery, *Phys. Rev.* **124**, 2000 (1961)

[de66] T. de Forest, Jr. and J. D. Walecka, *Adv. in Phys.* **15**, 1 (1966); (E) **15**, 491 (1966); (E) **17**, 479 (1968)

[de66a] T. de Forest, Jr., *Ph.D. Thesis, Stanford University* (1966)

[de67] T. de Forest, Jr., *Ann. of Phys.* **45**, 365 (1967)

[De73] V. De Alfaro, S. Fubini, G. Furlan, and C. Rossetti, *Currents in Hadron Physics*, North-Holland Publishing Company, Amsterdam (1973)

[de74] A. de Shalit and H. Feshbach, *Theoretical Nuclear Physics Vol. I, Nuclear Structure*, John Wiley and Sons, Inc., New York (1974)

[De75] T. DeGrand, R. L. Jaffe, K. Johnson, and J. Kiskis, *Phys. Rev.* **D12**, 2060 (1975)

[de83] T. de Forest, Jr., *Nucl. Phys.* **A392**, 232 (1983)

[De86] D. DeAngelis and H. Wegand, U. New Hampshire, *private communication*

[de86] P. K. A. de Witt Huberts, in *New Vistas in Electro-Nuclear Physics*, eds. E. L. Tomusiak *et al.*, *NATO ASI Series*, **B142**, Plenum, New York (1986) p. 331

[De99] F. J. Decker, Z. D. Farkas, and J. Turner, SLAC-PUB-8113 (1999)

[DE00] Website for DESY: *http://www.desy.de/*

[Dm90] V. Dmitrašinović, CEBAF Theory Study Group, CEBAF, Newport News, Virginia (1990), *unpublished*

[Dm92] V. Dmitrašinović, *Nucl. Phys.* **A537**, 551 (1992)

[Do69] T. W. Donnelly and G. E. Walker, *Phys. Rev. Lett.* **22**, 1121 (1969)

[Do73] T. W. Donnelly and J. D. Walecka, *Nucl. Phys.* **A201**, 81 (1973)

[Do75] T. W. Donnelly and J. D. Walecka, *Ann. Rev. Nucl. Sci.* **25**, 329 (1975)

[Do76] T. W. Donnelly and J. D. Walecka, *Nucl. Phys.* **A274**, 368 (1976)

[Do79] T. W. Donnelly and W. C. Haxton, *Atomic Data and Nuclear Data Tables* **23**, 103 (1979)

[Do79a] T. W. Donnelly and R. Peccei, *Phys. Rep.* **50**, 1 (1979)

[Do80] T. W. Donnelly and W. C. Haxton, *Atomic Data and Nuclear Data Tables* **25**, 1 (1980)

[Do93] J. F. Donoghue, E. Golowich, and B. R. Holstein, *Dynamics of the Standard Model*, Cambridge University Press, New York (1993)

[Do93a] J. D. Domingo, R. D. Carlini, B. A. Mecking, and J. Y. Mougey, *The CEBAF Experimental Equipment*, CEBAF Summer Workshop, June 15, 1992, *A.I.P. Conf. Proc.* **269**, eds. F. Gross and R. Holt, A.I.P., New York (1993) pp. 25–79

[Do99]   T. W. Donnelly and I. Sick, *Phys. Rev. Lett.* **82**, 3212 (1999)

[Dr61]   S. D. Drell and F. Zachariasen, *Electromagnetic Structure of Nucleons*, Oxford University Press, Oxford, England (1961)

[Dr64]   S. D. Drell and J. D. Walecka, *Ann. of Phys.* **28**, 18 (1964)

[Dr69]   D. Drechsel and H. Überall, *Phys. Rev.* **181**, 1383 (1969)

[Du76]   J. Dubach, J. H. Koch, and T. W. Donnelly, *Nucl. Phys.* **A271**, 279 (1976)

[Dy49]   F. J. Dyson, *Phys. Rev.* **75**, 1736 (1949)

[Dy89]   CEBAF PR 89-039, spks. S. Dytman

[Ed74]   A. R. Edmonds, *Angular Momentum in Quantum Mechanics*, 3rd Printing, Princeton University Press, Princeton, New Jersey (1974)

[Eg01]   H. Egiyan, *private communication*

[El55]   J. P. Elliot and T. H. R. Skyrme, *Proc Roy. Soc.* **A232**, 561 (1955)

[Fe49]   R. P. Feynman, *Phys. Rev.* **76**, 749 (1949)

[Fe49a] R. P. Feynman, *Phys. Rev.* **76**, 769 (1949)

[Fe51]   H. Feshbach, *Phys. Rev.* **84**, 1206 (1951)

[Fe58]   P. Federbush, M. L. Goldberger, and S. B. Treiman, *Phys. Rev.* **112**, 642 (1958)

[Fe65]   R. P. Feynman and A. R. Hibbs, *Quantum Mechanics and Path Integrals*, McGraw-Hill Book Co., Inc., New York (1965)

[Fe69]   R. P. Feynman, (quoted in [Bj69a])

[Fe71]   A. L. Fetter and J. D. Walecka, *Quantum Theory of Many-Particle Systems*, McGraw-Hill, New York (1971)

[Fe75]   G. Feinberg, *Phys. Rev.* **D12**, 3575 (1975)

[Fe80]   A. L. Fetter and J. D. Walecka, *Theoretical Mechanics of Particles and Continua*, McGraw-Hill, New York (1980)

[Fe91]   H. Feshbach, *Theoretical Nuclear Physics Vol. II, Nuclear Reactions*, John Wiley and Sons, Inc., New York (1991)

[Fe94]   T. C. Ferrée and D. S. Koltun, *Phys. Rev.* **C49**, 1961 (1994)

[Fi53]   V. L. Fitch and J. Rainwater, *Phys. Rev.* **92**, 789 (1953)

[Fi89]   CEBAF PR 89-028, spks. J. M. Finn and P. Ulmer; P. Ulmer and W. Van Orden, *private communication*

[Fi91]   CEBAF PR 91-010, spks. J. M. Finn and P. Souder

[Fo69]   L.L. Foldy and J. D. Walecka, *Ann. of Phys.* **54**, 447 (1969)

[Fo83]   F. Foster and G. Hughes, *Rep. Prog. Phys.* **40**, 1445 (1983)

[Fr60]   W. R. Frazer and J. R. Fulco, *Phys. Rev.* **117**, 1609 (1960)

[Fr60a] S. C. Frautschi and J. D. Walecka, *Phys. Rev.* **120**, 1486 (1960)

[Fr71]   H. Fritzsch and M. Gell-Mann, *Scale Invariance on the Light Cone*, talk given at the Conference on Fundamental Interactions at High Energy, Coral Gables, January 1971. Caltech Preprint 68-297

[Fr72]   J. I. Friedman and H. W. Kendall, *Ann. Rev. Nucl. Sci.*, **22**, 203 (1972)

[Fr72a] H. Fritzsch and M. Gell-Mann in *Proc. XVI Int. Conf. on High Energy Physics*, eds. J. D. Jackson and A. Roberts, Vol II, FNAL, Batavia, Illinois (1972), p. 135

[Fr72b] J. L. Friar and M. Rosen, *Phys. Lett.* **B39**, 615 (1972)

[Fr73]   J. L. Friar and J. W. Negele, *Nucl. Phys.* **A212**, 93 (1973)

[Fr73a] H. Fritzsch, M. Gell-Mann, and H. Leutwyler, *Phys. Lett.* **47B**, 365 (1973)

[Fr79]   B. Frois, *Lecture Notes in Phys.* **108**, Springer, Berlin (1979) p. 52

[Fu58]   S. Fubini, Y. Nambu, and V. Wataghin, *Phys. Rev.* **111**, 329 (1958)

[Fu91]   CEBAF PR 91-024, spks. H. Funsten

[Fu97]   R. J. Furnstahl, B. D. Serot, and H.-B. Tang, *Nucl. Phys.* **A618**, 446 (1997)

[Ge54]   M. Gell-Mann and F. E. Low, *Phys. Rev.* **95**, 1300 (1954)

[Ge64]   M. Gell-Mann, *Phys. Lett.* **8**, 214 (1964)

[Gl70]   S. L. Glashow, J. Iliopoulos, and L. Maiani, *Phys. Rev.* **D2**, 1285 (1970)

[Go61]   M. Gourdin, *Nuovo Cim.* **21**, 1094 (1961)

[Go79]   D. Gogney, *Nuclear Physics with Electromagnetic Interactions*, eds. H. Arenhövel and D. Drechsel, Lecture Notes in Physics **108**, Springer, Berlin (1979)

[Gr62]   T. A. Griffy, D. S. Onley, J. T. Reynolds, and L. C. Biedenharn, *Phys. Rev.* **128**, 833 (1962)

[Gr73a] D. J. Gross and F. Wilczek, *Phys. Rev. Lett.* **30**, 1343 (1973)

[Gr73b] D. J. Gross and F. Wilczek, *Phys. Rev.* **D8**, 3633 (1973)

[Gu34]   E. Guth, *Akad. Wiss. Wien, Math.-Naturw. Kl.* **24**, 299 (1934)

[Ha83]   E. Hadjimichael, B. Goulard, and R. Bornais, *Phys. Rev.* **C27**, 831 (1983)

[Ha84]   F. Halzen and A. D. Martin, *Quarks and Leptons*, John Wiley and Sons, New York (1984)

[HA96]   Hall A Collaboration, from experiment E89-033 (courtesy of C. Perdrisat, 1996)

[HB01]   Hall B Collaboration, TJNAF, *to be published*

[HC96]   Hall C Collaboration (1996)

[He69]   J. Heisenberg, R. Hofstadter, J. S. McCarthy, I. Sick, B. C. Clark, R. Herman, and D. G. Ravenhall, *Phys. Rev. Lett.* **23**, 1402 (1969)

[He83]   W. Hersman, *Phys. Lett.* **132B**, 47 (1983)

[He86] F. W. Hersman, W. Bertozzi, T. N. Buti, J. M. Finn, C. E. Hyde-Wright, M. V. Hynes, J. Kelly, M. A. Kovash, S. Kowalski, J. Lichtenstat, R. Lourie, B. Murdock, B. Pugh, F. N. Rad, C. P. Sargent, and J. P. Bellicard, *Phys. Rev.* **C33**, 1905 (1986)

[He89] CEBAF PR 89-031, spks. W. Hersman, R. Miskimen, and J. Lightbody

[He95] H. Hedayati-Poor, J. I. Johansson, and H. S. Sherif, *Phys. Rev.* **C51**, 2044 (1995)

[He00] jhh@pauli.unh.edu

[Ho56] R. Hofstadter *Rev. Mod. Phys.* **28**, 214 (1956)

[Ho63] R. Hofstadter, *Nuclear and Nucleon Structure*, Benjamin, NY (1963)

[Ho81] C. J. Horowitz and B. D. Serot, *Nucl. Phys.* **A368**, 503 (1981)

[Ho89] C. J. Horowitz and J. Piekarewicz, *Phys. Rev. Lett.* **62**, 391 (1989)

[Ho91] C. J. Horowitz, D. P. Murdoch, and B. D. Serot, in *Computational Nuclear Physics I: Nuclear Structure*, eds. K. Langanke, J. A. Marulin, and S. E. Koonin, Springer-Verlag, Berlin (1991) p. 128

[Hu83] V. W. Hughes and J. Kuti, *Ann. Rev. Nucl. Part. Sci.*, **33**, 611 (1983)

[Hu95] V. W. Hughes and J. D. Walecka, in *Symmetries and Fundamental Interactions in Nuclei*, eds. W. C. Haxton and E. M. Henley, World Scientific, Singapore (1995) p. 389

[Hy91] CEBAF PR 91-014, spks. C. Hyde-Wright

[Il87]   *Electron Scattering in Nuclear and Particle Science*, A.I.P. Conf. Proc. **161**, eds. C. N. Papanicolas, L. S. Cardman, and R. A. Eisenstein, A.I.P. New York (1987)

[Is77]   N. Isgur and G. Karl, *Phys. Lett.* **72B**, 109 (1977)

[Is80]   N. Isgur, in *The New Aspects of Subnuclear Physics*, ed. A. Zichichi, Plenum Press, New York (1980), p. 107

[Is81]   N. Isgur and G. Karl, *Phys. Rev.* **D23**, 817 (1981)

[Is85]   N. Isgur, *Act. Phys. Aust. Suppl. XXVII*, Springer-Verlag, Vienna (1985), p. 177

[Is85a] N. Isgur, G. Karl, and P. J. O'Donnell, *The Quark Structure of Matter*, World Scientific, Singapore (1985)

[Ja59]   M. Jacob and G. C. Wick, *Ann. of Phys.* **7**, 404 (1959)

[Ja62]   J. D. Jackson, *Classical Electrodynamics*, John Wiley and Sons, Inc., New York (1962)

[Ja66]   G. Jacob and T. A. J. Maris, *Rev. Mod. Phys.* **38**, 121 (1966)

[Ja73]   G. Jacob and T. A. J. Maris, *Rev. Mod. Phys.* **45**, 6 (1973)

[Ja76]   R. Jaffe and K. Johnson, *Phys. Lett.* **60B**, 201 (1976)

[Jo96]   J. Jourdan, *Nucl. Phys.* **A603**, 117 (1996)

[Jo00] M. K. Jones, K. A. Aniol, F. T. Baker, J. Berthot, P. Y. Bertin, W. Bertozzi, A. Besson, L. Bimbot, W. U. Boeglin, E. J. Brash, D. Brown, J. R. Calarco, L. S. Cardman, C.-C. Chang, J.-P. Chen, E. Chudakov, S. Churchwell, E. Cisbani. D. S. Dale, R. De Leo, A. Deur, B. Diederich, J. J. Domingo, M. B. Epstein, L. A. Ewell, K. G. Fissum, A. Fleck, H. Fonvieille, S. Frullani, J. Gao, F. Garibaldi, A. Gasparian, G. Gerstner, S. Gilad, R. Gilman, A. Glamazdin, C. Glashausser, J. Gomez, V. Gorbenko, A. Green, J.-O. Hansen, C. R. Howell, G. M. Huber, M. Iodice, C. W. de Jager, S. Jaminion, X. Jiang, W. Kahl, J. J. Kelly, M. Khayat, L. H. Kramer, G. Kumbartzki, M. Kuss, E. Lakuriki, G. Lavessiére, J. J. LeRose, M. Liang, R. A. Lindgren, N. Liyanage, G. J. Lolos, R. Macri, R. Madey, S. Malov, D. J. Margaziotis, P. Markowitz, K. McCormick, J. McIntyre, R. L. J. van der Meer, R. Michaels, B. D. Milbrath, J. Y. Mougey, S. K. Nanda, E. A. J. M Offerman, Z. Papandreo, C. F. Perdrisat, G. G. Petratos, N. M. Piskunov, R. I. Pomatsalyuk, D. L. Prout, V. Punjabi, G. Quéméner, R. E. Ransome, B. A. Raue, Y. Roblin, R. Roche, G. Rutledge, P. M. Rutt, A. Saha, T. Saito, A. J. Sarty, T. P. Smith, P. Sorokin, S. Strauch, R. Suleiman, K. Takahashi, J. A. Templon, L. Todor, P. E. Ulmer, G. M. Urciuoli, P. Vernin, B. Vlahovic, H. Voskanyan, K. Wijesooriya, B. B. Wojtsekhowski, R. Woo, F. Xiong, B. D. Zainea, and Z.-L. Zhou, *Phys. Rev. Lett.* **84**, 1398 (2000)

[Ju92]   K. Jungmann, V. W. Hughes, and G. zu Putlitz, eds., *The Future of Muon Physics*, Springer-Verlag, Berlin (1992)

[Ka78]   R. K. Kahn and F. J. Gilman, *Phys. Rev.* **D17**, 1313 (1978)

[Ka83]   G Kälbermann and J. M. Eisenberg, *Phys. Rev.* **D28**, 71 (1983)

[Ki75]   W. Kirk, *End Station A Spectrometers*, SLAC Beam Line, March 20, 1975

[Ki86]   E.-J. Kim, *Phys. Lett.* **B174**, 233 (1986)

[Ki87]   E.-J. Kim, *Ph.D. Thesis*, Stanford University (1987), *unpublished*

[Kl83]   W. E. Kleppinger and J. D. Walecka, *Ann. of Phys.* **146**, 349 (1983)

[Ko95]   D. S. Koltun and T. C. Ferrée, *Phys. Rev.* **C52**, 901 (1995)

[Ku01]   B. Kubis and U.-G. Meissner, *Nucl. Phys.* **A679**, 698 (2001)

[La80]   M. Lacombe, B. Loiseau, J. M. Richard, R. Vinh Mau, J. Côté, P. Pirès, and R. de Tourreil, *Phys. Rev.* **C21**, 861 (1980)

[La93]   L. Lapikás, *Nucl. Phys.* **A553**, 297 (1993)

[Le93]   C. W. Leemann, *CEBAF Progress Report*, CEBAF Summer Workshop, June 15, 1992, *A.I.P. Conf. Proc.* **269**, eds. F. Gross and R. Holt, A.I.P., New York (1993) pp. 11–24

[Le94]   M. B. Leuschner, J. R. Calarco, F. W. Hersman, E. Jans, G. J. Kramer, L. Lapikás, G. van der Steenhoven, P. K. A. de Witt Huberts, H. P. Blok, N. Kalantar-Nayestanaki, and J. Friedrich, *Phys. Rev.* **C49**, 955 (1994)

[Li79]   J. Lichtenstadt, J. Heisenberg, C. N. Papanicolas, C. P Sargent, A. N. Courtemanche, and J. S. McCarthy, *Phys. Rev.* **C20**, 497 (1979)

[Lo55] F. E. Low, *Phys. Rev.* **97**, 1392 (1955)

[Lo67] G. A. Loew, R.E. Neal, and E. Seppi, *The SLAC Accelerator: First Year of Operation*, talk presented at the 6th International Conference on High Energy Accelerators, Sept. 11-15, 1967, CEA, Boston. SLAC-PUB-351

[Ly51] E. M. Lyman, A. G. Hanson, and M. B. Scott, *Phys. Rev.* **84**, 626 (1951)

[Ma55] M. G. Mayer and J. H. D. Jensen, *Elementary Theory of Nuclear Shell Structure*, John Wiley and Sons, Inc., New York (1955)

[Ma69] L. C. Maximon, *Rev. Mod. Phys.* **41**, 193 (1969)

[Ma78] W. Marciano and H. Pagels, *Phys. Rep.* **36**, 137 (1978)

[Ma83] T. Matsui, *Phys. Lett.* **132B**, 260 (1983)

[Ma89] R. Machleidt, *The Meson Theory of Nuclear Forces and Nuclear Structure*, in *Advances in Nuclear Physics* **19**, eds. J. W. Negele and E. Vogt, Plenum Press, New York (1989), Chap. 2

[Ma89a] CEBAF PR 89-005, spks. R. Madey

[Ma90] K. Maung, CEBAF Theory Study Group, CEBAF, Newport News, Virginia (1990), *unpublished*

[Ma92] Bates Laboratory E 85-05, R. Madey, *private communication*

[Ma00] Website for Mainz Electron Accelerator: *http://www.kph.uni-mainz.de/*

[Ma00a] S. Malov, K. Wijesooriya, F. T. Baker, L. Bimbot, E. J. Brash, C. C. Chang, J. M. Finn, K. G. Fissum, J. Gao, R. Gilman, C. Glashausser, M. K. Jones, J. J. Kelly, G. Kumbartzki, N. Liyanage, J. McIntyre, S. Nanda, C. F. Perdrisat, V. A. Punjabi, G. Quéméner, R. D. Ransome, P. M. Rutt, D. G. Zainea, B. D. Anderson, K. A. Aniol, L Auerbach, J. Berthot, W. Bertozzi, P.-Y. Bertin, W. U. Boeglin, V. Breton, H. Breur, E. Burtin, J. R. Calarco. L. Cardman, G. D. Cates, C. Cavata, J.-P. Chen, E. Cisbani, D. S. Dale, R. De Leo, A. Deur, B. Diederich, P. Djawotho, J. Domingo, B. Doyle, J.-E. Ducret, M. B. Epstein, L. A. Ewell, J. Fleniken, H. Fonvieille, B. Frois, S. Frullani, F. Garibaldi, A. Gasparian, S. Gilad, A. Glamazdin. J. Gomez, V. Gorbenko, T. Gorringe, K. Griffioen, F. S. Hersman, J. Hines, R. Holmes, M. Holtrop, N. d'Hose, C. Howell, G. M. Huber, C. E. Hyde-Wright, M. Iodice, C. W. de Jager, S. Jaminion, K. Joo, C. Jutier, W. Kahl, S. Kato, S. Kerhoas, M. Khandaker, M. Khayat, K. Kino, W. Korsch, L. Kramer, K. S. Kumar, G. Laveissiére, A. Leone, J. J. LeRose, L. Levchuk, M. Liang, R. A. Lindgren, G. J. Lolos, R. W. Lourie, R. Madey, K. Maeda, D. M. Manley, D. J. Margaziotis, P. Markowitz, J. Marroncle, J. Martino, J. S. McCarthy, K. McCormick, R. L. J. van der Meer, Z.-E. Meziani, R. Michaels, J. Mougey, D. Neyret, E. A. J. M. Offermann, Z. Papandreou, R. Perrino, G. G. Petratos, S. Platchkov, R. Pomatsalyuk, D. L. Prout, T. Pussieux, O. Ravel, Y. Robin, R. Roche, D. Rowntree, G. A. Rutledge, A. Saha, T. Saito, A. J. Sarty, A. Serdarevic-Offermann, T. P. Smith, A. Soldi, P. Sorokin, P. Souder, R. Suleiman, J. A. Templon, T. Terasawa, L. Todor, H. Tsubota, H. Ueno,

P. E. Ulmer, G. M. Urciuoli, P. Vernin, S. van Verst, B. Vlahovic, H. Voskanyan, J. W. Watson, L. B. Weinstein, R. Wilson, B. Wojtsekhowski, V. Zeps, J. Zhao, and Z.-L. Zhou, *Phys. Rev.* **C62**, 057302 (2000)

[Mc62] K. W. McVoy and L. Van Hove, *Phys. Rev.* **125**, 1034

[Mc69] J. S. McCarthy and I. Sick, quoted in [Do69]

[Mc70] J. S. McCarthy, I. Sick, R. R. Whitney, and M. R. Yearian, *Phys. Rev. Lett.* **25**, 884 (1970)

[Mc77] J. S. McCarthy, I. Sick, and R. R. Whitney, *Phys. Rev.* **C15**, 1396 (1977)

[Me84] Z. E. Meziani, P. Barreau, M. Bernheim, J. Morgenstern, S. Turck-Chieze, R. Altemus, J. McCarthy, L. J. Orphanos, R. R. Whitney, G. P. Capitani, E. De Sanctis, S Frullani, and F. Garibaldi, *Phys. Rev. Lett.* **52**, 2130 (1984)

[Me89] CEBAF PR 89-045, spks. B. Mecking

[Mo69] E. J. Moniz, *Phys. Rev.* **184**, 1154 (1969)

[Mo69a] L. W. Mo and Y. S. Tsai, *Rev. Mod. Phys.* **41**, 205 (1969)

[Mo71] E. J. Moniz, I. Sick, R. R. Whitney, J. R. Ficenec, R. D. Kephart, and W. P. Towner, *Phys. Rev. Lett.* **26**, 445 (1971)

[Mo86] P. Morley and I. Schmidt, *Phys. Rev.* **D34**, 1305 (1986)

[Mo90] E. Moniz, *private communication* (1990)

[Mu89] CEBAF PR 89-024, spks. G. S. Mutchler

[Mu94] M. J. Musolf, T. W. Donnelly, J. Dubach, S. J. Pollock, S. Kowalski, and E. J. Beise, *Physics Reports*, **239**, 1 (1994)

[Mu97] A. H. Mueller, *Proceedings of the XXXVII Cracow School of Theoretical Physics*, Zakopane, Poland, May 30–June 10, 1997; *Acta Phys. Pol.* **28**, 2557 (1997)

[Na57] Y. Nambu, *Phys. Rev.* **106**, 1366 (1957)

[Ne82] J. W. Negele, *Rev. Mod. Phys.* **54**, 931 (1982)

[NS96] *Nuclear Science: A Long-Range Plan*, DOE/NSF Nuclear Science Advisory Committee (1996)

[Om59] R. Omnès, *Nuovo Cimento* **8**, 316 (1958)

[On63] D. S. Onley, T. A. Griffy, and J. T. Reynolds, *Phys. Rev.* **129**, 1689 (1963)

[Pa70] W. K. H. Panofsky, *Magnetic Spectrometers*, review talk presented at the High Energy Physics Instrumentation Conf. Sept. 8–12, 1970, Dubna USSR. SLAC-PUB-798

[Pa90] *Parity Violation in Electron Scattering*, Proc. Workshop at Cal. Inst. of Tech. Feb. 23–24, 1990, eds. E. J. Beise and R. D. McKeown, World Scientific, Singapore (1990)

[Pe89] CEBAF PR 89-014, spks. C. F. Perdrisat and V. Punjabi

[Po73] H. D. Politzer, *Phys. Rev. Lett.* **30**, 1346 (1973)

[Po74]   H. D. Politzer, *Phys. Rep.* **14**, 129 (1974)

[Po87]   S. J. Pollock, Ph.D. Thesis, Stanford University (1987), *unpublished*

[Po88]   S. J. Pollock, *Act. Phys. Pol.* **B19**, 419 (1988); (E) **B25**, 899 (1994)

[Pr65]   R. H. Pratt, J. D. Walecka, and T. A. Griffy, *Nucl. Phys.* **677** (1965)

[Pr69]   P. L. Pritchett, J. D. Walecka, and P. A. Zucker, *Phys. Rev.* **184**, 1825 (1969)

[Pr70]   P. L. Pritchett and P. A. Zucker, *Phys. Rev.* **D1**, 175 (1970)

[Pr75]   M. A. Preston and R. K. Bhaduri, *Structure of the Nucleus*, Addison-Wesley Publishing Company, Reading, Massachusetts (1975); 2nd printing (1982)

[Pr78]   C. Y. Prescott, W. B. Atwood, R. L. A. Cottrell, H. DeStaebler, E. L. Garwin, A. Gonidec, R. H. Miller, L. S. Rochester, T. Sato, D. J. Shereden, C. K. Sinclair, S. Stein, R. E. Taylor, J. E. Clendenin, V. W. Hughes, N. Sasao, K. P. Schüler, M. G. Borghini, L. Lübelsmeyer, and W. Jentschke, *Phys. Lett.* **77B**, 347 (1978)

[Pr79]   C. Y. Prescott, W. B. Atwood, R. L. A. Cottrell, H. DeStaebler, E. L. Garwin, A. Gonidec, R. H. Miller, L. S. Rochester, T. Sato, D. J. Shereden, C. K. Sinclair, S. Stein, R. E. Taylor, J. E. Clendenin, V. W. Hughes, N. Sasao, K. P. Schüler, M. G. Borghini, L. Lübelsmeyer, and W. Jentschke, with C. Young *Phys. Lett.* **84B**, 524 (1979)

[Qu83]   C. Quigg, *Gauge Theories of the Strong, Weak, and Electromagnetic Interactions*, Benjamin/Cummings, Reading, Massachusetts (1983)

[Ra54]   D. G. Ravenhall and D. R. Yennie, *Phys. Rev.* **96**, 239 (1954)

[Ra87]   D. G. Ravenhall, *A.I.P. Conf. Proc.* **161**, 153 (1987)

[Ra89]   A. S. Raskin and T. W. Donnelly, *Ann. of Phys.* **191**, 78 (1989)

[Re81]   E. Reya, *Phys. Rep.* **69**, 195 (1981)

[Ri80]   D. O. Riska, *Nucl. Phys.* **A350**, 227 (1980)

[Ri91]   CEBAF PR 91-008, spks. B. G. Ritchie

[Ro80]   R. Rosenfelder, *Ann. of Phys.* **128**, 188 (1980)

[Ro90]   R. G. Roberts, *The Structure of the Proton*, Cambridge University Press, Cambridge (1990)

[Ry96]   J. Ryckebusch, *Phys. Lett.* **B383**, 1 (1996)

[Ry97]   J. Ryckebusch, V. Van der Sluys, K. Heyde, H. Holvoet, W. Van Nespen, M. Waroquier, and M. Vanderhaegen, *Nucl. Phys.* **A624**, 581 (1997)

[Ry00]   J. Ryckebusch, S. Janssen, W. Van Nespen, and D. Debruyne, *Phys. Rev.* **C61**, 21603R (2000)

[Sa64]   A. Salam and J. C. Ward, *Phys. Lett.* **13**, 168 (1964)

[Sc49]   J. Schwinger, *Phys. Rev.* **75**, 790 (1949)

[Sc54]   L. I. Schiff, *Phys. Rev.* **96**, 765 (1954)

[Sc55] L. I. Schiff, *Phys. Rev.* **98**, 756 (1955)

[Sc58] *Selected Papers on Quantum Electrodynamics*, ed. J. Schwinger, Dover Publications, New York (1958)

[Sc68] L. I. Schiff, *Quantum Mechanics*, 3rd ed., McGraw-Hill Book Company, Inc., New York (1968)

[Sc89] CEBAF PR 89-004, spks. R. Schumacher

[Se79] B. D. Serot, *Nucl. Phys.* **A322**, 408 (1979)

[Se86] B. D. Serot and J. D. Walecka, *Adv. in Nucl. Phys.* **16**, eds. J. Negele and E. Vogt, Plenum, New York (1986)

[Se92] B. D. Serot, *Rep. Prog. Phys.* **55**, 1855 (1992)

[Se97] B. D. Serot and J. D. Walecka, *Int. J. Mod. Phys.* **E6**, 515 (1997)

[SL00] Website for SLAC: *http://www.slac.stanford.edu/*

[Sm72] R. A. Smith and E. J. Moniz, *Nucl. Phys.* **B43**, 605 (1972)

[So90] P. A. Souder, R. Holmes, D.-H. Kim, K. S. Kumar, M. E. Schulze, K. Isakovich, G. W. Dodson, K. A. Dow, M. Farkhondeh, S. Kowalski, M. S. Lubell, J. Bellanca, M. Goodman, S. Patch, R. Wilson, G. D. Cates, S. Dhawan, T. J. Gay, V. W. Hughes, A. Magnon, R. Michaels, and H. R. Schaefer, *Phys. Rev. Lett.* **65**, 694 (1990)

[St93] G. Sterman, *An Introduction to Quantum Field Theory*, Cambridge University Press, Cambridge (1993)

[Ta58] L. J. Tassie and F. C. Barker, *Phys. Rev.* **111**, 940 (1958)

[TJ00] Website for TJNAF(CEBAF): *http://www.jlab.org/*

[Tu68] S. T. Tuan, L. E. Wright, and D. S. Onley, *Nucl. Inst. and Meth.* **60**, 70 (1968)

[Tu00] W. Turchinetz, *private communication*

[Ub71] H. Überall, *Electron Scattering from Complex Nuclei*, 2 vols, Academic Press, New York (1971)

[Va78] J. W. Van Orden, *Ph.D. Thesis*, Stanford University (1978), *unpublished*

[Vi67] R. C. Vik, *Phys. Rev.* **163**, 1535 (1967)

[Vi77] R. D. Viollier and J. D. Walecka, *Acta Phys. Pol.* **B8**, 25 (1977)

[Vo60] R. Von Gehlen, *Phy. Rev.* **118**, 1455 (1960)

[Vo92] R. Voss in [Ju92], p. 169

[Wa50] J. C. Ward, *Phys. Rev.* **78**, 1824 (1950)

[Wa52] K. M. Watson, *Phys. Rev.* **88**, 1163 (1952).

[Wa59] J. D. Walecka, *Nuovo Cimento* **11**, 821 (1959)

[Wa68] J. D. Walecka and P. A. Zucker, *Phys. Rev.* **167**, 1479 (1968)

[Wa72] J. D. Walecka, *Acta Phys. Pol.* **B3**, 117 (1972)

[Wa75] J. D. Walecka, *Semileptonic Weak Interactions in Nuclei*, in *Muon Physics* Vol. II, eds V. W. Hughes and C. S. Wu, Academic Press, New York (1975), pp. 113–218

[Wa79] J. D. Walecka, *Lecture Notes in Physics* **108**, Springer-Verlag, N.Y. (1979) p. 484

[Wa83] J. D. Walecka, *Nucl. Phys.* **A399**, 387 (1983)

[Wa84] J. D. Walecka, *Lectures on Electron Scattering*, ANL-83-50, Argonne National Laboratory, Argonne, Illinois (1984); CEBAF (1987), *unpublished*

[Wa91] J. D. Walecka, *Lectures on Advanced Quantum Mechanics and Field Theory*, CEBAF, Newport News, VA (1991–92), *unpublished*

[Wa93] J. D. Walecka, *Overview of CEBAF Scientific Program*, CEBAF Summer Workshop, June 15, 1992, *A.I.P. Conf. Proc.* **269**, eds. F. Gross and R. Holt, A.I.P., New York (1993) pp. 87–136

[Wa94] J. D. Walecka, *Electron Scattering*, Conference on Perspectives in Nuclear Structure, the Niels Bohr Institute, Copenhagen, June 13–18, *Nucl. Phys.* **A574**, 271c (1994)

[Wa95] J. D. Walecka, *Theoretical Nuclear and Subnuclear Physics*, Oxford University Press, New York (1995)

[Wa97] J. D. Walecka, *Physics News in 1996*, A.I.P., College Park, MD (1997) p. 51

[We67] S. Weinberg, *Phys. Rev. Lett.* **19**, 1264 (1967)

[We72] S. Weinberg, *Phys. Rev.* **D5**, 1412 (1972)

[Wi37] E. P. Wigner, *Phys. Rev.* **51**, 106 (1937)

[Wi63] R. S. Willey, *Nucl. Phys.* **40**, 529 (1963)

[Wi69] K. G. Wilson, *Phys. Rev.* **179**, 1499 (1969)

[Wi74] K. G. Wilson, *Phys. Rev.* **D10**, 2445 (1974)

[Wi82] F. Wilczek, *Annu. Rev. Nucl. Part. Sci.* **32**, 177 (1982)

[Wi97] C. F. Williamson, T. C. Yates, W. M. Schmitt, M. Osborn, M. Deady, P. D. Zimmerman, C. C. Blatchley, K. K. Seth, M. Sarmiento, B. Parker, Y. Jin, L. E. Wright, and D. S. Onley, *Phys. Rev.* **C56**, 3152 (1997)

[Wo97] G. Wolf, *Proceedings of the XXXVII Cracow School of Theoretical Physics*, Zakopane, Poland, May 30–June 10, 1997; *Acta Phys. Pol.* **28**, 2587 (1997)

[Ya54] C. N. Yang and R. L. Mills, *Phys. Rev.* **96**, 191 (1954)

[Ye65] D. R. Yennie, F. L. Boos, and D. G. Ravenhall, *Phys. Rev.* **137B**, 882 (1965)

[Yo79] R. C. York and G. A. Peterson, *Phys. Rev.* **C19**, 574 (1979)

[Za66] N. Zagury, *Phys. Rev.* **145**, 1112 (1966)

[Za77]   H. Zarek, B. O. Pich, T. E. Drake, D. J. Rowe, W. Bertozzi, C. Creswell, A. Hirsch, M. V. Hynes, S. Kowalski, B. Norum, F. N. Rad, C. P. Sargent, C. F. Williamson, and R. A. Lindgren, *Phys. Rev. Lett.* **38**, 750 (1977)

[Ze91]   CEBAF PR 91-016, spks. B. Zeidman; B. Zeidman, *private communication*

[Zw65]   G. Zweig, *Symmetries in Elementary Particle Physics*, Proc. Int. School of Physics Ettore Majorana, 1964, ed. A. Zichichi, Academic Press, NY (1965) p. 192

Index

Printed in the United States
by Baker & Taylor Publisher Services